VARIOUS RESULTS, WITH REFERENCES TO THE SECTIONS WHERE THEY ARE DEVELOPED

$$\lambda = \frac{h}{p} = \frac{h}{\sqrt{2M(E-V)}}, \qquad E = \hbar\omega \tag{1.2}$$

$$H\Psi = i\hbar\frac{\partial\Psi}{\partial t}, \qquad H = -\frac{\hbar^2}{2M}\nabla^2 + V \tag{1.4}$$

$$\rho = |\Psi|^2, \qquad \vec{j} = \frac{\hbar}{2Mi}[\Psi^*(\vec{\nabla}\Psi) - (\vec{\nabla}\Psi^*)\Psi] \tag{1.5}$$

If $\Psi(\vec{r}, t) = \psi(\vec{r})e^{-iEt/\hbar}$, $\qquad H\psi = E\psi$ (2.1)

$a \times b \times c$ box, ∞ walls: $E_{n_x n_y n_z} = \dfrac{\pi^2\hbar^2}{2M}\left(\dfrac{n_x^2}{a^2} + \dfrac{n_y^2}{b^2} + \dfrac{n_z^2}{c^2}\right)$,

$$\psi = \sqrt{\frac{8}{abc}}\sin\left(\frac{\pi n_x x}{a}\right)\sin\left(\frac{\pi n_y y}{b}\right)\sin\left(\frac{\pi n_z z}{c}\right) \tag{2.4}$$

$V = \dfrac{Kx^2}{2}$: $E_n = (n + \frac{1}{2})\hbar\sqrt{\dfrac{K}{M}}$, $\qquad \beta^2 = \dfrac{\sqrt{KM}}{\hbar}$,

$$\psi_0 = \sqrt{\frac{\beta}{\sqrt{\pi}}}\,e^{-\beta^2x^2/2}, \qquad \psi_1 = \sqrt{\frac{\beta}{2\sqrt{\pi}}}\,2\beta x e^{-\beta^2x^2/2} \tag{2.7}$$

$V = gx,\ x \geq 0;\ V = \infty,\ x < 0$: $E_n = |\xi_n|\left(\dfrac{g^2\hbar^2}{2M}\right)^{1/3}$, $\qquad Ai(\xi_n) = 0$,

$$\xi_1 = -2.3381, \qquad \xi_2 = -4.0879, \qquad \xi_3 = -5.5206 \tag{2.9}$$

$$\langle\Omega\rangle = \int\Psi^*\Omega\Psi\,d\tau, \qquad \langle\vec{p}\rangle = \int\Psi^*\frac{\hbar}{i}\vec{\nabla}\Psi\,d\tau \tag{3.2}$$

If $\langle\Omega\rangle = \langle\Omega\rangle^*$ for all Ψ, $\displaystyle\int\Psi_1^*(\Omega\Psi_2)\,d\tau = \int(\Omega\Psi_1)^*\Psi_2\,d\tau$ (3.3)

$$xp_x - p_x x = [x, p_x] = i\hbar \tag{3.3}$$

Uncertainty principle: $\Delta P\Delta Q \geq \frac{1}{2}|\langle i[P, Q]\rangle|$ (3.4)

$$\Psi = \sum_n C_n\Psi_n, \qquad C_n = \int\Psi_n^*\Psi\,d\tau \tag{3.6}$$

$$\sum_n |C_n|^2 = 1, \qquad \langle\Omega\rangle = \sum_n |C_n|^2\omega_n \quad \text{if} \quad \Omega\Psi_n = \omega_n\Psi_n \tag{3.7}$$

$$\Psi_n \to |n\rangle, \qquad \Psi_m^* \to \langle m|, \qquad \int\Psi_m^*\Omega\Psi_n\,d\tau \to \langle m|\Omega|n\rangle \tag{3.8}$$

$$\frac{d}{dt}\langle\Omega\rangle = \left\langle\frac{i}{\hbar}\Big[H, \Omega\Big]\right\rangle + \left\langle\frac{\partial\Omega}{\partial t}\right\rangle \tag{3.10}$$

(*continued inside back cover*)

QUANTUM PHYSICS

QUANTUM PHYSICS

ROLF G. WINTER

College of William and Mary

WADSWORTH PUBLISHING COMPANY Belmont, California
A division of Wadsworth, Inc.

Physics Editor: H. Michael Snell
Production Editor: Greg Hubit Bookworks
Cover Design and Figure 1.2-6: Pat Winter

Printed in the United States of America

1 2 3 4 5 6 7 8 9 10—83 82 81 80 79

Library of Congress Cataloging in Publication Data

Winter, Rolf G
Quantum physics.

Includes bibliographical references and index.
1. Quantum theory. I. Title.
QC174.12.W55 530.1′2 78-25972
ISBN 0-534-00697-3

Preface

This book is a result of my thirty years' war in quantum physics. Through learning, teaching, and research, I have collected, mostly from the brains of others, ways of dealing with some quantal topics that seem to be useful to advanced undergraduates, and so I have gone down the usual route of increasingly voluminous handouts to publication.

My purpose is to introduce the basic structure and some applications of nonrelativistic quantum physics. The constraints of a course that I taught recently are reflected in the book. Most of my students had completed general physics in their first year and courses dealing with classical wave phenomena and qualitative modern physics in their second year. Only an elementary exposure to classical mechanics and electromagnetic theory is assumed. I have tried to select applications that both illustrate the general theory and also teach something about several fields of physics. Such an approach has intrinsic virtue, and my students having to choose their senior research area at the end of the year provided motivation. I have avoided topics that require much use of the phrase, "it can be shown that," and have concentrated on those that permit the calculation of something interesting from first principles. Formal matters are not stressed; starting the study of quantum physics is a hard business, and it is enough if the general structure of the theory begins to be absorbed osmotically.

Chapter 1 provides reminders of, but does not repeat, basic modern physics background, and it develops the Schrödinger equation. Chapter 2 solves the Schrödinger equation for some systems that can be handled in rectangular Cartesian coordinates. Chapter 3 contains the essential formal tools of the theory. I have tied it to the examples of Chapter 2, but the nature of the material guarantees that it will cause pain. The road then leads down from the glacial heights of Chapter 3 to pleasant and fertile valleys that both develop general methods and present applications in shorter chapters. The applications appear along with the methods. For example, calculations about the hydrogen atom, the deuteron, and quark-antiquark bound states are not concentrated in special sections, but are scattered throughout.

For a one-semester course I suggest either going straight through as far as time permits, or going through Section 3.8 without treating Sections 2.6, 2.9, and 3.4 in detail, and then making selections from later chapters.

For a two-semester course the entire book is either about right or somewhat long; there is a high density of things that should be chewed slowly and carefully. Some of the following sections might be omitted: 2.6, 3.4 except for the statement of the theorem (3.4-1), 3.9, 3.11, 5.2, 9.2, 9.3, 10.2, 10.4, 11.3, 12.3. These sections are more difficult than most, or a little off the main road of the argument, or both. Results of a few of these sections are mentioned subsequently, but it is not necessary to work through them to proceed.

At the risk of frightening some and insulting others, I will express the hope that occasionally this book will also be of use to graduate students, not for mathematical techniques, but for physical discussions. Not all advanced students are comfortable with questions about detecting particles in classically forbidden regions, about the meaning of the time-energy uncertainty principle, and so forth. I have put in some topics of this kind that seem to be considered, erroneously, too subtle for elementary courses and too trivial for advanced courses.

Books that try to be too self-contained, like overly solicitous parents, do a rotten job of raising the young, and numerous references to other texts and to journal articles are provided. It is important for undergraduates to learn that readable articles in *Physical Review Letters* do exist.

Most problems are imbedded in the text and should at least be read as they occur. To reduce the canonical I-don't-know-how-to-start-on-the-problems agony, there are worked-out examples, and many of the problems contain suggestions and parts of the answers. The long Chapters 1, 2, and 3 have additional problems at their ends to provide options and inspiration for review.

The notation is generally standard, and rococo symbols are avoided. A consequence is that sometimes the same symbol recurs with different uses in different places. The danger of confusion is negligible and the alternative of unfamiliar or highly decorated symbols can make easy things look hard.

Footnotes are numbered sequentially in each chapter. Equations and expressions that need labels are numbered sequentially in each section. Thus (3.2-1) is the first equation of the second section of the third chapter. Within each section, only the sequential number is used for reference. For example, in Section 3.2 the first equation is just called (1), while it is called by its full name (3.2-1) whenever it is referred to in another section.

I should like to thank those who helped me with this book. There are ways of looking at the subject that I trace to courses I took long ago, particularly with H. C. Corben, S. DeBenedetti, and F. Seitz. The William and Mary physics department has lots of people who know lots of things, and who like to talk; many of my colleagues made important contributions. H. C. von Baeyer and A. Sher were especially helpful and suggested many improvements. The students who used the preliminary versions took a lively interest in the project, caught many mistakes, and taught me how to explain things better. Mrs. S. Stout was indispensable while the manuscript was being prepared. Her ability to divine what I had meant to write was uncanny. The staff and the reviewers of Wadsworth Publishing Company deserve much of the credit for the final result. The following people reviewed the manuscript: Ronald A. Brown, SUNY, Oswego; Roland H. Good, Pennsylvania State University; B. J. Graham, U.S. Naval Academy; Arthur Hobson, University of Arkansas; Larry D. Johnson, Northeast Louisiana University; Mortimer N. Moore, California State University at Northridge; Vector J. Stenger, University of Hawaii, Manoa Campus; James V. Tucci, University of Bridgeport; Michael W. Webb, Alfred University.

Rolf G. Winter
Williamsburg, Virginia

Contents

0 | ***Exhortations*** ***1***

1 | ***Fundamentals*** ***5***

1.1 The Essence and Boundaries of Newtonian Mechanics 6
1.2 Wave-Particle Duality 8
1.3 Some Qualitative Quantum Physics 18
1.4 Development of the Schrödinger Equation 21
1.5 Probability Density, Current, and Conservation 30
1.6 Could Things Be Different? 36
1.7 Additional Chapter 1 Problems 38

2 | ***Simple Systems*** ***41***

2.1 Separation of the Schrödinger Equation 41
2.2 Step-Function Potentials 44
2.3 Rectangular and Delta Function Barriers 50

2.4 Particle in a Box. Degeneracy 56
2.5 One-Dimensional Rectangular and Delta Function Wells 60
2.6 Complex Potential Energies 76
2.7 The Harmonic Oscillator 80
2.8 Some Comments about Wave Functions 87
2.9 The Constant Force Problem 92
2.10 Additional Chapter 2 Problems 97

3 *Tools* ***101***

3.1 Expectation Values of $F(\vec{r})$ 101
3.2 Fourier Integrals and Expectation Values of $F(\vec{p})$ 105
3.3 Hermitian Operators 110
3.4 The Precise Uncertainty Principle 114
3.5 Orthogonality 118
3.6 Eigenfunction Expansions 122
3.7 Operators, Eigenvalues, and Measurements 127
3.8 Vectors, Matrices, and Dirac Bracket Notation 133
3.9 The Harmonic Oscillator Again 141
3.10 Time Derivatives of Expectation Values 148
3.11 Unstable States 153
3.12 Additional Chapter 3 Problems 162

4 *Systems of Several Particles* ***167***

4.1 Extension of the Theory 167
4.2 Identical Particles 174

5 *Electrons in Solids* ***187***

5.1 The Free Electron Model 189
5.2 Electrons in a Periodic Potential 202

6 *Angular Momentum* ***209***

6.1 General Properties of Angular Momentum 209
6.2 Orbital Angular Momentum Eigenfunctions 223

7 *Central Force Bound States* ***235***

7.1 Generalities and the Sphere of Constant Potential Energy 235
7.2 The Isotropic Harmonic Oscillator 242

7.3 The Coulomb Potential 246
7.4 Linear Potentials and Quarks 251

8 | ***Spin*** ***259***

8.1 Qualitative Background 259
8.2 Spin Eigenfunctions and Operators 266
8.3 Spinor Geometry 275
8.4 Magnetic Moments in Magnetic Fields 277

9 | ***The Addition of Angular Momenta*** ***287***

9.1 Spin $\frac{1}{2}$ Plus Spin $\frac{1}{2}$ 288
9.2 The General Problem of Angular Momentum Addition 295
9.3 The Magnetic Moment of the Deuteron 301

10 | ***Approximation Methods for Trapped Particles*** ***307***

10.1 Perturbation Theory for Nondegenerate States 308
10.2 Perturbation Theory for Degenerate States 317
10.3 The WKB Approximation 324
10.4 The Variation Method 332

11 | ***Scattering*** ***339***

11.1 Scattering Wave Functions and Cross Sections 340
11.2 The Born Approximation 344
11.3 Low-Energy Phase Shifts 356

12 | ***Transition Probabilities*** ***367***

12.1 Time-Dependent Perturbations 367
12.2 The Absorption and Emission of Light 370
12.3 Transitions to a Continuum 383

Index ***399***

R. H. FOWLER

A. PICCARD E. HENRIOT ED. HERZEN TH. DE DONDER E. SCHRÖDINGER W. PAULI W. HEISENBERG L. BRILLOUIN

P. EHRENFEST E. VERSCHAFFELT

P. DEBYE M. KNUDSEN W. L. BRAGG H. A. KRAMERS P. A. M. DIRAC A. H. COMPTON L. V. DE BROGLIE M. BORN N. BOHR

I. LANGMUIR M. PLANCK MADAME CURIE H. A. LORENTZ A. EINSTEIN P. LANGEVIN CH. E. GUYE C. T. R. WILSON

O. W. RICHARDSON

ABSENT: SIR W. H. BRAGG, MM. H. DESLANDRES AND E. VAN AUBEL

The fifth Solvay Conference on Physics, Brussels, October 1927. Every few years, the Solvay Conferences bring together small groups of physicists to discuss major unsolved problems. The 1927 meeting was pivotal in clarifying the content of the new quantum physics. The history of the conferences is described by Jagdish Mehra, The Solvay Conferences on Physics *(Dortrecht, Holland: R. Reidel Publishing Company, 1975). [Reproduced by permission of Instituts Internationaux de Physique et de Chimie (Solvay), Brussels, Belgium]*

0

Exhortations

The study of quantum physics leads one to pull together, sort, and clarify much of the physics that one has learned before, and it will pay to do a little reviewing at this time. Make sure that the basic ingredients of Newtonian mechanics are clear; many students discover in a quantum course that their classical concepts of kinetic, potential, and total energy are a little fuzzy. Review particularly your introduction to "modern physics" as in

K. W. Ford, *Classical and Modern Physics*
(Lexington, Mass.: Xerox College Publishing, 1972), Part VII

or similar books. We shall recapitulate many of the arguments there, but look now at the experiments and observations that led to the development of quantum ideas. Be sure that you have a feeling for magnitudes. What is the size of atoms? Of nuclei? What is the thermal energy at room temperature? At liquid helium temperature? What are the characteristic energies of chemistry, of nuclear physics, and of particle physics? What are the "classical radius of the electron," the Compton wavelength, the Bohr radius, and the fine structure constant?

Look over introductions to wave phenomena in a general physics book and in slightly more specialized texts such as

F. S. Crawford, *Waves, Berkeley Physics Course*, Vol. 3
(New York: McGraw-Hill Book Company, 1968),
A. P. French, *Vibrations and Waves*
(New York: W. W. Norton & Co., Inc., 1971),
J. R. Pierce, *Almost All About Waves*
(Cambridge, Mass: M.I.T. Press, 1974).

The basic nomenclature, and topics such as interference phenomena, the motion of wave packets, and so forth depend rather little on what is waving. If you know something about sound or other classical waves, you already know much of what follows.

As we proceed, keep on reviewing. A glance at previous encounters with our topics may prevent your getting lost in the underbrush. For example, the treatment of angular momentum will involve some serious mathematics, but many of the conclusions are described and partly justified in basic "modern physics" books.

Look at other quantum texts and find at least one that you will use regularly as collateral reading. The following range from a little more elementary to a little more advanced, depending on the particular topic, than this book:

E. E. Anderson, *Modern Physics and Quantum Mechanics*
(Philadelphia: W. B. Saunders Co., 1971),
R. Eisberg and R. Resnick, *Quantum Physics of Atoms, Molecules, Solids, Nuclei, and Particles*
(New York: John Wiley & Sons, Inc., 1974),
R. P. Feynman, R. B. Leighton, and M. Sands, *The Feynman Lectures on Physics*, Vol. III, *Quantum Mechanics*
(Reading, Mass.: Addison-Wesley Publishing Co., Inc., 1965),
A. P. French and E. F. Taylor, *An Introduction to Quantum Physics*
(New York: W. W. Norton & Co., Inc., 1978),
S. Gasiorowicz, *Quantum Physics*
(New York: John Wiley & Sons, Inc., 1974),
D. Park, *Introduction to the Quantum Theory*, 2d ed.
(New York: McGraw-Hill Book Company, 1974),
J. L. Powell and B. Crasemann, *Quantum Mechanics*
(Reading, Mass.: Addison-Wesley Publishing Co., Inc., 1961),
D. S. Saxon, *Elementary Quantum Mechanics*
(San Francisco: Holden-Day, Inc., 1968),
E. H. Wichman, *Quantum Physics, Berkeley Physics Course*, Vol. 4
(New York: McGraw-Hill Book Company, 1971),
K. Ziock, *Basic Quantum Mechanics*
(New York: John Wiley & Sons, Inc., 1969).

The easier parts of more advanced texts will be within your reach soon, and they are useful for extending some of our discussions. Such texts include

E. Merzbacher, *Quantum Mechanics*, 2d ed. (New York: John Wiley & Sons, Inc., 1970),
L. I. Schiff, *Quantum Mechanics*, 3d ed. (New York: McGraw-Hill Book Company, 1968),
A. Messiah, *Quantum Mechanics*, trans. G. M. Temmer (Amsterdam: North-Holland Publishing Company, 1962), particularly Vol. 1.

Keep reminding yourself that quantum physics deals with real things. Many of the experiments done in advanced undergraduate laboratories involve our topics. Look at books designed to help with the laboratories, particularly

A. C. Melissinos, *Experiments in Modern Physics* (New York: Academic Press, Inc., 1966);

the discussion of the underlying physics there complements our approach. Find out about research being done around you and try to connect it with your studies.

A knowledge of only the customary ingredients of freshman and sophomore mathematics is assumed. Through a cavalier nonrigorous use of physicists' mathematics, as distinguished from mathematicians' mathematics, specialized topics are developed as needed. These topics include Fourier series, Fourier integrals, tricks for dealing with differential equations, a few messy integrals, and some properties of matrices. If you find that the material is mathematically demanding, think about the reason. I believe that the difficulty has nothing to do with the particular techniques used, but comes from the fact that in the past you have been able to complete most calculations in five lines. Now you will have to do computations involving some pages of algebra, and the hard new thing that must be learned is calculational craftsmanship. You will need to organize your work so that you and others can verify it, to give short names to recurring long expressions, to check units and physical reasonableness at each step, and to reach for tables of formulas at the right moment, but these are obvious details. The key to it all is the realization that calculating, like carpentry, is indeed a craft, with tricks and habits that make the difference between success and failure.

Be critical. Keep asking yourself whether you really understand, and how you would explain it all to someone else. Devise better ways to present this material and make up other examples, for two reasons. First, that's the way to learn this kind of thing. Second, some of you will be writing physics books in a few years, and you should start practicing now.

Of bodies descending on inclined planes.

When a body descends on an inclined plane, its motion is continually accelerated by gravity, but in a less degree than when it descends perpendicularly; whence it follows, that what was said concerning the perpendicular descent of bodies is true of such as fall on oblique planes, allowance being made for the difference of acceleration.

The effect which gravity has on a body falling down an inclined plane, is to that which it exerts on another falling freely, as the perpendicular height of the plane, to its length.

Let A be a body upon an inclined plane, and from its centre let AG be drawn perpendicular to the horizon DC, and AF perpendicular to the inclined plane DB. Then, AE represents the whole force of gravity, and EF that part of it which urges the body down the plane. And the triangles AEF, DBC being similar, (Eucl. prop. 29 book 1) EF:AE::CB:BD, which makes the proposition abundantly manifest.

In classical mechanics, in the everyday events that form our prejudices, and in all physical experience accumulated until the twentieth century, the forces and initial conditions determine a unique outcome. From notes taken in 1809 by Robert D. Murchie, a student at the College of William and Mary, on lectures given by Bishop James Madison. (*College Archives, Earl Gregg Swem Library, College of William and Mary*)

1

Fundamentals

Newtonian mechanics works if the gravitational field is not much stronger than it is at the surface of our sun, if speeds lie below something like 0.1 c, and if distances are neither too large nor too small. If distances are large or gravity is strong, general relativity should be considered. If distances lie below 10^{-7} cm, we probably have to think about quantum mechanics. A more careful statement of the size boundary between Newtonian and quantum mechanics is the following: If the apertures in our system are very large compared to the de Broglie wavelength $\lambda = h/p$, we can be Newtonian. Thus even though our instinct might say "quantum mechanics" if someone says "proton," one can design a proton cyclotron with a knowledge of only classical physics; the protons have wavelengths $\lesssim 10^{-11}$ cm as soon as they get going, and the apertures in the device are of the order of a few centimeters.

We shall be concerned with nonrelativistic quantum mechanics, the domain where, as in classical physics, speeds are modest and gravitational fields are weak, but where sizes lie below those for which Newtonian mechanics works. Let us look first at some very simple, very profound comments about Newtonian mechanics that will help establish the boundaries.

1.1 The Essence and Boundaries of Newtonian Mechanics

All of us begin to use $\vec{F} = M\vec{a}$ at a tender age, but few of us think about the precise meanings of acceleration, mass, and force until we acquire some sophistication. Ontogeny recapitulates phylogeny: Newton's laws of motion were used successfully for two centuries before it became usual to examine what in them is definition and what can be checked experimentally. It is interesting that such discussions did not appear in the textbooks until the end of the nineteenth century, when relativity and quantum theory were imminent.[1] Perhaps the content of quantum theory will not be seen clearly until its successor becomes visible. It is hard to understand a physical theory unless one knows how it might be false.

The Newtonian structure rests on the assumptions that, for every particle, position as a function of time, $\vec{r}(t)$, and the derivatives $\dot{\vec{r}} \equiv \vec{v}, \ddot{\vec{r}} \equiv \vec{a}$, can be determined without any essential limitations on accuracy, and that there are inertial frames of reference in which $\vec{a} = 0$ for any particle that is very far from all other particles.

First, we develop the concept of *mass*. The qualitative concept involved is *difficulty of acceleration*. As the standard of difficulty of acceleration, take a block of platinum and call its mass one kilogram. The following comparison operation determines the mass of everything else. Take an unknown and the standard kilogram to an inertial frame in interstellar space and place them near each other. (If there are budgetary limitations, place the unknown and the standard on a smooth table and stretch a tiny spring between them.) Measure a_s, the magnitude of the acceleration of the standard, and a_1, the magnitude of the acceleration of the unknown, at the same time. Define M_1, the mass of the unknown, by

$$M_1 a_1 \equiv 1 \text{ kg} \cdot a_s.$$

Observe experimentally that the directions of $\vec{a}_1$ and $\vec{a}_s$ are opposite. Next, determine the masses $M_2, M_3, \ldots$ of other unknowns by the same process, and verify three properties of mass:

i. The mass of each object is a constant, independent of $\vec{r}$, t, $\vec{v}$, and $\vec{a}$.

ii. If the masses M_1 and M_2 of two objects have been determined, and these objects are placed in an inertial frame near each other and far from everything else, then $M_1 a_1 = M_2 a_2$. This observation assures that any mass can be used as a standard.

iii. Masses are additive. If two objects with masses M_1 and M_2 are combined, then the mass of the combination is $M_1 + M_2$.

[1] Ernst Mach, *The Science of Mechanics* 5th English ed. (LaSalle, Ill.: The Open Court Publishing Company, 1942), pp. 264–271, 303–306; R. B. Lindsay and Henry Margenau, *Foundations of Physics* (New York: John Wiley & Sons, Inc., 1936), pp. 79–98.

Finally, we complete the structure by introducing the force $\vec{F}$ in a shockingly simple way. By *definition*, the force acting on a mass is given by

$$\vec{F} \equiv M\vec{a},$$

where $\vec{a}$ is measured with respect to any inertial frame.

This development demotes $\vec{F} = M\vec{a}$ from the position of a basic law to that of a definition of force, a definition that is possible because of the characteristics of length, time, inertial frames, mass, and what can be measured about them. It has two related advantages over that originally presented by Newton. First, all quantities are defined through the use of clearly stated operations and, second, it is therefore particularly able to reveal the domain of validity of Newtonian mechanics. What happens when we leave the domain of our usual experience, and try to use these prescriptions to determine, for example, the mass of something that is large compared to our galaxy, or of an electron, or of something moving with half the speed of light?

Current estimates of the size of the universe, that is, of the size of the region from which one can in principle receive signals, are about 10^{10} light years.[2] It is not at all clear that one can find an approximately inertial frame for systems that extend over an appreciable fraction of this distance. Therefore general relativity is needed to discuss very large systems. Also, on a much smaller scale, the machinery of general relativity must be used to discuss strong gravitational fields.

If distances are modest and gravitational fields are weak, but speeds are not negligible compared to the speed of light, then special relativity is needed. The *difficulty of acceleration* of bodies becomes velocity dependent. The concept of simultaneity must be reinspected before one can discuss measuring the acceleration of two bodies "at the same time." If energy changes are not negligible compared to rest energies, masses are no longer additive: The sum of the mass of a neutron and the mass of a proton exceeds the mass of a deuteron.

In the domains of high speeds, of the very large, or of strong gravitational fields, coordinates have properties that violate our Newtonian prejudices, and laws of dynamics look very different from $\vec{F} = M\vec{a}$. However, the idea of definite and determined trajectories is still valid. The behavior of position and time may differ vastly from our everyday experience, but we retain the picture of particles moving along paths that can be predicted with certainty from a knowledge of the environment. In this sense, relativity is a part of classical physics. However, in the domain of the very small, of atomic and subatomic phenomena, something qualitatively different happens: The concept of the definite trajectory must be abandoned because particles are so easily disturbed that positions and momenta cannot be known simultaneously to useful accuracy. For an electron to be seen,

[2] A. Sandage and G. A. Tamman, *Astrophysical J.* **197**, 265 (1975) suggest 17.7 (+1.8, −1.5) × 10^9 years as the best estimate of the Hubble time.

it must scatter at least one photon, and that is enough to change its momentum appreciably. The idea of accelerations determined by a nearby mass, with only negligible interference by an observer, becomes useless, and the operations that lead to the definition of force cannot be performed.

Prequantum physics and everyday experience describe a deterministic universe. When we throw a ball or build a clock, we seem to prove that the future of a particle can be predicted from its initial conditions and its interactions. Quantum physics shows that, even for the most simple idealization of an isolated system, there is no way to predict precisely the results of all measurements. It may be that quantum theory is the most radical idea in the history of mankind.

1.2 *Wave-Particle Duality*

Around 1900 there began to accumulate experiments, and tentative explanations of these experiments, that led toward quantum theory as we know it. Review an elementary discussion of at least each of the following:

- Planck's black-body radiation law (1900).
- Einstein's interpretation of the photoelectric effect (1905).
- Einstein's discussion of specific heats (1907).
- The Ritz–Rydberg combination principle (1908).
- The Frank–Hertz experiment (1913).
- Bohr's model of the hydrogen atom (1913).
- The Stern–Gerlach experiment (1922).
- The Compton effect (1923).
- De Broglie's suggestion that electrons have a wave length $\lambda = h/p$ (1923).
- The Davisson–Germer electron diffraction experiment (1927).

PROBLEM 1.1

Take any one of the experiments and explanations mentioned above. Read about it in two or more "modern physics" books. Then look up the original paper on it. Summarize what you find in a couple of pages. Include the essential ideas and something interesting that emerged only from the original paper.

Although it was done after much of the theory had been developed, we shall use the Davisson–Germer experiment as our guide, because the essence of the quantal picture of electrons can be derived from it. Figure 1.2-1 shows the arrangement. The electrons arrive one by one at the photographic plate after being scattered by a crystal whose lattice parameters are known. Each electron

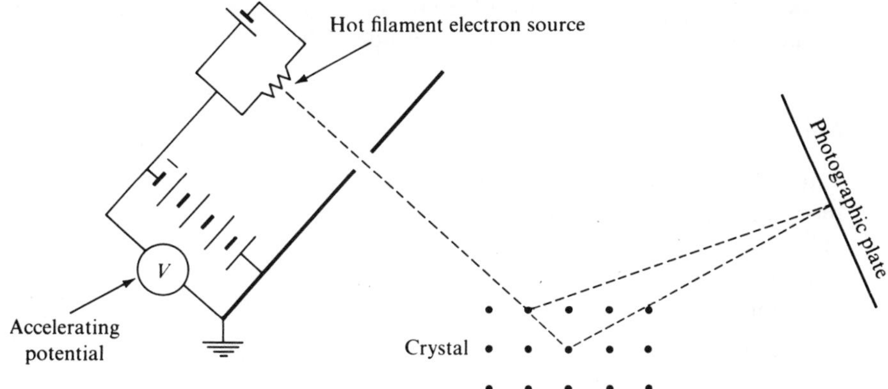

Fig. 1.2-1. The Davisson–Germer experiment.

blackens the emulsion at some point. As more and more electrons arrive, one sees that they have a high probability of landing in certain neighborhoods, so that the plate shows heavy blackening in some areas and not in others. The pattern is similar to that seen with x-ray diffraction, and it follows that there is something wavy about electrons.

Now think about the common elements in typical interference experiments with classical waves. There is generally a single source with a definite wavelength to give coherence. The locations of the scatterers or apertures are known, and the differences in path lengths to a detector can be calculated. The wavelets from the separate scatterers or apertures are added, with proper attention to their phases, to give the resultant wave at the detector. The square of the magnitude of this wave gives the observable quantity, the intensity. The precise nature of the wave is not important; these essentials are as applicable to a transverse electromagnetic wave as to a compressional wave in mint jelly. It *is* important what the wavelength is, and that the right way to calculate is to add the separate contributions and then to square the result.

Can the same things be said about the strange new electron waves? The experimental answer is *yes*, because the pattern on the photographic plate is predicted correctly if three assumptions are made:

i. The wavelength is given by $\lambda = h/p$ as proposed by de Broglie, with the momentum p determined by $p^2/2M = Ve$. Although other techniques are more accurate, the value of Planck's constant, $h = 6.63 \times 10^{-34}$ joule seconds, can be deduced from the diffraction pattern.

ii. The wave at the detector is the sum of the contributions scattered from the atoms in the crystal, with relative phases determined by the path-length differences.

iii. The observable quantity, the probability of finding an electron, is proportional to the square of the magnitude of the resultant wave calculated according to **(ii)**.

Some of the implications deserve special emphasis:

iv. There is a wavelength associated with electrons that have a definite momentum, but, when an electron is found, all of it appears at one point at one time.

v. If Ψ_1 is the wave scattered from one atom and Ψ_2 is the wave scattered from another atom, the resultant wave is $\Psi_1 + \Psi_2$. This conclusion seems trivial and obvious because it is so familiar from our experience with many classical wave phenomena, including light. One could, however, imagine a different world in which the presence of Ψ_1 changes the way Ψ_2 behaves; for example, shock waves in gases show such effects. The fact that Ψ_1 and Ψ_2 superposed give $\Psi_1 + \Psi_2$ means that the basic equations that govern the waves must be *linear*. This conclusion will be defined more carefully and exploited in Section 1.4.

vi. The measurable quantity in the experiment, the probability of finding an electron, is proportional to the square of the magnitude of the wave; it is not the wave itself. Chapter 3 shows how everything that can be known about a system can be calculated from the wave, although the wave itself cannot be detected directly.

PROBLEM 1.2

Consider the abstraction of an electron diffraction experiment shown in Figure 1.2-2. There are only two 0.1-cm-long slits, each 2.0 Å wide, with their centers separated by 5.0 Å. The current density of electrons incident on the slit system is 10^{-12} amp/cm^2. Choose a value of V which will give a nice diffraction pattern. Then calculate how long the experiment must be operated before the pattern on the plate begins to show a diffraction pattern that can be recognized readily.

For the time you have selected, make a quantitative graph of the expected number of electrons/cm^2 as a function of position along the photographic plate. Next, suppose that one of the slits is closed, and superpose on the same piece of graph paper a quantitative graph of the expected number of electrons/cm^2 as a function of position along the photographic plate.

It will help you to read Section 9.6 of F. S. Crawford, *Waves, Berkeley Physics Course*, Vol. 3 (New York: McGraw-Hill Book Company, 1968), and pp. 126–7 of E. E. Anderson, *Modern Physics and Quantum Mechanics* (Philadelphia: W. B. Saunders Co., 1971).

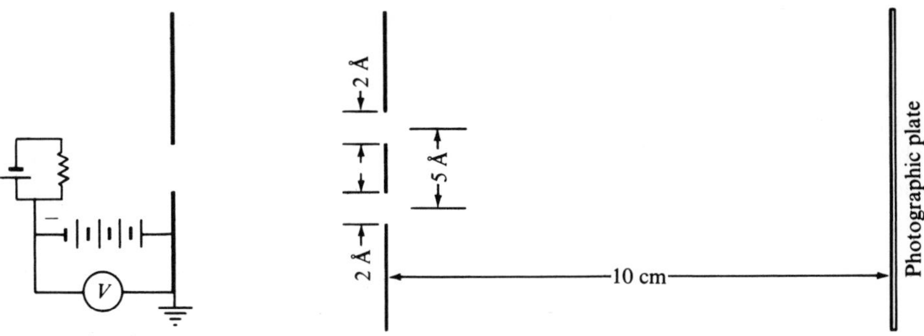

Fig. 1.2-2. A two-slit diffraction experiment with electrons.

The essence of these observations does not depend on the use of electrons. Soon after the Davisson–Germer experiment, diffraction experiments with helium atoms[3] showed that $\lambda = h/p$ for them also. Neutron diffraction[4] is a powerful tool in crystallography and other solid state research. All experience indicates that $\lambda = h/p$ is a universal relation, valid for elementary particles, atoms, locomotives, and planets, although macroscopic objects generally have wavelengths that are too small to be detected.

The next task is to develop a mathematical representation of the de Broglie waves. In one spatial dimension, a possible description of a wave with wavelength λ, angular frequency ω, and amplitude A is

$$Ae^{i(kx-\omega t)}. \tag{1.2-1}$$

To satisfy the de Broglie relation $p = h/\lambda$,

$$k \equiv \frac{2\pi}{\lambda} = \frac{p}{\hbar}, \tag{1.2-2}$$

$$\begin{aligned}\hbar \equiv \frac{h}{2\pi} &= 1.0546 \times 10^{-34} \text{ Joule sec} = 6.5822 \times 10^{-16} \text{ eV sec} \\ &= 1.9733 \times 10^{-5} \text{ eV cm}/c.\end{aligned}$$

Other choices include the sine or cosine of $(kx - \omega t)$. Since

$$\begin{aligned}e^{\pm i\phi} &= \cos\phi \pm i\sin\phi, \\ \cos\phi &= \frac{e^{+i\phi} + e^{-i\phi}}{2}, \\ \sin\phi &= \frac{e^{+i\phi} - e^{-i\phi}}{2i},\end{aligned}$$

[3] I. Estermann, R. Frisch, and O. Stern, *Z. Physik* **73**, 348 (1932).

[4] D. J. Hughes, *Pile Neutron Research* (Reading, Mass: Addison-Wesley Publishing Co., Inc., 1953), Ch. 10; G. E. Bacon, *Neutron Diffraction*, 2d ed. (London: Oxford University Press, 1962). With very slow neutrons, it is possible to observe diffraction by mechanically ruled gratings; see H. Scheckenhofer and A. Steyerl, *Phys. Rev. Lett.* **39**, 1310 (1977).

complex exponentials can always be expressed in terms of real trigonometric functions and vice versa, so that a decision to work in terms of functions like (1) is no restriction.[5] In many classical treatments of oscillations, complex exponentials are used to simplify the algebra, and the results can be written in terms of sines and cosines at the end of the calculation. However, in quantum physics, there are good physical reasons for the choice (1). The square of its magnitude, that is, (1) multiplied by

$$A^* e^{-i(kx-\omega t)},$$

where * denotes *complex conjugate*, is the constant $A^*A \equiv |A|^2$. If the square of the magnitude is proportional to the probability of finding the particle, then (1) implies that there is no preferred location, a reasonable result for a particle that is moving along with constant and definite momentum. On the other hand, $A \sin(kx - \omega t)$ gives a probability density $|A|^2 \sin^2(kx - \omega t)$, which oscillates and depends on the origins of x and t. A more solid reason for using complex waves emerges in Section 1.5. There it is shown that the wave must be complex if it is to describe the motion of a particle, subject to the condition that the total probability of finding the particle is conserved.

A particle of definite momentum is then represented by the plane wave (1). If several different momenta are present, the sum of several terms of the form (1) is needed. In typical experiments the momenta are, in fact, continuously distributed; there will generally be something that smears out the energy and the direction, and therefore the momentum. Expression (1) is a useful idealization, but a more realistic wave function is given by

$$\Psi(x, t) = \int_{-\infty}^{+\infty} A(k) e^{i[kx - \omega(k)t]}\, dk. \tag{1.2-3}$$

Each separate plane wave is labeled by its value of k and has its own amplitude $A(k)$. The angular frequency is called $\omega(k)$ to emphasize that different values of k may involve different frequencies. The integration over k adds all the contributions, and is shown as extending from $-\infty$ to $+\infty$ to announce that there is no restriction on the values of k that can occur.

Suppose that $A(k)$ is appreciable only for a small range of k, and that $\Psi(x, t)$ is, at any one time, appreciable only for a small range of x. Then Ψ is a *wave packet* that should behave, if we don't look too closely, like a classical particle, because position and momentum are roughly known. Each of the separate waves,

$$A(k) e^{i[kx - \omega(k)t]},$$

[5] Equations and expressions that need labels are numbered sequentially in each section. Thus (3.2-1) is the first equation of the second section of the third chapter. Within each section, only the sequential number is used for reference. For example, in this section the first expression is referred to simply as (1), while in other sections it is referred to by its full name (1.2-1).

has a phase velocity $\omega(k)/k$, but the motion of Ψ is determined by the *group velocity* v_g.[6] The separate waves combine to give a packet because they interfere constructively in one neighborhood and destructively elsewhere. Where they do interfere constructively, they must have nearly the same phase $kx - \omega(k)t$, and there will be some central value of k—call it k_0—at which a little change in k does not change the phase at all.

$$\left\{\frac{d}{dk}[kx - \omega(k)t]\right\}_{k_0} = x - \left[\frac{d\omega(k)}{dk}\right]_{k_0} t = 0.$$

The ratio of x to t that satisfies this condition is the velocity of the packet, so that

$$v_g = \left[\frac{d\omega(k)}{dk}\right]_{k_0}. \tag{1.2-4}$$

In the domain where Newtonian physics is valid, quantum calculations must reproduce the Newtonian results. This observation, often called the *correspondence principle*, was a powerful tool in the development of quantum theory. Here it requires that the center of a wave packet must follow the classical trajectory:

$$\frac{d\omega}{dk} = \frac{p}{M} = \frac{\hbar k}{M}.$$

Notice that the formalism is being told here that it is to describe a nonrelativistic theory because p/M, where M is the constant mass, is the nonrelativistic expression for the speed. Integration yields

$$\omega = \frac{\hbar k^2}{2M} + \alpha = \frac{p^2}{2M\hbar} + \alpha = \frac{E}{\hbar} + \alpha,$$

$$\hbar\omega \equiv h\nu = E + \alpha\hbar.$$

It is convenient to set the constant α equal to zero. Another choice would be equivalent to a shift in the origin of energy. The result, $E = h\nu$, is the relation originally proposed by Planck for photons, and (3) becomes

$$\Psi(x, t) = \int_{-\infty}^{+\infty} A\left(\frac{p}{\hbar}\right) e^{i(px - Et)/\hbar} \frac{dp}{\hbar}. \tag{1.2-5}$$

The Heisenberg uncertainty principle is already implied, as can be seen from a qualitative argument. Consider the wave packet shown in Figure 1.2-3 which represents a particle with uncertainty Δx in position. Let λ be an estimate

[6] F. S. Crawford, *Waves, Berkeley Physics Course*, Vol. 3 (New York: McGraw-Hill Book Company, 1968), Chap. 6, contains an excellent discussion of wave packets, phase velocity, and group velocity, and a more leisurely development of (4).

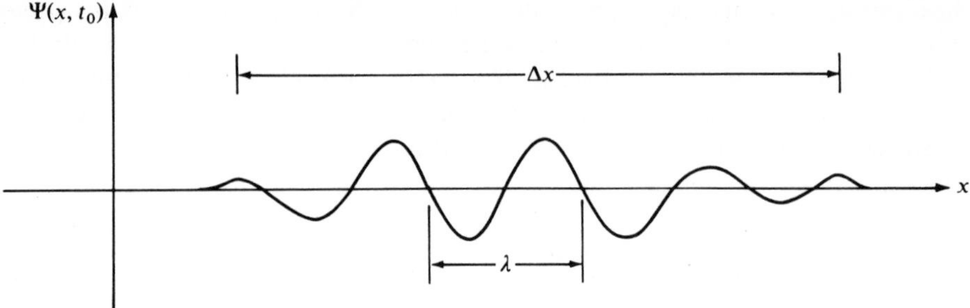

Fig. 1.2-3. A typical wave packet.

of wavelength obtained by looking at the center of the packet. If there is to be, through cancellation, a substantial reduction in Ψ near the ends of the packet, there must be an admixture of other wavelengths, around λ', which are in phase near the middle and out of phase near the ends. The length λ' must therefore fit into Δx at least once more (or once less) than λ; that is, if

$$\frac{\Delta x}{\lambda} = n,$$

then

$$\frac{\Delta x}{\lambda'} = n + 1.$$

Subtraction yields

$$\Delta x\left(\frac{1}{\lambda'} - \frac{1}{\lambda}\right) = 1,$$

which becomes, with $\lambda = h/p$ and $\lambda' = h/p'$,

$$\Delta x(p' - p) = h.$$

Since $p' - p$ is an estimate of the range of momenta present, it can be called Δp, the *momentum uncertainty*. We can always know less than the ultimate limits permit, and therefore turn the equality into an inequality:

$$\Delta x\, \Delta p \gtrsim h.$$

Note that Δx and Δp have not been well defined, and that numerical factors in this inequality depend on whether Δx and Δp are one or another measure of deviation from a central value. Section 3.4 shows that if Δx and Δp

are the root mean squared deviations of position and momentum from their means, then

$$\Delta x \, \Delta p \geq \frac{\hbar}{2}. \tag{1.2-6}$$

The uncertainty principle involving position and momentum is obtained by examining a packet as a function of position at fixed time. A similar argument regarding Ψ as a function of time at fixed x gives $\Delta E \, \Delta t \gtrsim h$.

PROBLEM 1.3

a. Carry out the argument suggested in the sentence above.

b. Obtain the $\Delta E \, \Delta t$ uncertainty principle from the $\Delta x \, \Delta p$ uncertainty principle by thinking about an observer watching a particle go past.

c. By thinking about a wave packet running around the circumference of a circle, obtain an uncertainty principle involving angle and angular momentum.

PROBLEM 1.4

What is the wavelength of a locomotive that is at rest? How well can the locomotive be localized? Discuss.

Many "thought experiments" have been used to examine the content of the uncertainty principles. For example, consider an attempt to measure the position of an electron with initial momentum p, in the x direction, by looking with a lens for a scattered photon of wavelength λ (Figure 1.2-4). Diffraction

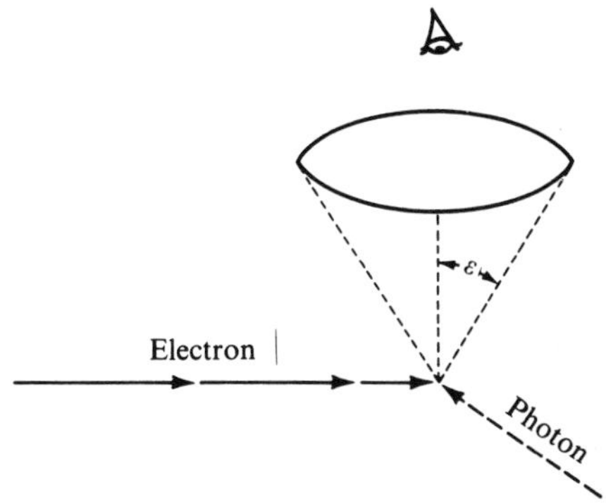

Fig. 1.2-4. An attempt to find an electron with a lens.

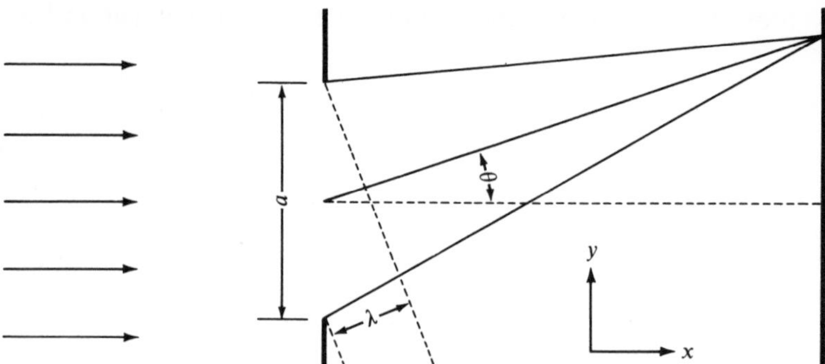

Fig. 1.2-5. Collimation of a beam of particles.

limits the resolution of this arrangement[7] to $\Delta x \simeq \lambda/\sin \varepsilon$. The momentum given to the electron by the photon is uncertain by $\Delta p = (h/\lambda) \sin \varepsilon$, since we know only that the photon went somewhere through the lens aperture. Therefore $\Delta x\, \Delta p = h$. Notice that the argument hinges on the fact that there must be at least one photon with momentum h/λ. The classical physicist would insist that the intensity could be reduced until the momentum carried by the light becomes negligible.

As another example, consider the collimation of a beam of particles by a slit of width a (Figure 1.2-5). For destructive interference, $\sin \theta = \lambda/a$. Since the trajectories of the particles can lie with large probability anywhere between the first interference minima, $\Delta p_y \simeq p \sin \theta = h/\lambda \sin \theta = (h/\lambda)(\lambda/a)$. Also, $\Delta y = a$. Therefore $\Delta p_y\, \Delta y \simeq h$. Note that p_x can remain essentially undisturbed; there is no limit on our simultaneous knowledge of, for example, y and p_x.

Look again at the interference experiment[8] examined in Problem 1.2. If either of the two slits is closed, electrons arrive at points where previously the two-slit interference caused the intensity to be zero. If both slits are opened again, the original pattern returns and the intensity again drops to zero at the interference minima. The classical mechanician will, of course, want to know, "How can the opening of a slit through which the electron does not go prevent it from arriving where it would have arrived if the slit were closed?" The quantum mechanician replies that if one does determine through which slit the electron goes, then the interference pattern is destroyed.

[7] F. A. Jenkins and H. E. White, *Fundamentals of Optics*, 4th ed. (New York: McGraw-Hill Book Company, 1976), Sec. 15.10.

[8] For similar discussions see L. I. Schiff, *Quantum Mechanics*, 3d ed. (New York: McGraw-Hill Book Company, 1968), Sec. 4; R. P. Feynman, R. B. Leighton, and M. Sands, *The Feynman Lectures on Physics*, Vol. III, *Quantum Mechanics* (Reading, Mass: Addison-Wesley Publishing Co. Inc., 1965) Chaps. 1, 2, 3; E. H. Wichmann, *Quantum Physics*, *Berkeley Physics Course*, Vol. 4 (New York: McGraw-Hill Book Company, 1974), Chaps. 5, 6.

Suppose that one mounts behind the screen any device that localizes the electron vertically to half the distance d between the slits. A less accurate measurement is uninteresting because it will not reveal through which slit the electron went. The uncertainty principle assures that $\Delta p_y \gtrsim 2h/d$, so that the measurement deflects the electrons through angles of the order of $2h/pd$. But the angle between a maximum in the interference pattern and an adjacent minimum is $\lambda/2d = h/2pd$. Therefore the measurement scatters the electrons about so that those initially heading for a maximum in the pattern can easily end up in a minimum.

It is therefore impossible to give operational meaning to talk about the electron going through this or that slit unless one introduces arrangements that destroy the interference pattern. All of us are tempted to mutter things like, "Well, I won't be able to check through which slit the electron goes without fouling things up, but surely it has to go either through one or the other." Anyone who yields to the temptation breaks the law: *Thou shalt not make any statement that can never be verified.* It is forbidden to say that the electron must go either through one or the other slit without accepting the consequences of determining that it does. This annoyingly dogmatic statement probably induces visions of stone tablets being lowered from above with piano wires. The justification is the experimental fact that nobody has yet made sense of quantum physics without accepting the law.

Niels Bohr made a crucial contribution to our understanding of these matters by formulating his *complementarity principle.*[9] He emphasized that there are properties that complement each other and that can be determined simultaneously to provide the complete classical description, but that cannot be so determined in the quantum domain. The pairs of variables that appear in the uncertainty principle are examples. Another example is the pair of properties, particleness and wavyness. An experiment, such as a diffraction study, which is designed to bring out the wave aspect of electrons will indeed show the wave aspect. We cannot, however, mix in procedures that involve the particle aspect of electrons without altering our wave aspect study.

It is worth noticing that the final stage of many experiments with electrons consists of detecting the electrons and thereby bringing out their particle properties. One therefore tends toward habits of thought that suggest that the electron is "really" a particle, and that the probability of finding this particle is to be calculated by doing tricky calculations about some unphysical wave. One can with this somewhat naive attitude do successfully lots of very complicated quantum calculations, but it is not an attitude that can withstand any profound and critical examination.

The electron is an electron. Be grateful that it shows some of the properties of a classical particle and some of the properties of a classical wave; you hardly

[9] N. Bohr, *Nature* **121**, 580 (1928); *Phys. Rev.* **48**, 696 (1935); *Atomic Theory and the Description of Nature* (London: Cambridge University Press, 1934); *Atomic Physics and Human Knowledge* (New York: John Wiley & Sons, Inc., 1958).

Fig. 1.2-6. Classical analog of an electron.

have the right to be annoyed at it for declining to accept neatly one or the other label. Perhaps a few hundred years ago, an explorer came back to his home town and tried to tell his neighbors about the duckbilled platypus. Perhaps the solid citizens, who hadn't done much traveling, demanded to know whether the explorer was talking about a duck or about a beaver; after all, they had seen all the kinds of animals that had been created. They asked the wrong question: we cannot leave behind the domain of experience in which our categories were developed and still expect all discoveries to fit neatly into these same categories.

1.3 *Some Qualitative Quantum Physics*

To understand, in the physicist's sense, what quantum phenomena are about, one must be able to do some reasonably complex calculations, but that is not sufficient. One must also be able to make rough pictures and estimates by using simple basic ideas. It is hard to make a real contribution without this knack. Furthermore, such a talent is good equipment for impressing colleagues at lunch and in the hallways at Physical Society meetings. The following is a sample of familiar arguments that involve little but the uncertainty principle.

To good approximation, the energy of a hydrogen atom is given by the kinetic energy plus the Coulomb potential energy of the electron;

$$E = \frac{p^2}{2M} - \frac{e^2}{4\pi\varepsilon_0 r}.$$

Since the electron is localized to an accuracy $\simeq r$, there will be fluctuations in momentum $\simeq \hbar/r$. In the ground state the electron will have a magnitude of momentum as small as possible, but this magnitude cannot be much smaller than the fluctuations. Therefore $p \simeq \hbar/r$, and

$$E \simeq \frac{\hbar^2}{2Mr^2} - \frac{e^2}{4\pi\varepsilon_0 r}.$$

The Coulomb attraction would like to make the atom indefinitely small and the potential energy indefinitely negative. The uncertainty principle, however, exacts a high cost in kinetic energy if r is reduced. Equilibrium between the two tendencies is established at the minimum E determined by $dE/dr = 0$:

$$-\frac{\hbar^2}{Mr^3} + \frac{e^2}{4\pi\varepsilon_0 r^2} = 0, \qquad r = \frac{4\pi\varepsilon_0 \hbar^2}{Me^2} = 0.528\,\text{Å},$$

$$E_{\min} = -\frac{M}{2\hbar^2}\left(\frac{e^2}{4\pi\varepsilon_0}\right)^2 = -13.6\,\text{eV}.$$

It is, of course, a contrivance that precisely the Bohr radius and the ground state energy have come out; for example, the choice $p \simeq \hbar/r$ is neither more nor less reasonable than $p \simeq \hbar/2r$. The essential ideas are, however, entirely justified.

The size of the deuteron is of the order of a fermi $\equiv f \equiv 10^{-13}$ cm. Therefore the fluctuations in momentum of, say, the neutron are around $\hbar/f$. The same argument that was used for the hydrogen atom then implies that the kinetic energy is roughly $\hbar^2/2M_n f^2 \simeq 30$ MeV. The binding energy of the deuteron is 2.2 MeV. There are no bound excited states. It follows that the deuteron is a loose structure in which the two nucleons see an average attractive potential energy of some tens of MeV, and in which the single bound state lies close to the top of the well. If nuclear forces were only slightly weaker, the deuteron would not be a stable structure, and you wouldn't be here.

PROBLEM 1.5

Why wouldn't you be here if the deuteron were unstable?

The exchange of pi mesons, or pions,[10] seems to be responsible for that part of the nucleon-nucleon interaction that has a range of about $1f$. One can think in terms of a virtual pion being emitted by one nucleon, living as long as the uncertainty principle lets it, and then being swallowed by another nucleon. If the range involved is $1f$, then the time that the pion exists must be at least $\Delta t = 1f/c = 3 \times 10^{-24}$ sec because the speed of the pion must be less than c, the speed of light. If there is a pion in transit, the energy of the system is wrong by at least $M_\pi c^2$, the rest energy of the pion. But that is allowed if the uncertainty in energy is around this amount, so that $M_\pi c^2 \simeq \Delta E \simeq \hbar/\Delta t$, or $M_\pi c^2 \simeq \hbar c/1f =$ 190 MeV. Actually $M_\pi c^2 = 140$ MeV for the charged pions $\pi^\pm$, and 135 MeV for the neutral π^0. It was a quantification of this argument that was used by H. Yukawa in 1935 to predict the existence of mesons.

[10] For a review of the properties of pions, muons, and other particles see, for example, K. W. Ford, *Classical and Modern Physics* (Lexington, Mass.: Xerox College Publishing, 1972), Chap. 27.

A generalization of these comments suggests that for any force with range R, there should be a quantum of the field that has a mass $\hbar/(Rc)$, that is, a particle whose Compton wavelength equals R. The nuclear force appears to have a "repulsive core," that is, an inner region within which the force prevents the nucleons from coming closer to each other, with a range of the order of a couple times 10^{-14} cm. A mass of $\hbar/0.2fc$ should be associated with this force, and the combined effects of things like nucleon-antinucleon pairs, various heavy mesons, and more complicated objects may be responsible. Whether it is right to think this way about the repulsive core is at present unclear. The other extreme of "infinite range" describes the electromagnetic and gravitational fields; that is, their potential falls off like $1/r$ without a characteristic length being involved. The masses of the quanta of these fields are then given by $\hbar/(\infty \cdot c)$. Therefore it makes sense that the electromagnetic quantum, the photon, has rest mass zero,[11] and the presently undetected[12] quantum of the gravitational field, the "graviton," should also have rest mass zero.

Suppose a decaying state has a mean life τ. Since one has only about τ to measure the energy of the state, the energy cannot be determined to an accuracy better than $\Delta E \simeq \hbar/\tau$. States indeed show a "natural width" of this size. Particles that can decay by uninhibited strong interactions have mean lives of the order of $1f/c \simeq 10^{-23}$ sec, so that $\Delta E \simeq \hbar/10^{-23}$ sec $\simeq 100$ MeV. For example, the ρ meson decays strongly into two pions, and its natural width is 150 MeV. In atomic physics, mean lives of around 10^{-8} sec are common, and the corresponding natural widths are $\hbar/10^{-8}$ sec $\simeq 10^{-7}$ eV. These widths are very difficult to see. One must eliminate the Doppler broadening caused by the random thermal motion of the emitting atoms toward and away from the spectrometer. Further, one must eliminate the *collision broadening*. In each collision the state of an atom is disturbed so much that we see a width $\hbar/t$, where t is the mean interval between collisions.

PROBLEM 1.6

a. For visible light from nitrogen, what is the gas temperature if Doppler broadening and the natural width are the same for states with a 10^{-8}-sec mean life?

b. What is the gas pressure at room temperature if collision broadening and the natural width are the same?

---

[11] For discussions of the upper limit on the rest mass of the photon, see A. S. Goldhaber and M. M. Nieto, "The Mass of the Photon," *Scientific American* **232** (5), 86 (1976). At present, a direct and relatively model independent upper limit on this rest mass is 10^{-48} gm, or 10^{-21} of the rest mass of the electron. Limits that depend on speculations regarding galactic magnetic fields range down to 10^{-10} of this value.

[12] Only static gravitational fields have been observed. Attempts to detect gravitational waves seem to be unsuccessful; see J. L. Levine and R. L. Garwin, *Phys. Rev. Lett.* **33**, 794 (1974). Individual graviton detection would be vastly more difficult.

1.4 *Development of the Schrödinger Equation*

Section 1.3 shows that rough estimates and qualitative pictures can be wrung out of the uncertainty principle. Further, we are already equipped to be more quantitative in a few particularly simple situations. If there are no forces, the wave packet discussed in Section 1.2 is a precise and general description of what the wave can do. The extremes of very slowly varying potential energy and of homogeneous regions bounded by impenetrable walls can also be studied without the development of more formal machinery.

If the potential energy changes slowly, in the sense that the change in wavelength per wavelength is small, it is useful to think in terms of a locally defined wavelength

$$\lambda(\vec{r}) = \frac{h}{p(\vec{r})} = \frac{h}{\sqrt{2M[E - V(\vec{r})]}} = \lambda_0\sqrt{\frac{E}{E - V(\vec{r})}}, \tag{1.4-1}$$

where

$$\lambda_0 = \frac{h}{\sqrt{2ME}}$$

is the wavelength that the particle of energy E would have if $V(\vec{r}) = 0$. Light going through the earth's atmosphere is a good analog. The index of refraction $n(\vec{r})$ is determined by the atmospheric density and depends on the altitude. In each neighborhood the wavelength is well defined, but to trace the light accurately through a few hundred miles of atmosphere, one must recognize that $\lambda(\vec{r}) = \lambda_0/n(\vec{r})$, where λ_0 is the wavelength in vacuum. For a slowly varying $n(\vec{r})$, the wave fronts will be nearly plane, and rays perpendicular to them give the path of the light. For the quantum waves, rays constructed in the same way give the trajectories of classical particles, as illustrated by the following example.

Suppose that the rays have some curvature. Look at a small neighborhood, in which the curvature of each ray and the wavelength along each ray are nearly constant. Let r be the radius of curvature and λ the wavelength on one ray, and let $r + dr$ and $\lambda + d\lambda$ be these quantities on an adjacent ray (Figure 1.4-1). From similar sectors,

$$\frac{r + dr}{r} = \frac{\lambda + d\lambda}{\lambda}.$$

Therefore since λ is h/p,

$$\frac{1}{r} = \frac{1}{\lambda}\frac{d\lambda}{dr} = \frac{\sqrt{2M[E - V(r)]}}{h}\frac{d}{dr}\frac{h}{\sqrt{2M[E - V(r)]}}$$

$$= \frac{1}{2[E - V(r)]}\frac{dV}{dr} = \frac{1}{Mv^2}\frac{dV}{dr},$$

$$\frac{Mv^2}{r} = \frac{dV}{dr}.$$

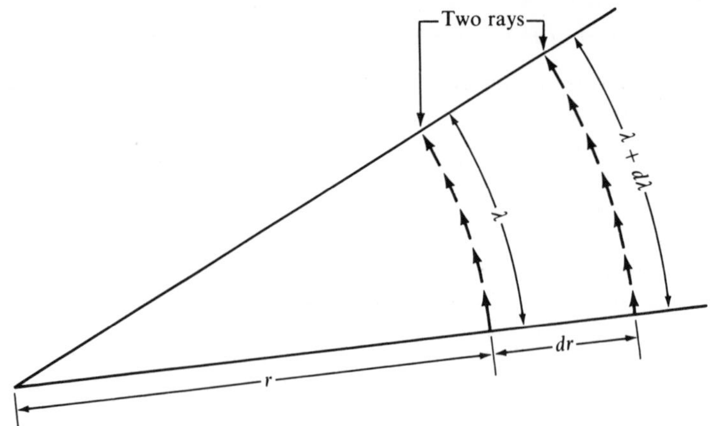

Fig. 1.4-1. Curvature of wavefronts in a region of slowly varying potential energy.

Since dV/dr is the magnitude of the centripetal force, the result is the Newtonian answer. This conclusion is no great miracle. Classical physics is the domain in which the de Broglie wavelength is very small compared to distances over which things change appreciably.

Next, consider a particle trapped by rigid walls in a region of zero potential energy (Figure 1.4-2). The de Broglie wave must go to zero at the impenetrable walls, and so there is a discrete spectrum of possible wavelengths. The picture is like that of sound waves in a box or of the vibrations of a string stretched between fixed supports. If in a one-dimensional problem the rigid walls are at $x = 0$ and $x = a$, then

$$n_x \frac{\lambda}{2} = a, \qquad n_x = 1, 2, 3, \ldots .$$

The magnitude of the momentum is h/λ, and the energy is given by

$$\frac{p^2}{2M} = \frac{h^2}{8Ma^2} n_x^2. \tag{1.4-2}$$

The requirement that the waves must fit into the region produces a discrete spectrum of energies

$$\frac{h^2}{8Ma^2}, \qquad \frac{4h^2}{8Ma^2}, \qquad \frac{9h^2}{8Ma^2}, \ldots$$

that can be expressed in terms of a *quantum number* n_x. Section 2.4 examines this situation more carefully.

These examples show that in special cases one can get some information without developing the theory further. However, better tools are needed to deal

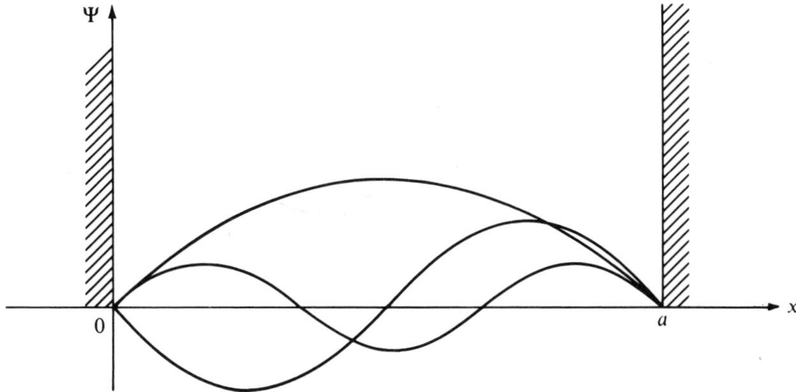

Fig. 1.4-2. Standing waves between impenetrable walls.

with most interesting problems, and it will help to have a *wave equation* whose solution $\Psi(\vec{r}, t)$ describes the wave aspect of particles.[13] Integral equations cannot be excluded; such equations are, in fact, used for some very important formulations of quantum theory. However, most of us have been raised to feel more comfortable with differential equations, and there is much useful experience with the partial differential equations that describe classical sound and electromagnetic waves. We therefore look for something that can be described symbolically by

$$F(\Psi, \text{derivatives of } \Psi, V, M, \hbar) = 0, \tag{1.4-3}$$

that is, something that connects Ψ and its derivatives with the potential energy function $V(\vec{r}, t)$ and the fundamental constants that should enter. The procedure will be to think about general physical observations and to deduce the most simple form of (3) that is in accord with these observations. This approach is distantly related in spirit to that used to develop $\vec{F} = M\vec{a}$ in Section 1.1.

Section 1.2 lists some assumptions and implications that come out of the Davisson–Germer experiment. Statement (v) there emphasizes that the linear superposition of waves is valid; it requires that if Ψ_1 is a solution to (3), and Ψ_2 is another solution to (3), then $\Psi_1 + \Psi_2$ is also a solution to (3). It follows that (3) must be *linear* and *homogeneous*, that is, that its form must be

$$\Omega\Psi = 0, \tag{1.4-4}$$

[13] I have a revealing language problem here. The word *particle* is misleading because the objects of our studies have both particle and wave properties. Some people have semiseriously proposed using the word *wavicle*. However, it is customary to use the word *particle* for electrons, protons, neutrons, neutrinos, and so forth, and also photons, without there being any implication of classical particle behavior.

where Ω is a *linear operator*. An operator is *linear* if, for any Ψ_1 and Ψ_2,

$$\Omega(\Psi_1 + \Psi_2) = \Omega\Psi_1 + \Omega\Psi_2, \tag{1.4-5}$$

and, for any constant b,

$$\Omega(b\Psi) = b\Omega\Psi. \tag{1.4-6}$$

That the equation is *homogeneous* means that all its terms do involve Ψ, and that the right side is zero rather than some constant or some function independent of Ψ. For example, for any functions $u(\vec{r}, t)$, $v(\vec{r}, t)$ and $w(\vec{r}, t)$,

$$\left[v(\vec{r}, t) + u(\vec{r}, t)\frac{\partial}{\partial t} + w(\vec{r}, t)\frac{\partial^2}{\partial x^2}\right]\Psi = 0$$

is linear and homogeneous, while any equation containing Ψ^2 or $\sqrt{\Psi}$ or $\Psi(\partial\Psi/\partial x)$ or $(\partial\Psi/\partial t)^2$ is nonlinear. Equation (4) is just what is needed to guarantee linear superposition; *any* linear combination of solutions is another solution. If $\Omega\Psi_1 = 0$ and $\Omega\Psi_2 = 0$, then, for any constants b_1 and b_2,

$$\begin{aligned}\Omega(b_1\Psi_1 + b_2\Psi_2) &= \Omega(b_1\Psi_1) + \Omega(b_2\Psi_2)\\ &= b_1\Omega\Psi_1 + b_2\Omega\Psi_2 = 0.\end{aligned}$$

If nature had been less kind and had asked for a *non*linear structure, the mathematics would be much harder. The theory of linear differential equations is well developed and not too difficult, but little is known about nonlinear equations. Even if there were no compelling physical reasons, we would certainly try to find a linear equation that works, and give up the search only with mutters, curses, and great reluctance.

For one spatial dimension x, a general trial form of (4) is

$$\left[f_{00} + f_{10}\frac{\partial}{\partial t} + f_{01}\frac{\partial}{\partial x} + f_{11}\frac{\partial^2}{\partial t\,\partial x} + f_{02}\frac{\partial^2}{\partial x^2} + f_{20}\frac{\partial^2}{\partial t^2} + \cdots\right]\Psi = 0,$$

or

$$\left[\sum_{i,j} f_{ij}\frac{\partial^{i+j}}{\partial t^i\,\partial x^j}\right]\Psi = 0, \tag{1.4-7}$$

where

$$f_{ij} = f_{ij}(V, M, \hbar).$$

If space and time are *homogeneous*, that is, if the origins of x and t are arbitrary, then the f_{ij} can depend on x and t only through their dependence on $V(x, t)$. For example, if one of the f_{ij} were something like $\alpha t + \beta x^2$ in a region of zero potential energy, then the behavior of the system would change even though nothing interacts with it, and experiments done at different times and positions in interstellar space would give different answers. The assumption being made here is that inertial frames exist and that (7) is valid in them; it is really the same as the assumption made about inertial frames in Section 1.1.

The generality of (7) is much reduced because, if $\Psi^*\Psi$ is the probability density, then there should not be time derivatives higher than the first. Suppose there were a $\partial^2\Psi/\partial t^2$. Then $\partial\Psi/\partial t$ at $t = 0$, as well as Ψ, could be specified freely everywhere in space as an initial condition, and that would make trouble for the *correspondence principle* and *conservation of particles.* To understand the correspondence principle difficulty, look at a typical wave packet, such as the one sketched in Figure 1.2-3. The position of the particle is somewhere within the extent of the packet, and the momentum is around h divided by a wavelength estimated near the center of the packet. The initial wave function $\Psi(x, 0)$ describes, as well as the uncertainty principle permits, the initial position and the initial momentum of the particle. These are the initial conditions that can be specified in classical mechanics; the initial acceleration can never be specified in the same way because it is determined by the equation of motion. The function $\partial\Psi/\partial t$, however, implies the average acceleration. For example, if the packet were being compressed so that the wavelengths are decreasing, then the average speed would be increasing and there would be an acceleration. It follows that the possibility of specifying the initial $\partial\Psi/\partial t$ corresponds to the possibility of specifying the initial acceleration, and should be ruled out.

To see the trouble with conservation, recall that the probability of finding a particle in any neighborhood is $\Psi^*\Psi$ evaluated there, so that the time derivative of the probability is

$$\Psi^* \frac{\partial\Psi}{\partial t} + \frac{\partial\Psi^*}{\partial t}\Psi$$

If both Ψ and $\partial\Psi/\partial t$ could be given any value one pleased, this expression could be made positive *everywhere*, and the total probability of finding the particle somewhere in the world would increase. Such a result would be silly, because the *total* probability of finding the particle must be constant; if the local probability increases in one neighborhood, it must decrease in another. This argument and other aspects of particle conservation are examined more carefully in the next section, but it is already clear that freedom to specify $\partial\Psi/\partial t$ and particle conservation will generally be inconsistent. It follows then that, in the trial form (7), $f_{ij} = 0$ for $i \geq 2$.

Next, observe that in a region where $V = 0$, a general plane wave

$$\Psi = Ae^{i(kx - \hbar k^2 t/2M)}$$

must be a solution for *any* $k = p/\hbar$.

Now write some derivatives of this special solution:

$$\frac{\partial \Psi}{\partial x} = ik\Psi, \qquad \frac{\partial \Psi}{\partial t} = -\frac{i\hbar k^2}{2M}\Psi, \qquad \frac{\partial^2 \Psi}{\partial x^2} = -k^2\Psi, \qquad \frac{\partial^2 \Psi}{\partial t\, \partial x} = \frac{\hbar k^3}{2M}\Psi,$$

$$\frac{\partial^3 \Psi}{\partial x^3} = -ik^3\Psi, \qquad \frac{\partial^3 \Psi}{\partial t\, \partial x^2} = \frac{i\hbar k^4}{2M}\Psi, \qquad \frac{\partial^4 \Psi}{\partial x^4} = k^4\Psi, \ldots.$$

If (7) contains derivatives that bring out different powers of k, then substitution of the plane wave solution yields an algebraic equation for k, and only the discrete values of $p = \hbar k$ so determined would be possible. Our experience, however, is that an electron gun can be adjusted to give any momentum. Further, if only discrete values of momentum were possible, Galilean invariance would be lost: Given any value of momentum, any other value can be generated by looking at the electron from the point of view of a moving observer. We conclude that the equation may contain only derivatives that all bring out of a plane wave the same power of k; it must be either

$$f_{01}\frac{\partial \Psi}{\partial x} = 0, \qquad \text{or}$$

$$f_{10}\frac{\partial \Psi}{\partial t} + f_{02}\frac{\partial^2 \Psi}{\partial x^2} = 0, \qquad \text{or}$$

$$f_{11}\frac{\partial^2 \Psi}{\partial t\, \partial x} + f_{03}\frac{\partial^3 \Psi}{\partial x^3} = 0, \qquad \text{or} \ldots.$$

The first try, $\partial\Psi/\partial x = 0$, is useless; $\Psi = \Psi(t)$ only is not acceptable. The others can all be summarized by

$$\frac{\partial^n}{\partial x^n}\left\{f_{1n}\frac{\partial}{\partial t} + f_{0n+2}\frac{\partial^2}{\partial x^2}\right\}\Psi = 0.$$

The ratio of f_{1n} to f_{0n+2} is fixed by requiring that the plane wave be a solution:

$$\frac{\partial^n}{\partial x^n}\left\{f_{1n}\left(\frac{-i\hbar k^2}{2M}\right) + f_{0n+2}(-k^2)\right\}Ae^{i[kx-(\hbar k^2/2M)t]} = 0,$$

$$\frac{f_{0n+2}}{f_{1n}} = \frac{-i\hbar}{2M}.$$

Because the equation is linear and homogeneous, only the ratio can be determined, and it will turn out to be convenient to write

$$\frac{\partial^n}{\partial x^n}\left\{i\hbar\frac{\partial}{\partial t} + \frac{\hbar^2}{2M}\frac{\partial^2}{\partial x^2}\right\}\Psi = 0.$$

The $n = 0$ choice,

$$-\frac{\hbar^2}{2M}\frac{\partial^2\Psi}{\partial x^2} = i\hbar\frac{\partial\Psi}{\partial t},$$

gives an equation satisfied by any wave packet in free space. It is the simplest choice and easiest to use; it is, in one dimension, the Schrödinger equation for zero potential energy.

PROBLEM 1.7

Study the $n \neq 0$ equations. Do they make sense? Can you exclude them on the basis of any experiment or reasonableness requirement? Notice that any solution of the $n = 0$ equation is also a solution of the $n \neq 0$ equations, but not vice versa. You should therefore think about the additional solutions that $n \neq 0$ permits.

In three dimensions a plane wave describing definite momentum and definite energy is given by

$$\Psi(\vec{r}, t) = Ae^{i(\vec{k}\cdot\vec{r}-\omega t)} = Ae^{i(\vec{p}\cdot\vec{r}-Et)/\hbar}, \tag{1.4-8}$$

where

$$E = \frac{p^2}{2M} = \frac{p_x^2 + p_y^2 + p_z^2}{2M} = \frac{\hbar^2}{2M}(k_x^2 + k_y^2 + k_z^2).$$

For this wave function,

$$\frac{\hbar}{i}\frac{\partial}{\partial x}\Psi = p_x\Psi, \qquad \frac{\hbar}{i}\vec{\nabla}\Psi = \vec{p}\Psi,$$

$$-\frac{\hbar^2}{2M}\nabla^2\Psi = \frac{p^2}{2M}\Psi, \qquad i\hbar\frac{\partial\Psi}{\partial t} = E\Psi.$$

The obvious three-dimensional statement of the free-particle Schrödinger equation is

$$-\frac{\hbar^2}{2M}\nabla^2\Psi = i\hbar\frac{\partial\Psi}{\partial t}. \tag{1.4-9}$$

This equation says, "The $p^2/2M$ operator, $-(\hbar^2/2M)\nabla^2$, operates on the wave-function Ψ and produces the same thing as $i\hbar\,\partial/\partial t$ operating on Ψ."

The free-particle Schrödinger equation and a general wave packet

$$\Psi(\vec{r}, t) = \int_{-\infty}^{+\infty}\int_{-\infty}^{+\infty}\int_{-\infty}^{+\infty} dk_x\,dk_y\,dk_z\,A(\vec{k})e^{i(\vec{k}\cdot\vec{r}-\hbar k^2 t/2M)} \tag{1.4-10}$$

contain the same information. To construct a useful theory, we must now introduce interactions.

PROBLEM 1.8

Verify formally that (10) satisfies Schrödinger's equation (9). Be pedantic; quote relevant theorems about things like exchange of order of differentiation and integration.

Suppose that the interactions can be described by a scalar potential energy $V(\vec{r})$. Look first at a region in which the potential energy is a constant V_0. There the kinetic energy is

$$\frac{p^2}{2M} = E - V_0. \tag{1.4-11}$$

The origin of potential energy is arbitrary; only the difference between the total energy E and the potential energy is measurable. In the region of constant potential, (8) should still be a solution if the magnitude of momentum p is given not by $\sqrt{2ME}$, but by

$$\sqrt{2M(E - V_0)}.$$

Just as above,

$$-\frac{\hbar^2}{2M}\nabla^2 A e^{i(\vec{p}\cdot\vec{r}-Et)/\hbar} = \frac{p^2}{2M} A e^{i(\vec{p}\cdot\vec{r}-Et)/\hbar},$$

$$i\hbar\frac{\partial}{\partial t} A e^{i(\vec{p}\cdot\vec{r}-Et)/\hbar} = E A e^{i(\vec{p}\cdot\vec{r}-Et)/\hbar},$$

but $E \neq p^2/2M$; the wave no longer satisfies (9). The repair is easy. Instead of (9), write

$$-\frac{\hbar^2}{2M}\nabla^2\Psi = i\hbar\frac{\partial\Psi}{\partial t} - V_0\Psi, \tag{1.4-12}$$

and the connection (11) between kinetic, total, and potential energies is obeyed. Suppose next that the potential is not quite constant, but that it varies very slowly, in the sense discussed at the beginning of this section. Then the wavelength $\lambda(\vec{r})$ and the magnitude of momentum $p(\vec{r})$ are given by (1). In each neighborhood the difference between a truly constant potential energy V_0 and a sufficiently slowly varying $V(\vec{r})$ must be undetectable, and (12) can safely be replaced by

$$-\frac{\hbar^2}{2M}\nabla^2\Psi + V(\vec{r})\Psi = i\hbar\frac{\partial\Psi}{\partial t}. \tag{1.4-13}$$

Now the question of rapidly varying potential energies must be faced. It is such potentials that made quantum theory necessary; classical mechanics is usually adequate if $V(\vec{r})$ varies slowly. There are many things that one might try at this point, such as various fancy operators that reduce to $V(\vec{r})$ in the classical limit. Once more, we choose the road of greatest simplicity, and *assume that* (13) *is generally valid, regardless of the behavior of* $V(\vec{r})$. "Assume," of course, means that we shall try it, and, if it gives useful answers, retain it.

It does give useful answers, and is the celebrated Schrödinger equation. It is the most simple construction that fits the observations that have been adduced in this chapter. It says, "The kinetic energy operator $-(\hbar^2/2M)\nabla^2$ plus the potential energy operator $V(\vec{r})$ applied to Ψ equals $i\hbar\, \partial/\partial t$ applied to Ψ." Note that it is acceptable to call $V(\vec{r})$ an operator. A multiplying function is a perfectly good operator; it just happens to be a very simple one. Anything that, when applied to a Ψ, produces another function is called an operator. A compact way of writing (13) is

$$H\Psi = i\hbar\frac{\partial}{\partial t}\Psi, \tag{1.4-14}$$

where

$$H = -\frac{\hbar^2}{2M}\nabla^2 + V(\vec{r}) \tag{1.4-15}$$

represents $p^2/2M + V(\vec{r})$, the sum of the kinetic and potential energies. In classical mechanics this expression is often called the *Hamiltonian*, and the name has been taken over into quantum physics.

Everything in this section should be seen not as derivation, but as reasonable guides. Note particularly the way the potential energy function $V(\vec{r})$ was introduced. Classically $V(\vec{r})$ is defined as the work needed to bring the particle from a reference position, usually infinite distance, to $\vec{r}$, with momentum equal to zero initially and finally. The uncertainty principle forbids such a determination of $V(\vec{r})$ in the quantum domain. Nevertheless, one studies the hydrogen atom by writing the Schrödinger equation with $V = -e^2/4\pi\varepsilon_0 r$. Where did that really come from? An adequately sophisticated view might be the following.

There is a function $V(\vec{r})$ whose classical limit is the potential energy. We have to use our experience and our cunning to guess $V(\vec{r})$, and we have guessed right when the Schrödinger equation gives right answers. It is very interesting that the classical potential energy applicable to a large-scale model of the quantum system works for a variety of problems.

The Coulomb potential gives accurate results for the hydrogen atom. If magnetic effects are important, a suitable introduction of the classical formalism for magnetic fields gives the right answers. It follows that, for nonrelativistic systems, one should insert the classical descriptions of electromagnetic interactions in the Schrödinger equation.

There is not much experience regarding the effect of gravity on quantum phenomena. The gravitational interaction is much weaker than the other interactions between particles. It is seldom significant except for large and electrically nearly neutral systems. It has however been verified[14] that the wavelength of neutrons is given by the expression (1) when $V(\vec{r})$ is the classical gravitational potential energy.

What about the nuclear domain? Since the range of $V(\vec{r})$ is only about 10^{-13} cm, there is no classical guidance in selecting the form of the potential. For nuclear problems we can only choose functional forms and adjust parameters until the Schrödinger equation yields correct answers, and $V(\vec{r})$ cannot be deduced from or interpreted as a classical potential energy.

1.5 *Probability Density, Current, and Conservation*

In the quantum domain the uncertainty principle makes it impossible to know the precise trajectories of particles. What can be known is the *probability* that a particle is in one or another region, and it is useful to define a *probability density* $\rho(\vec{r}, t)$. Look at an infinitesimal box with volume $dx\,dy\,dz \equiv d\tau$ that lies between x and $x + dx$, y and $y + dy$, z and $z + dz$.

$$\rho(x, y, z, t)dx\,dy\,dz \equiv \rho(\vec{r}, t)d\tau$$

is the probability that the particle is in the little box at time t. In other words, $\rho(\vec{r}, t)$ is the probability per unit volume evaluated at $\vec{r}$ at the time t. It satisfies the condition

$$\iiint \rho(\vec{r}, t)d\tau = 1 \tag{1.5-1}$$

if at time t there definitely is one particle somewhere. The integral must be taken at least over all regions where the particle might be. Since ρ is zero wherever the particle cannot go, it is correct, and often useful, to integrate between infinite limits.

PROBLEM 1.9

A train goes back and forth between two towns separated by 60 km. It spends negligible time in the two stations. It accelerates uniformly until it is

[14] R. Colella, A. W. Overhauser, and S. A. Werner, *Phys. Rev. Lett.* **34**, 1472 (1975).

halfway between the towns and then decelerates uniformly until it reaches its destination.

a. Plot as a function of distance the time-averaged probability density $\rho(x)$ for finding the front of the train at x. Since there is always one train, the solution must satisfy

$$\int_0^{60\,\mathrm{km}} \rho(x)dx = 1.$$

b. At an unspecified time, 10 km from one of the towns, a 0.020-km-long bridge collapses. What is the probability that the front of the train is on the bridge at the moment of collapse?

c. There is a causeway between the $x = 15$ km and $x = 30$ km points of the trip. At an unspecified time the entire causeway collapses. What is the probability that the front of the train is on it at the moment of collapse?

Section 1.2 presents evidence that the probability density for finding the particle is proportional to $\Psi^*\Psi \equiv |\Psi|^2$, and we take

$$\rho(\vec{r}, t) = |\Psi(\vec{r}, t)|^2. \tag{1.5-2}$$

In going from the observation that $|\Psi|^2$ is *proportional* to ρ to the requirement that it is *equal* to ρ, something additional has, of course, been imposed. If there is one particle in the apparatus at time t_0, then (2) is consistent with (1) if

$$\iiint |\Psi(\vec{r}, t_0)|^2\, d\tau = 1. \tag{1.5-3}$$

This requirement can easily be met. If a wave function $\Psi_u(\vec{r}, t_0)$ does *not* satisfy (3),

$$\iiint d\tau |\Psi_u(\vec{r}, t_0)|^2 = K^2(t_0) \neq 1,$$

replace it by

$$\Psi(\vec{r}, t_0) \equiv \frac{e^{i\delta}}{K(t_0)} \Psi_u(\vec{r}, t_0),$$

where δ is any real constant. The function Ψ then does satisfy (3), and it is *normalized*, while Ψ_u is *unnormalized.* The Schrödinger equation is linear and homogeneous so that, from condition (1.4-6), Ψ is a solution if Ψ_u is a solution.

Suppose that Ψ is normalized at t_0, what will happen later? If particles are neither added nor removed, then

$$\iiint d\tau |\Psi(\vec{r}, t)|^2 = 1 \tag{1.5-4}$$

must be true at *all* times, not just at $t = t_0$; *probability must be conserved.* To see that it is, examine the time rate of change of $|\Psi(\vec{r}, t)|^2 \equiv \rho(\vec{r}, t)$.

$$\frac{\partial \rho}{\partial t} = \left(\frac{\partial \Psi^*}{\partial t}\right)\Psi + \Psi^*\left(\frac{\partial \Psi}{\partial t}\right).$$

Use the Schrödinger equation (1.4-13),

$$\frac{\partial \Psi}{\partial t} = \frac{1}{i\hbar}\left(-\frac{\hbar^2}{2M}\nabla^2\Psi + V\Psi\right).$$

$$\frac{\partial \rho}{\partial t} = -\frac{\hbar}{2iM}[\Psi^*(\nabla^2\Psi) - (\nabla^2\Psi^*)\Psi] + \frac{\Psi^*(V\Psi)}{i\hbar} - \frac{(V^*\Psi^*)\Psi}{i\hbar}. \tag{1.5-5}$$

The usual potential energy V is a real multiplying function and gives cancellation of the last two terms. The first two terms can be rewritten to yield

$$\frac{\partial \rho}{\partial t} = -\vec{\nabla}\cdot\frac{\hbar}{2iM}[\Psi^*(\vec{\nabla}\Psi) - (\vec{\nabla}\Psi^*)\Psi].$$

Therefore if the *current density* is defined by

$$\vec{j} \equiv \frac{\hbar}{2iM}[\Psi^*(\vec{\nabla}\Psi) - (\vec{\nabla}\Psi^*)\Psi], \tag{1.5-6}$$

there results a *conservation equation*

$$\frac{\partial \rho}{\partial t} + \vec{\nabla}\cdot\vec{j} = 0. \tag{1.5-7}$$

This equation appears whenever something is described by a density ρ (so much per unit volume), a current density $\vec{j}$ (so much per unit time per unit area perpendicular to the flow), and conservation, in the sense that our something is neither created nor destroyed. Equation (7) applies not only to quantum phenomena, but also to the flow of electric charge,[15] to gas flow,[16,17] and to traffic flow.[18] If you have not done so, derive this equation by thinking about a small volume $\Delta x\ \Delta y\ \Delta z$ and equating the time rate of increase of stuff inside to the net flow across the six faces.[16]

Look at the structure of the current as given by (6). If Ψ is a real function, then $\vec{j}$ must be zero. As anticipated in Section 1.2, the introduction of a complex

[15] E. M. Purcell, *Electricity and Magnetism, Berkeley Physics Course*, Vol. 2 (New York: McGraw-Hill Book Company, 1965), Sec. 4.2.

[16] T. C. Bradbury, *Theoretical Mechanics* (New York: John Wiley & Sons, Inc., 1968), Secs. 2.7, 2.8.

[17] K. R. Symon, *Mechanics* 3d ed. (Reading, Mass,: Addison-Wesley Publishing Co., Inc., 1971), Secs. 8.6–8.9.

[18] D. C. Gazis, "Traffic Flow and Control: Theory and Applications," *American Scientist* **60**, 414–424 (1972).

Ψ is not gratuitous cruelty, but is necessary: A net transport of probability cannot be described by a real Ψ.

If (7) is satisfied, then the preservation of normalization is assured.

$$\frac{d}{dt}\iiint \Psi^*\Psi\, d\tau = \frac{d}{dt}\iiint \rho\, d\tau = \iiint \frac{\partial \rho}{\partial t}\, d\tau$$

because the limits of integration are held fixed. Therefore

$$\frac{d}{dt}\iiint \Psi^*\Psi\, d\tau = -\iiint \vec{\nabla}\cdot\vec{j}\, d\tau = -\iint \vec{j}\cdot d\vec{S},$$

where in the last step Gauss' divergence theorem[15] has been used to express the result as an integral over the surface bounding the volume, with surface element $d\vec{S}$. If this surface is taken so large that it encloses all of the apparatus, then particles cannot reach the surface, $\vec{j}$ is zero there, and

$$\frac{d}{dt}\iiint \Psi^*\Psi\, d\tau = 0. \tag{1.5-8}$$

If the normalization integral is unity at one time, it will remain unity at all times.

Section 1.4 gives arguments against the Schrödinger equation having time derivatives higher than the first. The discussion there regarding conservation can now be made more concrete. As is pointed out in Section 1.4, the occurrence of a $\partial^2\Psi/\partial t^2$ would permit the specification of $\partial\Psi/\partial t$ as an initial condition, so that $\partial\rho/\partial t$ at $t = 0$ could be given any value. It is now clear that such a freedom would permit violation of the conservation equation (7) no matter what functional form is used for $\vec{j}$.

EXAMPLE

Calculate $\vec{j}$ for $\Psi = Ae^{i(\vec{p}\cdot\vec{r}-Et)/\hbar}$. Evaluate all constants in Ψ if it describes 1.0 milliamp/cm^2 of 2.0 keV electrons that are traveling in the $x - y$ plane at $+30°$ with respect to the x axis.

SOLUTION

$$\vec{\nabla}e^{i(\vec{p}\cdot\vec{r}-Et)/\hbar} = \left(\hat{e}_x\frac{\partial}{\partial x} + \hat{e}_y\frac{\partial}{\partial y} + \hat{e}_z\frac{\partial}{\partial z}\right)e^{i(p_x x + p_y y + p_z z - Et)/\hbar}$$

$$= \frac{i}{\hbar}(\hat{e}_x p_x + \hat{e}_y p_y + \hat{e}_z p_z)e^{i(p_x x + p_y y + p_z z - Et)/\hbar} = \frac{i\vec{p}}{\hbar}e^{i(\vec{p}\cdot\vec{r}-Et)/\hbar}.$$

$$\therefore\quad \vec{j} = \frac{\hbar}{2Mi}[\Psi^*\vec{\nabla}\Psi - (\vec{\nabla}\Psi^*)\Psi]$$

$$= \frac{\hbar}{2Mi}\left[\Psi^*\left(\frac{i\vec{p}}{\hbar}\right)\Psi - \left(-\frac{i\vec{p}}{\hbar}\right)\Psi^*\Psi\right] = \frac{\vec{p}}{M}\Psi^*\Psi = \vec{v}\rho,$$

where $\vec{v} \equiv \vec{p}/M$, and $\rho = \Psi^*\Psi = A^*A$. Next, choose E, $\vec{p}$, and A to match the given conditions. Since 2.0 keV/Mc^2 = 2.0 keV/511 keV $\simeq$ 0.004, nonrelativistic mechanics can be used if about 0.4% accuracy is sufficient.

$$p = \sqrt{2ME} = \sqrt{2Mc^2E}/c$$

$$= \sqrt{2 \times 511{,}000 \text{ eV} \times 2000 \text{ eV}}/c = 0.0452 \text{ MeV}/c.$$

$$\therefore \quad \frac{p}{\hbar} = \frac{0.0452 \text{ MeV}/c}{1.973 \times 10^{-11} \text{ MeV cm}/c} = 2.29 \times 10^9/\text{cm}.$$

$$\therefore \quad \frac{\vec{p}\cdot\vec{r}}{\hbar} = \frac{p}{\hbar}(\cos 30^\circ x + \cos 60^\circ y + \cos 90^\circ z)$$

$$= 2.29 \times 10^9/\text{cm}\ (0.866x + 0.500y).$$

$$\frac{E}{\hbar} = \frac{2000 \text{ eV}}{6.58 \times 10^{-16} \text{ eV sec}} = 3.04 \times 10^{18}/\text{sec}.$$

$$v = \frac{p}{M} = \frac{0.0452 \text{ MeV}/c}{0.511 \text{ MeV}/c^2} = 0.0885c = 2.65 \times 10^9 \text{ cm/sec}.$$

The current density

$$j = \frac{10^{-3} \text{ coul/sec cm}^2}{1.602 \times 10^{-19} \text{ coul/electron}} = 6.24 \times 10^{15} \text{ electrons/sec cm}^2.$$

Since

$$j = \rho v = |A|^2 v, \qquad |A| = \sqrt{\frac{j}{v}} = 1.53 \times 10^3/\text{cm}^{3/2}.$$

$$\therefore \quad \Psi(\vec{r}, t) = 1.5 \times 10^3/\text{cm}^{3/2} \exp i[2.3 \times 10^9/\text{cm}\ (0.87x + 0.50y) - 3.0 \times 10^{18} t/\text{sec}]$$

times any arbitrary phase factor $e^{i\delta}$, δ any real number.

PROBLEM 1.10

Protons with 1.0 MeV kinetic energy are traveling in the $+x$ direction. A 20-cm^2 counter, with its face perpendicular to the x axis, gives 5000 counts/sec. Obtain a wave function that describes these circumstances.

We have here a formalism in which particles are neither created nor destroyed. Such a formalism is entirely appropriate for electrons in the domain of atomic physics, where they may make transitions from one atomic state to another, and where they may be transferred from one atom to another, but where the total number of electrons remains constant. Also, in the domain of low-energy nuclear physics, there is conservation of nucleons; the total number of neutrons and protons is conserved. However, there are processes such as

$$\gamma \to e^+ + e^-, \tag{1.5-9}$$

$$\pi^0 \to 2\gamma, \tag{1.5-10}$$

$$n \to p^+ + e^- + \bar{\nu}_e, \tag{1.5-11}$$

and hundreds of other reactions in which conservation in any direct sense is violated. How do such particle reactions fit into our story? They don't: We are looking at nonrelativistic quantum physics, a domain restricted to energy exchanges that are small compared to the rest energies involved. This is precisely the restriction that is violated when particle transmutation occurs. In reaction (9), the gamma ray must have an energy greater than twice the rest energy of an electron. The two gamma rays in (10) carry off all of the rest energy of the neutral pion. In (11) the energy change is small compared to the rest energy of the nucleon, which is conserved, but exceeds that of the electron, which is not conserved. Notice that the theory of photons has to invoke much more than some modest modification of the Schrödinger equation. Since photons can have arbitrarily small energies, they can be created and destroyed in arbitrarily low energy processes, and they are not conserved if they interact at all.

Although the topic is outside the principal concerns of this text, it is important to realize that, even for relativistic particle reactions, various conservation laws are obeyed. For example, with suitably defined bookkeeping, there appears to be conservation of the total number of particles in each of several broad categories. Another conservation law that seems always to be obeyed is that for electric charge; no one has found reactions in which the total charge before some collision differs from the total after the collision. For example, for center-of-mass energies $> M_\pi c^2$, there is a good yield of

$$n + p \to n + n + \pi^+, \qquad n + p \to n + p + \pi^0, \qquad n + p \to p + p + \pi^-,$$
$$n + n \to n + n + \pi^0, \qquad n + n \to n + p + \pi^-,$$
$$p + p \to n + p + \pi^+, \qquad p + p \to p + p + \pi^0,$$

but none for two neutrons going to two neutrons and a positive pion;

$$n + n \not\to n + n + \pi^+.$$

If word arrives tomorrow that a charge-disconserving particle reaction has been discovered, a profound modification of physics would have to be made. Here is a way of indicating the extent of the upset.

Suppose that the particle reaction $n + n \to n + n + \pi^+$ is found. We can then build a perpetual motion machine (Figure 1.5-1). Enclose the particle

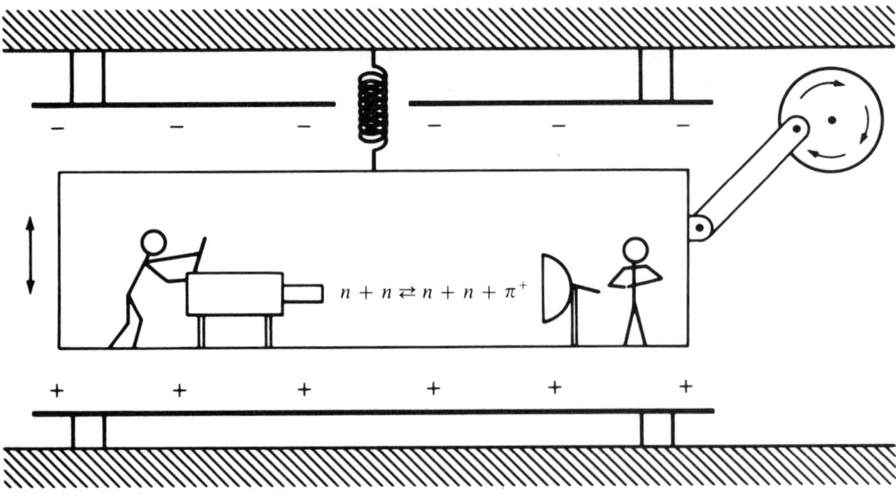

Fig. 1.5-1. A perpetual motion machine that works if charge is not conserved.

accelerator, its power source, and surrounding laboratories in a big Faraday cage. Make the reaction, catch the products, bring them back together, and make the inverse of the reaction. By being careful and efficient, return the entire system to its initial state without importing any additional energy into the Faraday cage. Of course, we have to be careful of everything, and, for example, collect and reuse the kinetic energy of the reaction products. Now go back and forth repeatedly; $n + n \rightleftarrows n + n + \pi^+$. Since we are very efficient, it costs nothing. Next, suspend the Faraday cage from a spring between charged capacitor plates. Since the charge inside the cage changes, the charge on the outside of the cage changes and the cage is pushed up and down. Attach a crankshaft made of nonconducting material and extract the energy. In other words, if we believe in microscopic reversibility (as much energy is needed to produce $n + n \to n + n + \pi^+$ as is released by $n + n + \pi^+ \to n + n$) and gauge invariance (the events in the lab are not influenced by the absolute potential of the cage), then nonconservation of charge implies nonconservation of energy.[19] It follows that (7) is strictly valid for electric charge in all domains of physics—at least until someone does an experiment that proves that it is wrong.

1.6 *Could Things Be Different?*

Section 1.1 shows that very general considerations require the structure of Newtonian mechanics to be what it is. Given the existence of inertial frames

[19] P. Morrison, *Am. J. Phys.* **26**, 358 (1958).

and the arbitrarily accurate measurability of positions and times, one can define mass, observe some general properties of masses, and complete the structure with a definition of force.

Is the structure of nonrelativistic quantum theory determined uniquely by the general characteristics of its domain? The answer depends on the rules that we set for ourselves, and many modifications have been explored. There is, for example, a large and swampy literature on attempts to make nonlinear theories consistent with what is known. However, if we demand the most simple theory that fits the observations, it is hard to avoid the Schrödinger equation. Superposition is guaranteed only by linear and homogeneous equations. The requirement that $|\Psi|^2$ shall be the probability density in the spirit of wave-particle duality, and that the most simple story that satisfies the correspondence principle should be used, led directly to the Schrödinger equation.

PROBLEM 1.11

Here is a rainy-Sunday-afternoon entertainment masquerading as a problem: Accept the need for a linear homogeneous equation and for the correspondence principle. Cook up an additional term for the Schrödinger equation that is dimensionally right, that doesn't spoil things we don't want spoiled, and that gives different results far from the classical limit.

A fundamentally different alternative to the usual form of quantum theory is to be found in the so-called *hidden-variables theories*.[20] In these theories the apparent indeterminacy is the result of our failure to deal with as-yet-undiscovered internal coordinates. A good analogy is the classical physicists' view of pressure fluctuations on a small area in the wall of a bottle filled with gas. There the Newtonian view is that one could in principle predict how many molecules will bounce how hard against the area, but that the complexity of the problem makes it useful to discuss only statistical quantities such as means and root-mean-squared deviations from these means: The statistical view is a matter of practicality, not of fundamentals. Similarly, it is suggested that there are variables which might or might not become measurable some day, which do determine exactly where each electron goes in a diffraction experiment, and that quantum fluctuations are merely the result of our ignorance of these variables.

At present there are no logical or experimental reasons for adopting or rejecting hidden-variables theories, and attitudes toward these theories are often influenced substantially by individual tastes and ideologies.[21] The efforts

[20] F. J. Belinfante, *A Survey of Hidden Variables Theories* (Oxford: Pergamon Press, 1973).

[21] It would be an interesting sociology research project to determine whether, for example, Marxists and Royalists tend to prefer deterministic hidden-variables theories.

of the hidden-variables partisans have forced the clarification of the foundations of quantum theory. However, the majority view is that until an experiment changes our minds, it is best to view quantum indeterminacy as ultimate and intrinsic.

1.7 *Additional Chapter 1 Problems*

PROBLEM 1.12

Section 1.2 mentions diffraction experiments that were done with He atoms in 1932. The "diffraction grating" was a crystal. Roughly what energy, in eV, should the He atoms have to give a good pattern? At what temperature is this energy the thermal energy? Design a device that will give a He atom beam of the desired energy, and a device that will detect the scattered He atoms. Then look up designs that have been used, in Reference 3 or in more recent discussions of atomic beam techniques.

PROBLEM 1.13

Figure 1.2-4 shows an arrangement for localizing an electron. Suppose that ε, the half-angle of the aperture, is 30°, that the photons are x-rays with $\lambda = 2.0$ Å, and that the kinetic energy of the electrons before detection is 100 eV. Sketch roughly but with some attention to scale the wave function of an electron

a. *before* it has been detected, and

b. immediately *after* it has been detected.

PROBLEM 1.14

An electron counter of area a is set up with its face parallel to the z axis, and with the perpendicular to its face making an angle ϕ with the x axis. The electron wave function is given by $\Psi = A \exp i(\vec{k} \cdot \vec{r} - \omega t)$. Obtain the average number of counts/sec.

PROBLEM 1.15

In Section 1.3 the energy of the hydrogen atom ground state is estimated through an application of the uncertainty principle. Use the same idea to estimate the ground state energy for a spherically symmetric harmonic oscillator, that is, for $V = Kr^2/2$.

PROBLEM 1.16

A beam of 1.0 eV electrons is collimated by a slit and strikes a screen 1.0 meter beyond the slit. Estimate roughly the smallest width that can be achieved for the strip within which most of the electrons strike the screen.

PROBLEM 1.17

Equation (1.5-8) shows that the normalization integral is a constant for any solution to the Schrödinger equation. Suppose now that Ψ_1 and Ψ_2 are two *different* solutions to the same Schrödinger equation. What can be said about

$$\frac{d}{dt}\iiint \Psi_1^*\Psi_2 \, d\tau?$$

A clever way to deal with this question is to recognize that any linear combination of Ψ_1 and Ψ_2 is a Ψ that satisfies (1.5-8).

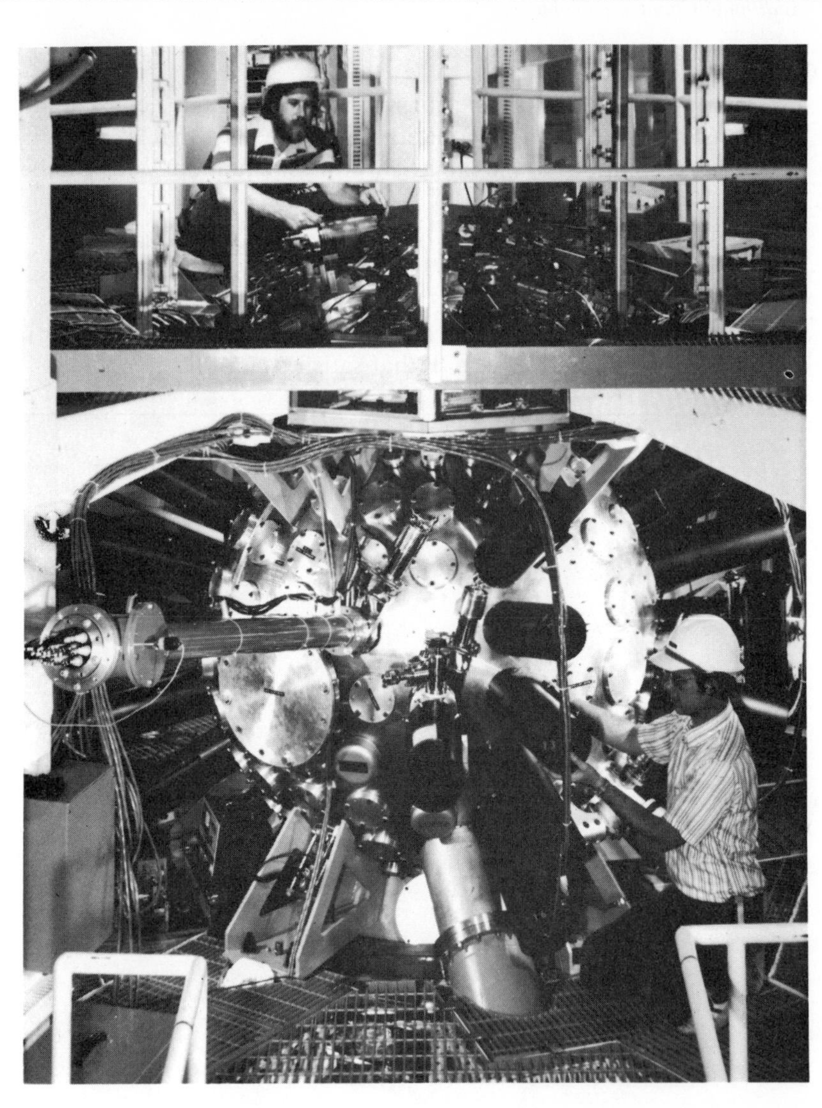

A laser fusion experiment. Twenty large lasers, located in long halls on two sides of the target chamber region shown, are designed to deliver more than 10^{13} watts in short pulses to the center of the chamber. The power is focused on a small deuterium–tritium fuel pellet, and compresses and heats it so that fusion can occur. The system is described in Physics Today, *April 1978, pp. 17–21. A high temperature is needed to give the deuterons and tritons an appreciable chance of coming within each others' nuclear force ranges. The Coulomb repulsion acts as a barrier between the deuterons and tritons, and the probability of penetration of this barrier is a rapidly varying, roughly exponential, function of the energy; see Sections 2.3 and 10.3.* (*Courtesy of Lawrence Livermore Laboratory*)

2

Simple Systems

Whether you know it or not, the first chapter is very hard. It is qualitative, but it is loaded with general arguments and hints that become easier to appreciate after one has lived with quantum ideas for a while and has done some detailed calculations. Promise yourself that you will read it again in a few months.

There are more generalities and formalism ahead. However, what is needed now is a look at a few simple solutions to the Schrödinger equation to give concreteness to these discussions.

2.1 *Separation of the Schrödinger Equation*

Whenever possible, partial differential equations are subdued by separating them into ordinary differential equations, one for each of the variables. Let us begin by looking for a solution of the form

$$\Psi(\vec{r}, t) = \psi(\vec{r}) f(t). \tag{2.1-1}$$

Capital Ψ will be used for wave functions that do include time dependence, and lowercase ψ will be used for those that do not. Assume $V = V(\vec{r})$ only. Substitute into the Schrödinger equation and divide by Ψ:

$$\frac{1}{\psi(\vec{r})}\left[-\frac{\hbar^2}{2M}\nabla^2\psi(\vec{r}) + V(\vec{r})\psi(\vec{r})\right] = \frac{1}{f(t)}\, i\hbar\frac{df(t)}{dt}.$$

The left side depends only on $\vec{r}$ and the right side depends only on t. Since $\vec{r}$ and t are independent variables, the two sides can be equal only if they are both equal to a constant, here called E with forethought. There are then two equations,

$$i\hbar\frac{df(t)}{dt} = Ef(t), \tag{2.1-2}$$

$$-\frac{\hbar^2}{2M}\nabla^2\psi(\vec{r}) + V(\vec{r})\psi(\vec{r}) = E\psi(\vec{r}). \tag{2.1-3}$$

Calling the separation constant E was a good idea; (2) is solved by

$$f(t) = Ae^{-iEt/\hbar}, \tag{2.1-4}$$

which again yields the time variation obtained in Section 1.2, with E identified as the total energy.

Equation (3) is called the time-independent Schrödinger equation and says that the kinetic plus potential energies operating on ψ produce the total energy multiplying ψ. Both (2) and (3) are *eigenvalue* equations of the form:

Operator operating on a function = constant multiplying the function. (2.1-5)

The constant, in this case E, is the *eigenvalue*, and the function that satisfies the equation is the *eigenfunction.*

The separation turns up a definite constant E and produces a state of definite energy. In general, one may well be dealing with a state, such as the wave packet (1.4-10), which involves the superposition of many energies. It is important to remember that asking for the form (1) means asking for a very special case. However, it turns out that the most general solution can be written as the sum of terms like (1). For example, the expression (1.4-10) for a three-dimensional wave packet in a $V(\vec{r}) = 0$ region, written in terms of $\vec{p}$ rather than $\vec{k}$,

$$\Psi(\vec{r}, t) = \iiint g(\vec{p})e^{i(\vec{p}\cdot\vec{r} - Et)/\hbar}\, d^3p, \tag{2.1-6}$$

can be read in this spirit. To each $\psi(\vec{r}) = e^{i\vec{p}\cdot\vec{r}/\hbar}$, there belongs a time dependence

$$f(t) = e^{-iEt/\hbar} = e^{-i(p^2/2M)t/\hbar}.$$

Each such pair is multiplied by a constant coefficient $g(\vec{p})$, and the integration adds all contributions to give the total wave function.

Whenever the energy is definite, so that

$$\Psi(\vec{r}, t) = \psi(\vec{r})e^{-iEt/\hbar},$$

the function of position $\psi(\vec{r})$ is the envelope for the complex oscillation $e^{-iEt/\hbar}$. The result is a standing wave, or normal mode. The probability density,

$$|\Psi(\vec{r}, t)|^2 = |\psi(\vec{r})|^2 e^{+iEt/\hbar} e^{-iEt/\hbar} = |\psi(\vec{r})|^2,$$

is independent of the time, and such states are often called *stationary states.* The conservation equation (1.5-7) becomes

$$\vec{\nabla} \cdot \vec{j} = 0,$$

that is, in Cartesian coordinates,

$$\frac{\partial j_x}{\partial x} + \frac{\partial j_y}{\partial y} + \frac{\partial j_z}{\partial z} = 0.$$

Suppose, for example, that $\vec{j}$ is everywhere in the x direction; $j_y = j_z = 0$. Then $\partial j_x/\partial x = 0$ and j_x is constant. If it were true that $j_x(x_1) > j_x(x_2)$, $x_1 < x_2$, then the probability of finding the particle between x_1 and x_2 would increase in time, $\partial\rho/\partial t$ would be positive there, and the state would not be stationary.

Now look at the time-independent Schrödinger equation (3). It still involves three spatial dimensions, and we want to break it down into three ordinary differential equations. The form of $V(\vec{r})$ determines whether this second separation is possible. If $V(\vec{r})$ can be written as the sum of three terms, each of which depends on only one spatial coordinate, we have a chance; if not, we don't. The first step is therefore to look for a coordinate system which fits the problem. For example, the Coulomb potential can be written in spherical polar coordinates as $V_r(r) + V_\theta(\theta) + V_\phi(\phi)$, with $V_r(r) = q_1q_2/4\pi\varepsilon_0 r$, $V_\theta(\theta) = 0$, and $V_\phi(\phi) = 0$. With Cartesian coordinates, $V(\vec{r}) \sim q_1q_2(x^2 + y^2 + z^2)^{-1/2}$, and separation is impossible. The coordinate system must match the symmetry of the problem, even though there is often the price, as in going from Cartesians to spherical polars, of a more complicated ∇^2.

Consider first a potential that can be separated in Cartesians:

$$V(x, y, z) = V_x(x) + V_y(y) + V_z(z).$$

Substitute into (3) the trial form $\psi(x, y, z) = \psi_x(x)\psi_y(y)\psi_z(z)$, and then divide by ψ:

$$\frac{1}{\psi_x}\left[-\frac{\hbar^2}{2M}\frac{d^2\psi_x}{dx^2} + V_x\psi_x\right] + \frac{1}{\psi_y}\left[-\frac{\hbar^2}{2M}\frac{d^2\psi_y}{dy^2} + V_y\psi_y\right] + \frac{1}{\psi_z}\left[-\frac{\hbar^2}{2M}\frac{d^2\psi_z}{dz^2} + V_z\psi_z\right] = E.$$

The first term depends on only x, the second on only y, and the third on only z. Their sum equals the constant E. Therefore since x, y, and z are independent variables, each must equal a constant. Call the three constants E_x, E_y, and E_z. There are then three equations

$$
\begin{aligned}
-\frac{\hbar^2}{2M}\frac{d^2\psi_x}{dx^2} + V_x\psi_x &= E_x\psi_x, \\
-\frac{\hbar^2}{2M}\frac{d^2\psi_y}{dy^2} + V_y\psi_y &= E_y\psi_y, \\
-\frac{\hbar^2}{2M}\frac{d^2\psi_z}{dz^2} + V_z\psi_z &= E_z\psi_z,
\end{aligned}
\tag{2.1-7}
$$

subject to the condition

$$E_x + E_y + E_z = E. \tag{2.1-8}$$

Look at the trivial case $V_x = V_y = V_z = 0$. The x equation is solved by

$$\psi_x = Ae^{+ip_x x/\hbar} + Be^{-ip_x x/\hbar},$$

subject to $p_x^2/2M = E_x$. Since p_x can be positive or negative, all possibilities are given by just the first term. Similarly, $\psi_y \sim e^{+ip_y y/\hbar}$ with $E_y = p_y^2/2M$ and $\psi_z \sim e^{+ip_z z/\hbar}$ with $E_z = p_z^2/2M$. The condition (8) says that

$$\frac{p_x^2 + p_y^2 + p_z^2}{2M} = \frac{p^2}{2M} = E.$$

Now reassemble the pieces.

$$
\begin{aligned}
\psi = \psi_x\psi_y\psi_z &\sim e^{ip_x x/\hbar}e^{ip_y y/\hbar}e^{ip_z z/\hbar} = e^{i\vec{p}\cdot\vec{r}/\hbar}, \\
\Psi(\vec{r}, t) = \psi(\vec{r})f(t) &= g(\vec{p})e^{i(\mathbf{p}\cdot\mathbf{r} - Et)/\hbar},
\end{aligned}
$$

where $g(\vec{p})$ is anything that does not depend on $\vec{r}$ or t. Finally, one comes full circle by adding lots of waves through an integration over $\vec{p}$; once more, the general free-particle wave packet (6) appears.

2.2 *Step-Function Potentials*

Suppose that $V = V(x)$ only. Then we can separate (2.1-3) in Cartesian coordinates, write free waves $\psi_y \sim e^{ip_y y/\hbar}$ and $\psi_z \sim e^{ip_z z/\hbar}$ for the y and z dependence, and concentrate on the x behavior. Suppose further that (Figure 2.2-1)

$$
\begin{aligned}
V(x) &= 0, \qquad x < 0 \\
&= V_0, \qquad x \geq 0.
\end{aligned}
\tag{2.2-1}
$$

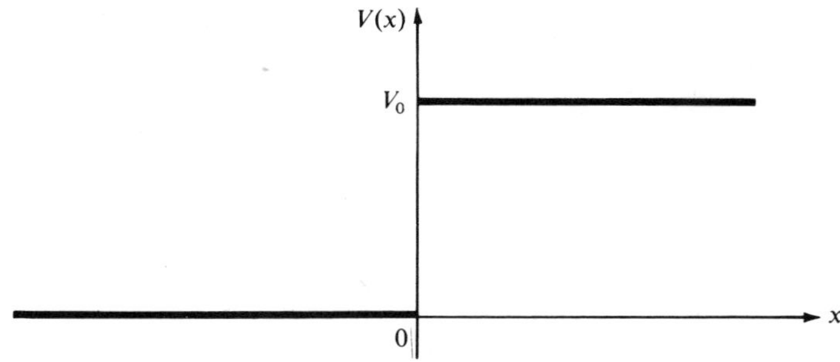

Fig. 2.2-1. Step-function potential energy.

This form implies infinite force at the origin and zero force everywhere else, and is a model of a potential energy function that rises over a distance short compared to the wavelength. It might, for example, describe the potential energy of an electron near the edge of a large block of metal.

What are the boundary conditions at $x = 0$? Integrate

$$-\frac{\hbar^2}{2M}\frac{d^2\psi_x}{dx^2} + V(x)\psi_x = E_x\psi_x$$

from $-\varepsilon$ to $+\varepsilon$:

$$-\frac{\hbar^2}{2M}\left[\left(\frac{d\psi_x}{dx}\right)_{+\varepsilon} - \left(\frac{d\psi_x}{dx}\right)_{-\varepsilon}\right] + \int_{-\varepsilon}^{+\varepsilon} V\psi_x\,dx = E_x\int_{-\varepsilon}^{+\varepsilon}\psi_x\,dx,$$

or

$$\left(\frac{d\psi_x}{dx}\right)_{+\varepsilon} - \left(\frac{d\psi_x}{dx}\right)_{-\varepsilon} = \left(\frac{2M}{\hbar^2}\right)2\varepsilon(\overline{V\psi_x} - E_x\overline{\psi}_x),$$

where $\overline{}$ denotes mean value over the interval $-\varepsilon \leq x \leq +\varepsilon$. If the potential jump is finite, the right side goes to zero as $\varepsilon \to 0$, so that $d\psi_x/dx$ is continuous. The wave function itself must also be continuous, as one might expect because the measurable $|\psi_x|^2$ should have a unique value. If V_0 is increased, there will be increasing curvature in ψ_x at $x = 0$; as $V_0 \to \infty$, ψ_x remains continuous but $d\psi_x/dx$ becomes discontinuous.

Consider first $E_x > V_0$. With momentum along the x axis only, $E = E_x$, $k \equiv \sqrt{2ME}/\hbar$, and $K \equiv \sqrt{2M(E - V_0)}/\hbar$,

$$\begin{aligned}\psi = \psi_x &= Ae^{+ikx} + Be^{-ikx}, \qquad x < 0,\\ &= Ce^{+iKx} + De^{-iKx}, \qquad x \geq 0.\end{aligned}$$

The expression (1.5-6) for the current and, explicitly, Problem 1.10, show that Ae^{+ikx} and Ce^{+iKx} describe positive x components of momentum, while

Be^{-ikx} and De^{-iKx} describe negative x components of momentum. Suppose that there is an electron gun somewhere near $x = -\infty$, and that there are no other sources of electrons, and no changes in the potential except the one at $x = 0$. Then $D = 0$ because e^{-iKx} represents electrons traveling from right to left in the $+x$ region; there is nothing that can give rise to such electrons. The constant A is determined by the requirement that the current density, the number of electrons arriving per m^2 per sec, equals $|A|^2 v = |A|^2 k\hbar/M$; see Section 1.5. The remaining constants B and C are fixed in terms of A by the requirements of continuity and continuity of the derivative;

$$A + B = C,$$
$$ikA - ikB = iKC,$$

so that

$$B = \frac{k - K}{k + K} A, \qquad C = \frac{2k}{k + K} A.$$

The transmission coefficient is the ratio of the current density computed from Ce^{iKx} to that computed from the incident wave Ae^{ikx}:

$$T = \frac{|C|^2 \dfrac{\hbar K}{M}}{|A|^2 \dfrac{\hbar k}{M}} = \frac{4kK}{(k + K)^2}. \tag{2.2-2}$$

The reflection coefficient is given by

$$R = \frac{|B|^2 \dfrac{\hbar k}{M}}{|A|^2 \dfrac{\hbar k}{M}} = \left(\frac{k - K}{k + K}\right)^2. \tag{2.2-3}$$

Note that, happily,

$$T + R = 1. \tag{2.2-4}$$

Classically $E_x > V_0$ would give $T = 1$, $R = 0$. It is the wave aspect of particles that produces reflection here. A sudden change in potential energy gives a sudden change in wavelength, and any wave is at least partially reflected at a surface where there is such a sudden change.

Next, suppose that the particles are again incident along the x axis, but that $E < V_0$. The form of the solution is unchanged for $x < 0$, but for $x \geq 0$,

$$\frac{d^2\psi}{dx^2} = \alpha^2 \psi, \qquad \alpha \equiv \frac{\sqrt{2M(V_0 - E)}}{\hbar},$$

and

$$\psi = Fe^{-\alpha x} + Ge^{+\alpha x}.$$

Since $e^{+\alpha x}$ diverges as $x \to +\infty$, G must be zero. Continuity and continuity of derivative require

$$A + B = F, \qquad ikA - ikB = -\alpha F,$$

$$B = \frac{ik + \alpha}{ik - \alpha} A, \qquad F = \frac{2ik}{ik - \alpha} A.$$

Since $|B|^2 = |A|^2$, the reflection coefficient $R = 1$. Since the current carried by $Fe^{-\alpha x}$ is zero, as it is for any function that can be written as (constant) · (real function), the transmission coefficient is zero. The *net* current in the $x < 0$ region is also zero because as much is reflected as is incident, and there also the wave function can be written as (constant) · (real function).

$$\begin{aligned} Ae^{ikx} + Be^{-ikx} &= A\left[e^{ikx} + \frac{ik + \alpha}{ik - \alpha} e^{-ikx}\right] \\ &= \frac{A}{ik - \alpha}[ik(e^{ikx} + e^{-ikx}) - \alpha(e^{ikx} - e^{-ikx})] \\ &= \frac{2iA}{ik - \alpha}[k \cos kx - \alpha \sin kx]. \end{aligned}$$

This result is a special case of a general fact: If there is no net current in, say, the x direction, then the x dependence of ψ is described by (constant) · (real function). Look at the expression (1.5-6) for the current $\vec{j}$. If $j_x = 0$,

$$\psi^* \frac{\partial \psi}{\partial x} = \psi \frac{\partial \psi^*}{\partial x}.$$

Let $\psi = u + iv$, u and v real. Then

$$(u - iv)(u' + iv') = (u + iv)(u' - iv'),$$

$$\frac{u'}{u} = \frac{v'}{v},$$

$$u = \text{constant} \cdot v,$$

and ψ can be written as a possibly complex constant times a real function of x. Conversely, if there is to be a nonzero current, then the real and imaginary parts of ψ must have different dependence on position.

The appearance of a nonzero probability $|\psi|^2$ in the "classically forbidden" region is an immediate consequence of the wave aspect of particles; one finds analogous situations with classical waves. In the example of total internal reflection of light at a glass-to-air interface, there are fields in the air that decrease exponentially with distance away from the interface. The $1/e$ distance is typically a fraction of a wavelength. "Mechanical" waves propagating through arrays of coupled pendulums driven below the cutoff frequency of the system[1] show similar behavior.

What can one say about $|\psi|^2 \neq 0$ in classically forbidden regions when the particle aspect of electrons is brought out? "Bringing out the particle aspect" really means doing a localization experiment, so we next inspect what happens in an attempt to find an electron in $x > 0$ (Figure 2.2-2) by, for example, illuminating the region and detecting scattered photons.[2]

One can see quite generally that whenever an electron is localized, it must have enough energy to be in the region classically. *Localization* means transition to a state represented by a wave packet, construction of a packet requires oscillating functions, and oscillating functions imply real momenta and positive kinetic energies. For example, suppose that there is an attempt to find the electron in the region where $|\psi|^2 \sim e^{-2\alpha x}$ is appreciable, that is, within about $1/(2\alpha)$ of $x = 0$. Unless $\Delta x < 1/(2\alpha)$, the photons might have scattered from the electron in $x < 0$. Therefore $\Delta p \gtrsim \hbar\alpha$, and an energy bigger than

$$\frac{\hbar^2\alpha^2}{2M} = V_0 - E$$

can easily have been given to the electron if it is found. But $V_0 - E$ plus the original energy E is just what is needed for an electron to acquire classical rights to the $x > 0$ region.

Trying to be more subtle is of no avail. One might go very far into the $x > 0$ region in the hopes of using long-wavelength, very low-energy photons to find the electron. In the $x = +1$-mile neighborhood, can the electron be found by illuminating the region with photons of 0.01-mile wavelength and negligible energy that are traveling in, say, the y direction? If detection inside the barrier is to be claimed, the intensity of the light, $I(x)$, must increase with x faster than the probability density of the electron, $|\psi(x)|^2$, decreases. Suppose that one looks for photons scattered through an angle θ with respect to the y axis. If they are to provide information about the electron, then these photons must be really scattered, not diffracted by a collimating system. Inspection of diffraction at a straightedge shows that if there is not much diffraction through θ, and if the

[1] F. S. Crawford, Jr., *Waves, Berkeley Physics Course*, Vol. 3 (New York: McGraw-Hill Book Company, 1968), Sec. 3.5, particularly pp. 134–135.

[2] The variation of electron density in the classically forbidden region outside a solid surface has been explored by H. D. Hagstrum and T. Sakurai, *Phys. Rev. Lett.* **37**, 615 (1976). They shot positive ions at a positively charged surface and saw the neutralization of the ions by the electrons. Different ion energies gave different distances of closest approach to the surface, and the spatial variation of the electron $|\psi|^2$ was extracted. The result is not a simple exponential because the solid contains electrons with many different energies. The measurements were actually used to learn something about the electron energy distribution.

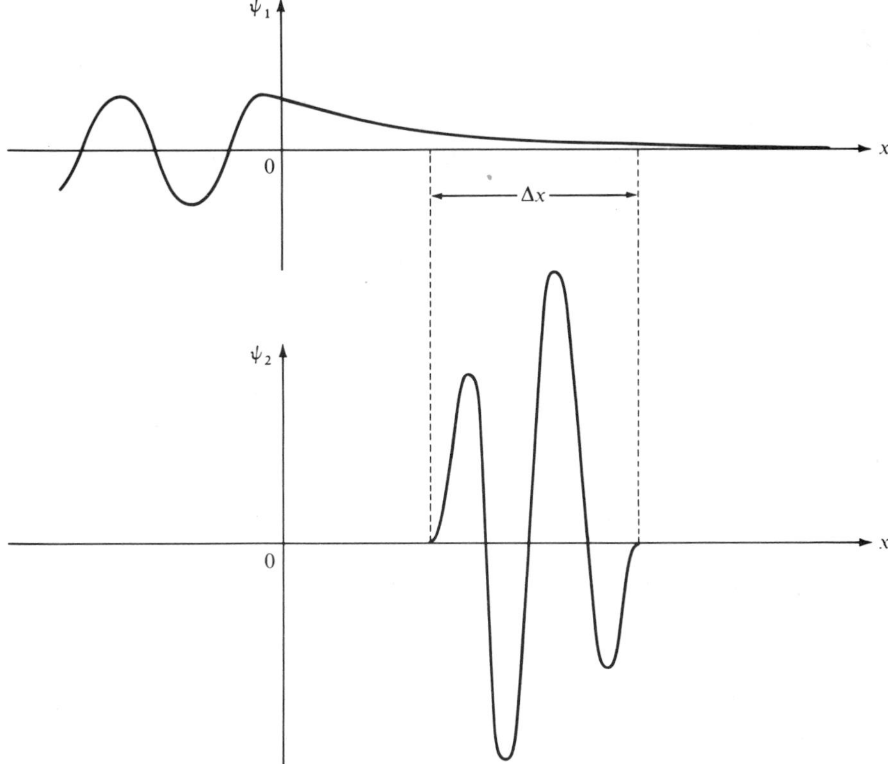

Fig. 2.2-2. The wave function ψ_1 is a decreasing exponential, $Fe^{-\alpha x}$, in the classically inaccessible region. After a localization experiment that finds the electron in Δx, the wave function becomes the packet ψ_2.

shadow intensity is to rise faster than $e^{2\alpha x}$, then $\lambda \lesssim \sin\theta/\alpha$;[3] arbitrarily long wavelength photons can *not* be used. The photons therefore have momentum at least equal to $h\alpha/\sin\theta$, and transfer momentum of at least $h\alpha$ in scattering through θ. The uncertain energy given to the electron is then of the order of $(h\alpha)^2/2M = (2\pi)^2(V_0 - E)$, which is more than enough to raise the energy above V_0.

Superficially, it may seem like convenient shorthand to say, "In $x > 0$ the electrons have imaginary momentum and negative kinetic energy," but this statement is profoundly misleading. We should not say what is true in $x > 0$ without including the consequences of determining that the electrons are indeed there. Here also, the law proclaimed in Section 1.2 is applicable: *Thou shalt not make any statement that can never be verified.*

[3] F. A. Jenkins and H. E. White, *Fundamentals of Optics*, 4th ed. (New York: McGraw-Hill Book Company, 1976), Secs. 18.11, 18.12.

PROBLEM 2.1

Suppose that particles of energy E are incident on the potential step at an angle ϕ with respect to the normal; $p_y/p_x = \tan\phi$. At what angle does the transmitted beam go? Compare your result with Snell's law in optics. What is the transmission coefficient? What does total internal reflection mean here?

2.3 Rectangular and Delta Function Barriers

We continue to suppose that $V = V(x)$ only, and next look at the example shown in Figure 2.3-1,

$$V = V_0, \qquad 0 \le x \le a,$$
$$V = 0, \qquad x < 0 \text{ or } x > a.$$

Suppose that particles with energy $\hbar^2 k^2/2M$ are incident from a source at large negative x. For $x < 0$,

$$\psi = \psi_x = Ae^{ikx} + Be^{-ikx}.$$

For $x > a$, since there is no negative momentum wave, $\psi = Ce^{ikx}$. Within the barrier, for $0 \le x \le a$,

$$\begin{aligned} \psi &= Fe^{+iKx} + Ge^{-iKx} \qquad &\text{if } E > V_0, \\ &= Ie^{-\alpha x} + Je^{+\alpha x} \qquad &\text{if } E < V_0. \end{aligned}$$

The constant A is again determined by the requirement that the incident flux density shall be $v|A|^2$. Continuity and continuity of the derivatives at 0 and at a

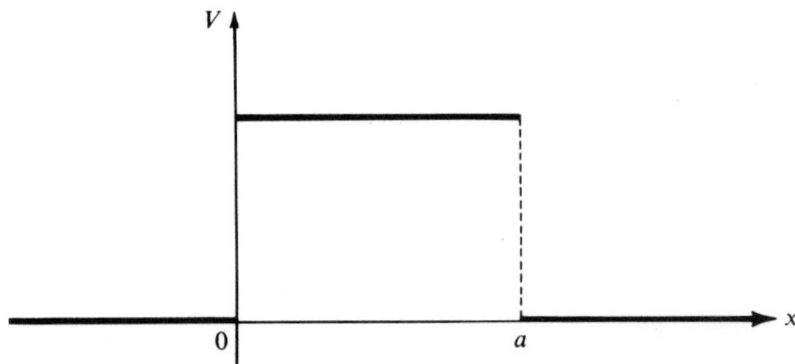

Fig. 2.3-1. A rectangular potential energy barrier.

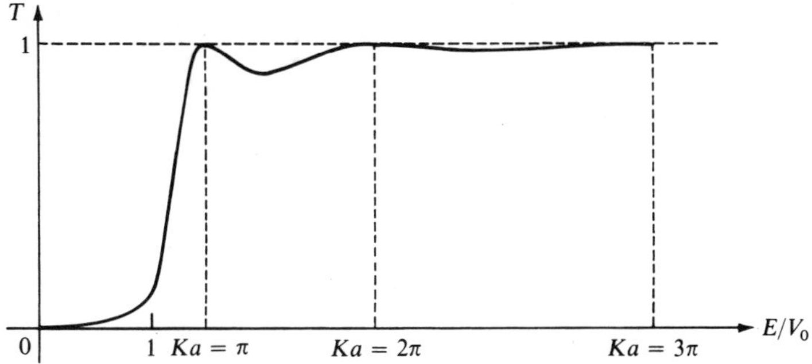

Fig. 2.3-2. Transmission coefficient of a rectangular barrier. Details depend on the choice of parameters; here, $a^2 V_0 M = 10\hbar^2$.

provide four more conditions, and some algebra gives the transmission and reflection coefficients. The transmission coefficient is

$$T = \left[1 + \frac{V_0^2 \sin^2 Ka}{4E(E - V_0)}\right]^{-1}, \qquad E > V_0, \tag{2.3-1}$$

$$T = \left[1 + \frac{V_0^2 \sinh^2 \alpha a}{4E(V_0 - E)}\right]^{-1}, \qquad E < V_0, \tag{2.3-2}$$

and is graphed in Figure 2.3-2. Note that $T = 1$ if $Ka = n\pi$. If you understand reflection in thin films,[4] you understand this fact.

PROBLEM 2.2

For $E = V_0/2$ and $\alpha a = 1$, obtain the constants B, C, I, and J in terms of A. Sketch graphs of the wave function in the three regions, $x < 0$, $0 \leq x \leq a$, and $a < x$. Note that you will have to graph separately the real and imaginary parts, and that both are continuous and have continuous first derivatives. Sketch graphs of the probability density in each of the three regions. Calculate the current density in each of the three regions. Obtain the transmission coefficient and check it against (2).

PROBLEM 2.3

Obtain T of the rectangular barrier for the special case $E = V_0$

a. by working through the problem with $E = V_0$ from the beginning,
b. by taking the appropriate limit of (2.3-1), and
c. by taking the appropriate limit of (2.3-2).

---

[4] F. S. Crawford, Jr., *Waves*, *Berkeley Physics Course*, Vol. 3 (New York: McGraw-Hill Book Company, 1968), Sec. 5.5.

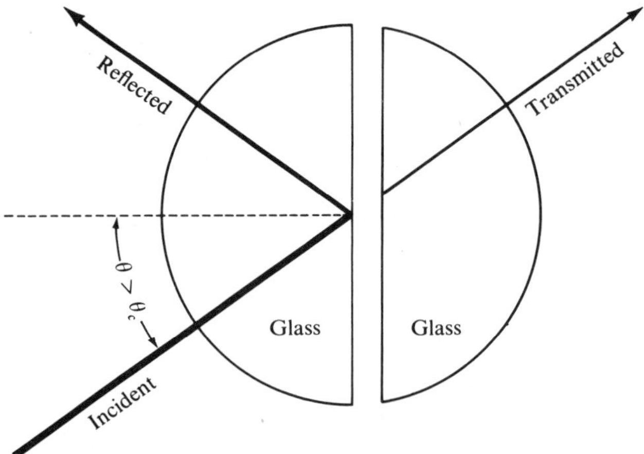

Fig. 2.3-3. Transmission of light through an "inaccessible" region.

Classical particle mechanics provides quite a different answer: All particles get through if $E > V_0$, and none get through if $E < V_0$. Nonzero transmission through classically forbidden regions is a characteristic quantum result. It is often called *tunneling*; during the early days of quantum theory, it was seen as a surprising jailbreak through potential energy walls that should have been escape proof. Classical *waves* do show behavior that is really the same as tunneling. Suppose a second piece of glass is brought near a piece of glass in which light is incident on an internal surface at greater than the critical angle (Figure 2.3-3). A faint transmitted beam becomes visible when the gap is of the order of a wavelength, and becomes more intense as the gap is decreased. If the gap is made negligible compared to a wavelength, the transmitted intensity becomes equal to the incident intensity, and the reflected beam disappears.[5]

The end of Section 2.2 deals with attempts to localize a particle where $V > E$. The same arguments and conclusions are right here. If a particle is localized between $x = 0$ and $x = a$, it must be kicked hard enough so that its total energy exceeds V_0. Figure 2.2-2 is applicable, except for the immaterial modification that the exponential behavior stops at $x = a$. However, the particle can be detected in the $x < 0$ and $x > a$ regions with arbitrarily small disturbance; these two regions are indefinitely large, and the probability density is constant in them. Much the same can be said about the experiment with light sketched in Figure 2.3-3. There the incident, reflected, and transmitted beams can be observed without any appreciable influence on their intensities. If, on the other

[5] F. S. Crawford, *Waves*, *Berkeley Physics Course*, Vol. 3 (New York: McGraw-Hill Book Company, 1968), Sec. 7.2, particularly pp. 345–346; D. D. Coon, *Am. J. Phys.* **34**, 240 (1966).

hand, one tries to detect the light in the gap between the two blocks of glass, something with an index of refraction different from that of air must be inserted, and everything changes.

In the quantum domain, the penetration of classically inaccessible regions is an important and familiar process. The ideas of this section are applicable to tunneling in electronic phenomena such as field emission from metal surfaces and the operation of tunnel diodes, and nuclear phenomena such as alpha decay. However, the rectangular one-dimensional barrier is a little too simple a model, and it will pay to study methods, as in Chapter 10, that can handle more complicated potential energy functions. You should, however, review at this time an elementary treatment of alpha decay.[6]

Now we shall look at an even simpler model of a barrier. Suppose that the barrier thickness is small compared to the wavelength and that the barrier height is large compared to the energy. Such a very narrow tall spike can be described by the *Dirac delta function* $\delta(x - b)$ defined by

$$\begin{aligned} \delta(x - b) &= 0 \qquad \text{for } x \neq b, \\ \int \delta(x - b)dx &= 1, \end{aligned} \tag{2.3-3}$$

where the integral is taken over any interval that includes b. In other words, $\delta(x - b)$ gets indefinitely large at b in just such a way as to keep the area under the spike equal to unity, even though the width of the spike is made negligible.

If $f(x)$ is continuous at b, then

$$\int f(x)\delta(x - b)dx = f(b). \tag{2.3-4}$$

This statement is often taken as a definition of $\delta(x - b)$. A good way to understand the delta function is to think about a narrow peak $\Delta(x - b)$ which is not quite a delta function because it has some width and finite height, but which does cover unit area. In Figure 2.3-4 a continuous $f(x)$ passes through the neighborhood of b. Because $f(x)$ varies rather little in the small interval about b in which $\Delta(x - b)$ is appreciable, and because $\int \Delta(x - b)dx = 1$,

$$\int f(x)\Delta(x - b)dx \simeq f(b) \int \Delta(x - b)dx = f(b).$$

If the width of Δ approaches zero, then $\Delta(x - b)$ becomes the delta function $\delta(x - b)$, and this equation becomes (4).

[6] K. W. Ford, *Classical and Modern Physics* (Lexington, Mass.: Xerox College Publishing, 1972), pp. 1318–1324. Some historical background is given by E. U. Condon, *Am. J. Phys.* **46**, 319 (1978).

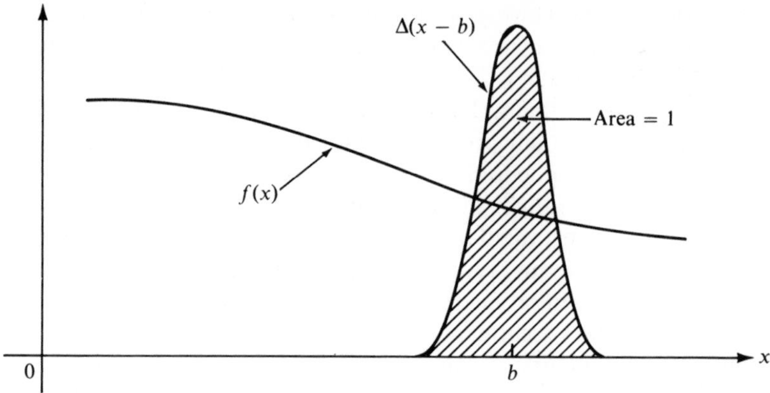

Fig. 2.3-4. A peak Δ that is almost narrow enough to behave like a delta function.

EXAMPLE

a. Invent some functional forms that have the properties of $\Delta(x - b)$.

b. The *three-dimensional* Dirac delta function $\delta(\vec{r} - \vec{r}')$ is zero if $\vec{r} \neq \vec{r}'$, and $\iiint \delta(\vec{r} - \vec{r}')d\tau = 1$ if the volume of integration includes $\vec{r}'$. Express $\delta(\vec{r} - \vec{r}')$ in Cartesian coordinates and in spherical polar coordinates.

SOLUTION

a.

$$\frac{e^{-(x-b)^2/\Gamma^2}}{\Gamma\sqrt{\pi}}; \qquad \frac{e^{-|x-b|/\Gamma}}{2\Gamma};$$

$$\frac{\dfrac{\Gamma}{2\pi}}{(x-b)^2 + \dfrac{\Gamma^2}{4}}; \qquad \frac{\sin^2\left(\dfrac{x-b}{\Gamma}\right)}{\pi\Gamma\left(\dfrac{x-b}{\Gamma}\right)^2};$$

$$\frac{1}{\Gamma} \text{ if } b - \frac{\Gamma}{2} < x < b + \frac{\Gamma}{2} \text{ and zero otherwise}; \ldots.$$

Each of these possibilities satisfies $\int_{-\infty}^{+\infty} \Delta(x - b) = 1$, and becomes $\delta(x - b)$ if $\Gamma \to 0$.

b. In Cartesian coordinates, if $\vec{r} = (x, y, z)$ and $\vec{r}' = (x', y', z')$,

$$\delta(\vec{r} - \vec{r}') = \delta(x - x')\delta(y - y')\delta(z - z').$$

This expression is zero unless $x = x'$, $y = y'$, and $z = z'$. Also,

$$\iiint \delta(x - x')\delta(y - y')\delta(z - z')dx\,dy\,dz$$

$$= \int \delta(x - x')dx \int \delta(y - y')dy \int \delta(z - z')dz = 1 \times 1 \times 1 = 1.$$

In spherical polars, if $\vec{r} = (r, \theta, \varphi)$ and $\vec{r}' = (r', \theta', \varphi')$,

$$\delta(\vec{r} - \vec{r}') = \frac{1}{r^2}\delta(r - r')\delta(\cos\theta - \cos\theta')\delta(\varphi - \varphi').$$

This expression is zero unless $r = r'$, $\theta = \theta'$, and $\varphi = \varphi'$. Also, since the volume element is $r^2\,dr \sin\theta\,d\theta\,d\varphi$,

$$\int_0^\infty \int_0^\pi \int_0^{2\pi} \frac{1}{r^2}\delta(r - r')\delta(\cos\theta - \cos\theta')\delta(\varphi - \varphi')r^2\,dr\sin\theta\,d\theta\,d\varphi$$

$$= \int_0^\infty \delta(r - r')dr \int_{-1}^{+1} \delta(u - u')du \int_0^{2\pi} \delta(\varphi - \varphi')d\varphi = 1 \times 1 \times 1 = 1.$$

Suppose there is a barrier of large height V_0 and small width a at the origin. Let $V_0 a = U$. Then

$$V(x) = U\delta(x)$$

approximates such a barrier. Notice that the parameter U has the dimensions (energy) · (length). Since the dimensions of $\delta(x)$ are *per length*, $V(x)$ has the dimensions of energy, as it should. Now consider the Schrödinger equation with such a potential:

$$-\frac{\hbar^2}{2M}\frac{d^2\psi}{dx^2} + U\delta(x)\psi = E\psi. \tag{2.3-5}$$

One could deal with this problem by taking suitable limits in the rectangular barrier problem, but it is easier to attack this equation directly. Integrate it from $-\varepsilon$ to $+\varepsilon$:

$$-\frac{\hbar^2}{2M}\left[\left(\frac{d\psi}{dx}\right)_{+\varepsilon} - \left(\frac{d\psi}{dx}\right)_{-\varepsilon}\right] + U\int_{-\varepsilon}^{+\varepsilon} \delta(x)\psi(x)dx = E\int_{-\varepsilon}^{+\varepsilon} \psi(x)dx.$$

The term involving the potential energy becomes, from (4), simply $U\psi(0)$. The term on the right is $2\varepsilon E\bar{\psi}$, where $\bar{\psi}$ is the mean value of ψ in the interval $-\varepsilon \le x \le +\varepsilon$. Since $\bar{\psi}$ and E are finite, this term vanishes as $\varepsilon \to 0$. The result is then

$$\left(\frac{d\psi}{dx}\right)_{+\varepsilon} - \left(\frac{d\psi}{dx}\right)_{-\varepsilon} = \frac{2M}{\hbar^2}U\psi(0), \tag{2.3-6}$$

a condition on the discontinuity of the derivative at the potential energy spike. The wave function ψ itself is continuous at $x = 0$. If it were not, then measurable physical quantities would be discontinuous there.

PROBLEM 2.4

Prove that ψ is continuous at the location of the delta function potential by examining the suitable limit of the rectangular barrier problem.

Since $V = 0$ for $x < 0$ and $x > 0$, then, if the source of electrons is at $-\infty$,

$$\begin{aligned} \psi &= Ae^{ikx} + Be^{-ikx}, \qquad x < 0, \\ &= Ce^{ikx}, \qquad x > 0. \end{aligned}$$

Continuity gives $A + B = C$, and (6) gives

$$ikC - (ikA - ikB) = \frac{2mUC}{\hbar^2}.$$

Therefore

$$T = \left|\frac{C}{A}\right|^2 = \frac{1}{1 + \dfrac{M^2U^2}{k^2\hbar^4}} = \frac{1}{1 + \dfrac{MU^2}{2\hbar^2 E}}. \tag{2.3-7}$$

PROBLEM 2.5

Consider two delta function barriers separated by a distance l;

$$V(x) = U\delta(x + l) + U\delta(x).$$

Obtain, as a function of energy, the transmission coefficient of the two barriers together, that is, of the probability that an electron of energy E gets from $x < -l$ to $x > 0$. (*Note:* Organize your algebra carefully for this one. It can get messy.)

2.4 *Particle in a Box. Degeneracy*

Suppose that an electron is inside a box-shaped force-free region, and is constrained to remain inside that region by a very steep and high potential increase at the boundaries:

$$\begin{aligned} V(x, y, z) &= 0 \qquad \text{if } 0 \le x \le a \text{ and } 0 \le y \le b \text{ and } 0 \le z \le c, \\ &= \infty \qquad \text{otherwise.} \end{aligned}$$

This potential can be written in the form

$$V(x, y, z) = V_x(x) + V_y(y) + V_z(z),$$

where

$$\begin{aligned} V_x &= 0, & 0 \le x \le a, \\ &= \infty, & x < 0 \text{ or } a < x; \\ V_y &= 0, & 0 \le y \le b, \\ &= \infty, & y < 0 \text{ or } b < y; \\ V_z &= 0, & 0 \le z \le c \\ &= \infty, & z < 0 \text{ or } c < z. \end{aligned}$$

The time-independent Schrödinger equation then separates into the three equations (2.1-7). For the x behavior,

$$\frac{d^2\psi_x}{dx^2} = -k_x^2\psi_x, \qquad k_x \equiv +\frac{\sqrt{2ME_x}}{\hbar},$$

and the general solution is

$$\psi_x(x) = A \sin k_x x + B \cos k_x x. \tag{2.4-1}$$

Since $\psi_x(0) = 0$, $B = 0$. To have $\psi(a) = 0$, one must require that

$$k_x = \frac{\pi n_x}{a}, \qquad n_x = 1, 2, 3, \ldots,$$

$$E_x = \frac{\pi^2 n_x^2 \hbar^2}{2Ma^2}.$$

The value $n_x = 0$ would make ψ_x, and therefore $\psi = \psi_x\psi_y\psi_z$, identically zero everywhere, and must be excluded. The choice $A = \sqrt{2/a}$ yields a normalized wave function,

$$\int_0^a |\psi_x|^2\, dx = 1.$$

PROBLEM 2.6

Rather than (1), one could write

$$\psi_x(x) = Ce^{ik_x x} + De^{-ik_x x}. \tag{2.4-2}$$

Use the boundary conditions and the normalization requirement to evaluate C and D and to obtain the energy, and show that the two approaches are equivalent.

The requirement that the wave must fit into the box has produced discrete energy eigenvalues. The particular result, $E_x = \pi^2 n_x^2 \hbar^2/2Ma^2$, is of course special to this model. We will, however, see repeatedly how the imposition of boundary conditions limits the possible wave functions and corresponding energies to certain discrete possibilities. Messy potential energy functions may generate messy mathematics, but the general scheme has now been illustrated. If an electron is described by a wave, and if the electron is confined to some region, then the wave has to fit into that region. Through the connection between the kinetic energy and the curvature of ψ, the fitting requirement leads to the energy levels of the system.

Also, notice that the result illustrates the uncertainty principle. Since the electron is somewhere in the box, $\Delta x \simeq a$. The magnitude of p_x is $\hbar k_x = \pi n_x \hbar/a$, so that $\Delta p_x \simeq 2p_x = 2\pi n_x \hbar/a$. Therefore

$$\Delta x\, \Delta p_x = 2\pi n_x \hbar,$$

or h for the lowest state, $2h$ for the first excited state, and so forth. The approximate equality in $\Delta x\, \Delta p_x \gtrsim h$ holds for the ground state, while the inequality holds for the excited states.

Similarly,

$$\psi_y = \sqrt{\frac{2}{b}} \sin \frac{\pi n_y y}{b}, \qquad E_y = \frac{\pi^2 n_y^2 \hbar^2}{2Mb^2},$$

$$\psi_z = \sqrt{\frac{2}{c}} \sin \frac{\pi n_z z}{c}, \qquad E_z = \frac{\pi^2 n_z^2 \hbar^2}{2Mc^2}.$$

The wave function is then

$$\psi = \psi_x \psi_y \psi_z = \sqrt{\frac{8}{abc}} \sin\left(\frac{\pi n_x x}{a}\right) \sin\left(\frac{\pi n_y y}{b}\right) \sin\left(\frac{\pi n_z z}{c}\right), \tag{2.4-3}$$

and the energy

$$E = E_x + E_y + E_z = \frac{\pi^2 \hbar^2}{2M}\left(\frac{n_x^2}{a^2} + \frac{n_y^2}{b^2} + \frac{n_z^2}{c^2}\right). \tag{2.4-4}$$

If a, b, and c are all different, then, unless there is some special coincidence, only one set of the three quantum numbers (n_x, n_y, n_z), and the one wave function corresponding to these numbers, will give each of the possible energy levels. If however two or all three of the lengths are equal, something quite different happens: There are many more physically distinguishable states than there are energy levels. Suppose that $a = b = c$, so that

$$E = \frac{\pi^2 \hbar^2}{2Ma^2}(n_x^2 + n_y^2 + n_z^2). \tag{2.4-5}$$

Then $(n_x, n_y, n_z) = (1, 1, 2)$, and $(1, 2, 1)$, and $(2, 1, 1)$ all produce the same $E = (\pi^2 \hbar^2/2Ma^2) \times 6$, but the three wave functions look quite different and can be distinguished from each other by, for example, measurements of the probability distribution. These three wave functions are said to be *degenerate.* Since the symmetry of the Hamiltonian is the cause, the situation can be labeled *symmetry degeneracy*. The states (1, 1, 1), (2, 2, 2), (3, 3, 3), . . . are not degenerate. The (1, 1, 2), (1, 2, 1), (2, 1, 1) situation is one of threefold degeneracy. Since (1, 2, 3), (1, 3, 2), (2, 1, 3), (2, 3, 1), (3, 1, 2), (3, 2, 1) represent six linearly independent, physically distinguishable states that all have energy $(\pi\hbar^2/2Ma^2) \times 14$, they are an example of sixfold degeneracy, again caused by symmetry.

Now consider the two states (1, 1, 7) and (1, 5, 5). Both have an energy of $(\pi^2\hbar^2/2Ma^2) \times 51$. Because one cannot go from one state to the other by relabeling the axes or by rotation, they are an example of *accidental* degeneracy.

The terms *linearly independent* and *physically distinguishable* recur in this discussion and deserve more emphasis. Is $-\psi$, or $i\psi$, physically distinguishable from ψ? The answer is no. The probability density and the current density are the same. Chapter 3 shows that *all* measurables are the same. Multiplying ψ by a constant of course produces a different *function*, but since there is no change in the physics, one says that the *state* is the same. Next, consider three degenerate states, such as those schematically called (1, 1, 2), (1, 2, 1), and (2, 1, 1) above. They are truly different. How about a linear combination of the three, $A(1, 1, 2)+ B(1, 2, 1) + C(2, 1, 1)$? It has a different probability density and is a different state. Since there are infinitely many possibilities for A, B, and C, is this really an example of infinite, rather than threefold degeneracy? Such a description is not useful. The way of counting that makes sense, and that was implicitly invoked above, is the following: If N *linearly independent* states all have the same energy, then there is N-fold degeneracy.

Absolute symmetry is hard to achieve. For example, a box might be designed so that $a = b = c$, but the construction may not be perfect. Perhaps the gravitational field, or a gentle breeze blowing through the lab, makes the different directions look a little different and breaks the symmetry. How nearly the same must the different energies be if it is useful to call the corresponding states *degenerate*? This question will reappear in Section 10.2, but here is a preliminary and partial answer. The ground state, in this example (1, 1, 1), is not degenerate. Every excited state, perhaps because of electromagnetic interactions, actually decays. Every excited state therefore has a natural width $\Delta E = \hbar/\tau$. It is useful to call two states *degenerate* if there is appreciable overlap of their natural widths, or if other small effects shift their energies about by amounts comparable to their separation.

PROBLEM 2.7

A particle of mass M is in an $a \times a \times a$ rigid-walled box. Calculate and plot, as a function of the energy E, the total number of states with energy *less* than E, for $0 \leq E \leq 20\pi^2\hbar^2/2Ma^2$. Your graph will be a series of steps, with a discontinuous jump at every energy that permits additional states. (This problem is not as artificial as it looks. For example, Chapter 5 deals with the number of states, below some maximum energy, that are available for electrons in a piece of metal.)

2.5 One-Dimensional Rectangular and Delta Function Wells

Consider now the example

$$\begin{aligned} V(x) &= -V_0, \qquad |x| \le \frac{a}{2}, \\ &= 0, \qquad |x| > \frac{a}{2}, \end{aligned} \tag{2.5-1}$$

with $V_0 > 0$ (Figure 2.5-1). For an electron, this form of potential could in principle be produced by four large parallel conducting sheets and a battery as shown in Figure 2.5-2. A 1-eV electron has $\lambda = 1.23 \times 10^{-7}$ cm; mounting sheets with a separation less than the wavelength of an electron of reasonable energy is beyond even the capabilities of the physics department machine shop. Real situations that can approximate such a potential are, for example, those involving a metal film supported between two pieces of insulator.

Since $V = V(x)$ only, the Schrödinger equation separates again into free-particle equations for the y and z motion, and an x equation in which all the interesting things happen. For the x behavior, two profoundly different situations are now possible, depending on whether the energy associated with the x motion is greater than zero or less than zero. Since only the x motion is of interest, call this energy E rather than E_x, and the wave function ψ rather than ψ_x.

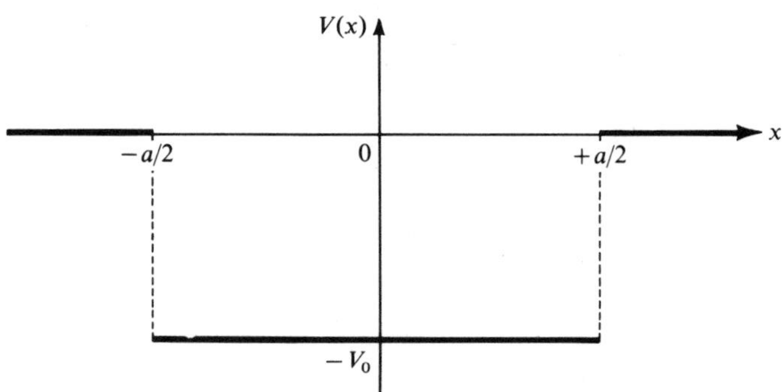

Fig. 2.5-1. A one-dimensional rectangular well.

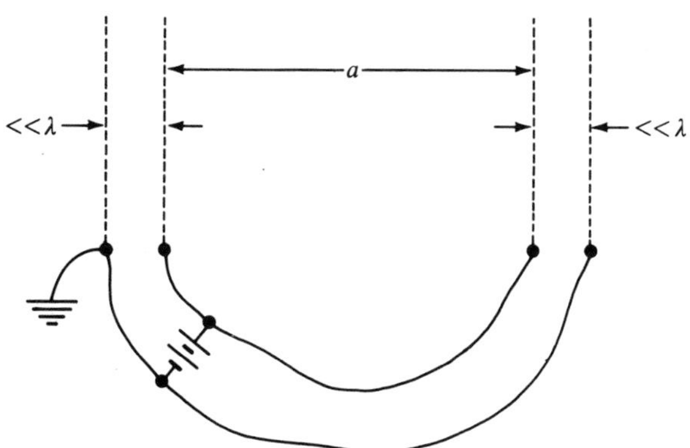

Fig. 2.5-2. Four parallel conducting sheets and a battery wired to give the potential described by (1).

If $E > 0$, then the problem is much like that of Section 2.3. If the electrons are incident from the left,

$$\begin{aligned}
\psi &= Ae^{ikx} + Be^{-ikx}, \qquad k \equiv \frac{\sqrt{2ME}}{\hbar}, \qquad x < -\frac{a}{2}, \\
&= Fe^{iKx} + Ge^{-iKx}, \qquad K \equiv \frac{\sqrt{2M(E+V_0)}}{\hbar}, \qquad -\frac{a}{2} \leq x \leq +\frac{a}{2}, \\
&= Ce^{ikx}, \qquad +\frac{a}{2} < x.
\end{aligned}$$

The constant A is determined by the incident flux, and B, F, G, and C are determined from continuity and continuity of derivatives at $-a/2$ and at $+a/2$. The transmission coefficient is $|C/A|^2$, and is given by (2.3-1) if V_0 is replaced by $-V_0$. The electrons come from and escape to indefinitely large distances, their wavelengths are not constrained to fit into some region, and their energy can therefore be adjusted to any value. The Schrödinger equation has solutions for any energy greater than zero.

Although all values of $E > 0$ are possible, certain energies in this range have something special about them. Look at the probability of finding an electron inside the well,

$$P(\text{inside}) = \int_{-a/2}^{+a/2} |Fe^{iKx} + Ge^{-iKx}|^2\, dx.$$

Divide this quantity by the incident flux density $j_i = |A|^2\hbar k/M$. The result is

$$\frac{P(\text{inside})}{j_i} = \frac{2kMa}{\hbar} \frac{(K^2+k^2) + (K^2-k^2)\dfrac{\sin(2Ka)}{2Ka}}{4K^2k^2 + (K^2-k^2)^2 \sin^2 Ka}. \tag{2.5-2}$$

PROBLEM 2.8

Verify (2), and use this opportunity to outgrow your caveperson approach to calculations. Here are some suggestions. You will need to obtain some of the constants A, B, F, G, and C by imposing continuity and continuity of derivatives at the boundaries of the potential well. The well is given as extending from $-a/2$ to $+a/2$, and that will clarify the symmetry of the $E < 0$ states. However, for this problem things are easier if the origin is shifted so that the well extends from $x = -a$ to $x = 0$. Since (2) is a physical measurable that does not involve the choice of origin, we can change to

$$\begin{aligned} V &= -V_0, \qquad & -a \le x \le 0, \\ &= 0, & \text{otherwise,} \end{aligned}$$

without affecting the result. Next, note that only the *ratios* of the coefficients are interesting, so that any one of them can be set equal to unity. This time, choose $C = 1$. Since one tends to think of the transmitted current, $|C|^2\hbar k/M$, as an effect rather than a cause, there is a psychological barrier against this choice. Once you have made it, however, continuity and continuity of derivative at the right-hand edge of the well ($x = 0$) become simply

$$F + G = 1, \qquad iKF - iKG = ik.$$

These equations immediately give F and G. Look next at the left-hand edge of the well, $x = -a$, and use continuity and continuity of derivative to determine A. You might like to redo Problems 2.2, 2.3, and 2.5 with this approach.

Expression (2) has some interesting physics in it. As is pointed out in the beginning of Section 2.3, the transmission coefficient becomes unity whenever $Ka = \pi, 2\pi, 3\pi, \ldots$, that is, whenever the wave fits across the well in this example, or across the barrier of Section 2.3. Expression (2) says that the probability of finding electrons inside the well is relatively large whenever this condition is satisfied. The effect is most pronounced if the energy is only slightly above zero and the well is deep, in the sense that $K^2 \gg k^2$. Then (2) becomes

$$\frac{2kMa}{\hbar}\,\frac{1 + \dfrac{\sin(2Ka)}{2Ka}}{4k^2 + K^2\sin^2 Ka},$$

which equals

$$\frac{2Ma}{\hbar}\frac{1}{4k}$$

when $Ka = n\pi$, and

$$\frac{2Ma}{\hbar}\frac{k}{K^2}$$

when $Ka = (n + \frac{1}{2})\pi$. The ratio of these values, $K^2/4k^2$, can be huge. We are therefore looking at a typical resonance phenomenon.

Now examine $E < 0$. The solution to the Schrödinger equation is

$$\begin{aligned}\psi &= Ae^{-\alpha x} + Be^{+\alpha x}, \qquad \alpha \equiv \frac{\sqrt{2M(-E)}}{\hbar}, \qquad x < \frac{-a}{2}\\ &= Fe^{iKx} + Ge^{-iKx} = I\sin Kx + J\cos Kx, \qquad \frac{-a}{2} \le x \le +\frac{a}{2},\\ &= Ce^{-\alpha x} + De^{+\alpha x}, \qquad x > +\frac{a}{2}.\end{aligned} \tag{2.5-3}$$

The definition of K is the same as in the $E > 0$ case. Inside the well $e^{\pm iKx}$ or $\sin Kx$ and $\cos Kx$ can be used; the two forms of solution are completely equivalent, as Problem 2.6 shows. If we were feeling singularly silly, we could even use things like $Y^{+iKx} + Z\cos Kx$. All measurable quantities would come out the same regardless of the form in which ψ is written, but the form $I\sin Kx + J\cos Kx$ is most convenient here. Solutions of the form e^{+iKx} and e^{-iKx} are convenient when the probability current is the interesting quantity because they display directly the contributions to this current. If however the energy lies below the energy of an electron that is far away and at rest, below zero for the potential energy given by (1), then there is a *bound state*: The electron is trapped and must wander around inside and in the neighborhood of the well, and the net current must be zero. Since $\sin Kx$ and $\cos Kx$ are proportional to $e^{iKx} \mp e^{-iKX}$, they represent a superposition of equal amounts of positive and negative currents.

There is an additional reason for using $\sin Kx$ and $\cos Kx$. Whenever $V(\vec{r})$ is an even function, in the sense that $V(\vec{r}) = V(-\vec{r})$, then, if there is no degeneracy, $\psi(\vec{r})$ must be either a purely even function or a purely odd function. The Schrödinger equation is

$$-\frac{\hbar^2}{2M}\nabla^2\psi(\vec{r}) + V(\vec{r})\psi(\vec{r}) = E\psi(\vec{r}).$$

Change $\vec{r}$ into $-\vec{r}$. Since ∇^2 and $V(\vec{r})$ are unchanged,

$$-\frac{\hbar^2}{2M}\nabla^2\psi(-\vec{r}) + V(\vec{r})\psi(-\vec{r}) = E\psi(-\vec{r}).$$

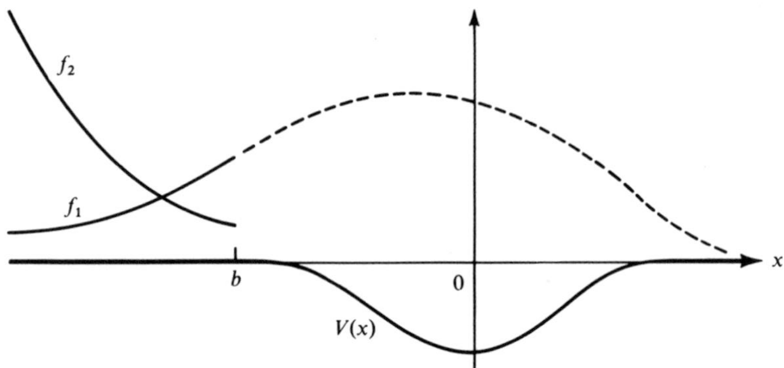

Fig. 2.5-3. Construction of a unique bound-state solution.

The two wave functions $\psi(\vec{r})$ and $\psi(-\vec{r})$ therefore satisfy the same Schrödinger equation and have the same energy. If there is no degeneracy, then $\psi(+\vec{r})$ and $\psi(-\vec{r})$ must be the same except for a multiplying constant:

$$\psi(+\vec{r}) = c\psi(-\vec{r}).$$

Change $\vec{r}$ into $-\vec{r}$:

$$\psi(-\vec{r}) = c\psi(+\vec{r}).$$

Therefore $c^2 = 1$,

$$c = \pm 1.$$

If $c = +1$, then $\psi(\vec{r})$ is said to have *even parity*. If $c = -1$, then $\psi(\vec{r})$ is said to have *odd parity*.

Notice the assumption that there is no degeneracy. In one-dimensional problems, for bound states with well-behaved potential energy functions, it is true that no degeneracy occurs. This fact is a statement regarding the uniqueness of solutions of ordinary linear differential equations.[7] Suppose we go far to the negative x side of the potential well, where $V(x)$ is zero, say to $x < b$ (Figure 2.5-3). For bound states $E < 0$, and the two linearly independent solutions there are $f_1 = Ae^{+\alpha x}$ and $f_2 = Be^{-\alpha x}$. For given energy E there are no other solutions, because a second-order ordinary differential equation has two and only two. Beginning at $x = b$, the rest of the two solutions that begin with f_1 and f_2 can be constructed by some analytic or numerical procedure. The answers are unique because both the values and the derivatives of f_1 and f_2 are known at $x = b$. Thinking like a not very bright digital computer may clarify this point: Evaluate

[7] R. P. Agnew, *Differential Equations* (New York: McGraw-Hill Book Company, 1968), Chap. 6; W. Leighton, *An Introduction to the Theory of Ordinary Differential Equations* (Belmont, Calif.: Wadsworth Publishing Co., Inc., 1976), Chap. 4.

f_1 and df_1/dx at b. Obtain d^2f_1/dx^2 from the Schrödinger equation. Go to $x = b + \Delta x$. Calculate $f_1(b + \Delta x) = f_1(b) + (df_1/dx)_b\,\Delta x$. Calculate

$$\left(\frac{df_1}{dx}\right)_{b+\Delta x} = \left(\frac{df_1}{dx}\right)_b + \left(\frac{d^2f_1}{dx^2}\right)_b \Delta x = \left(\frac{df_1}{dx}\right)_b + \frac{2M}{\hbar^2}[V(b) - E]f_1(b)\Delta x.$$

Go to $b + 2\,\Delta x$. Calculate $f_1(b + 2\,\Delta x) = f_1(b + \Delta x) + (df_1/dx)_{b+\Delta x}\,\Delta x$. Calculate

$$\left(\frac{df_1}{dx}\right)_{b+2\,\Delta x} = \left(\frac{df_1}{dx}\right)_{b+\Delta x} + \left(\frac{d^2f_1}{dx^2}\right)_{b+\Delta x} \Delta x$$

$$= \left(\frac{df_1}{dx}\right)_{b+\Delta x} + \frac{2M}{\hbar^2}[V(b + \Delta x) - E]f_1(b + \Delta x)\Delta x.$$

Go to $b + 3\,\Delta x$. Calculate .... This procedure is unique at each stage. There are then two linearly independent solutions, one of which begins with f_1 in $x < b$ and the other of which begins with f_2 in $x < b$. The one that begins with f_2 must be rejected because it diverges as $x \to -\infty$. The result is that there is only one possible solution, the one that begins with f_1, and so there is no degeneracy.

PROBLEM 2.9

Think about the three conditions buried in the sentence, "In one-dimensional problems, for bound states with well-behaved potential energy functions, it is in fact true that no degeneracy occurs."

a. "In one-dimensional problems ..." Suppose that it is a two- or three-dimensional problem. Describe where the argument fails. Give an example of two different wave functions that correspond to the same energy.

b. "..., for bound states ..." Suppose that it is a one-dimensional problem about unbound, $E > 0$, states. Describe where the argument fails. Give an example of two different wave functions that correspond to the same energy.

c. "... with well-behaved potential energy functions, ..." Suppose $V(x)$ behaves as shown in Figure 2.5-4, so that $V(x) = \infty$ between $-l$ and $+l$. Describe where the argument fails. Give an example of two different wave functions that correspond to the same energy.

This digression has been useful because we know now that there cannot be more than one solution for a given negative energy, and that each solution must be either a definitely even function of x, $\psi(-x) = +\psi(x)$, or a definitely odd function of x, $\psi(-x) = -\psi(x)$. With this information look again at the solution (3).

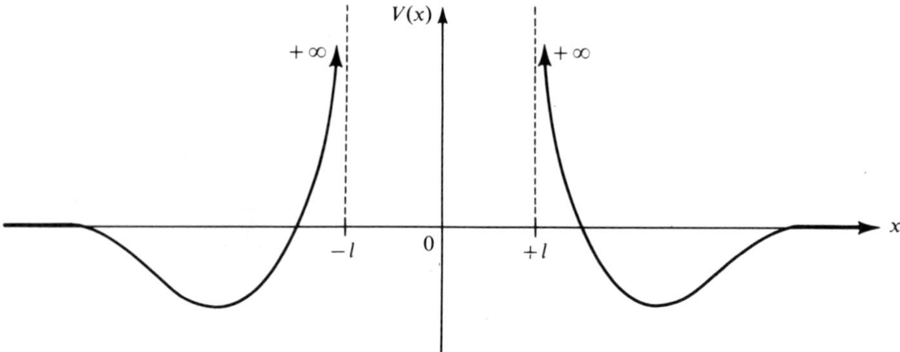

Fig. 2.5-4. A highly singular potential.

Let us first find the *even* solutions. These must be described in $-a/2 \leq x \leq +a/2$ by $J \cos Kx$; the constant I must be set equal to zero because it multiplies an odd function. Further, A and D must be zero to keep ψ finite as $x \to \pm\infty$. Also, since the function is even, $B = C$. The set of equations (3) then becomes, with $\psi = \psi_e$ for *even*,

$$\begin{aligned} \psi_e &= Be^{\alpha x}, && x < \frac{-a}{2}, \\ &= J\cos Kx, && \frac{-a}{2} \leq x \leq \frac{+a}{2}, \\ &= Be^{-\alpha x}, && \frac{+a}{2} < x. \end{aligned} \tag{2.5-4}$$

Continuity of ψ_e at $a/2$ yields

$$J \cos \frac{Ka}{2} = Be^{-\alpha a/2}. \tag{2.5-5}$$

Continuity of $d\psi_e/dx$ at $a/2$ yields

$$-KJ \sin\left(\frac{Ka}{2}\right) = -\alpha Be^{-\alpha a/2}. \tag{2.5-6}$$

Divide the second of these equations by the first:

$$K \tan\left(\frac{Ka}{2}\right) = \alpha. \tag{2.5-7}$$

Since K and α depend on E, this equation determines the possible values of the energy. It is a transcendental equation that cannot be solved for E in closed form and must be solved numerically. We shall look in detail at (7) later, but first we want to get a qualitative picture of the basic physical situation that selects certain allowed energies.

Inside the well $\psi_e \sim \cos Kx$:

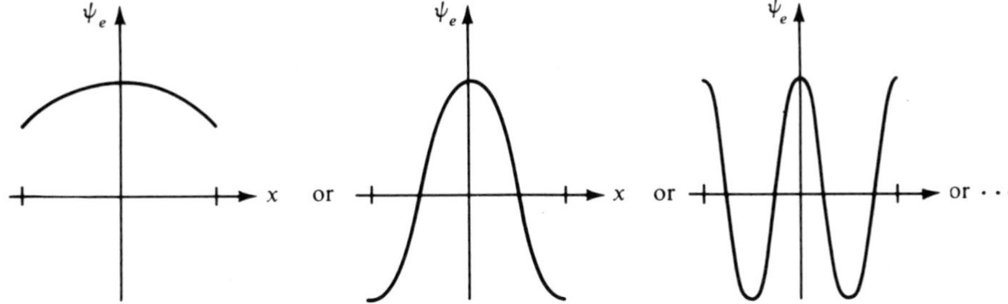

Outside, there are decreasing exponentials, $\psi_e \sim e^{-\alpha|x|}$:

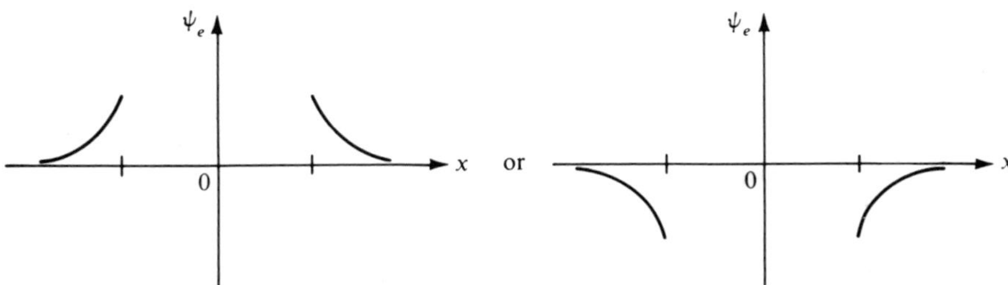

There is no difficulty in making ψ_e continuous at $\pm a/2$ because both $\cos Kx$ and $e^{-\alpha|x|}$ can be multiplied by any constant, so that (5) can be satisfied for any E, and therefore for any K and α. In general, however, continuity of the derivative will not be simultaneously achieved. For example, let us explore energies in the neighborhood of the lowest possible energy. If the energy is too low, the slope of the exponential is too steep at the edge of the well, or, if you prefer, the slope of the cosine is not steep enough:

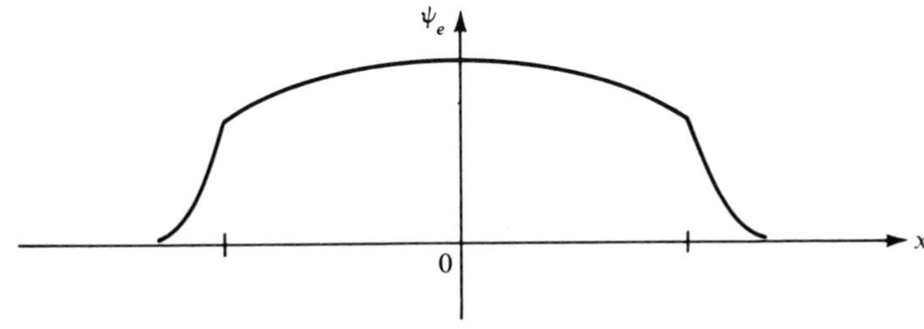

If the energy is too high, the picture looks like this:

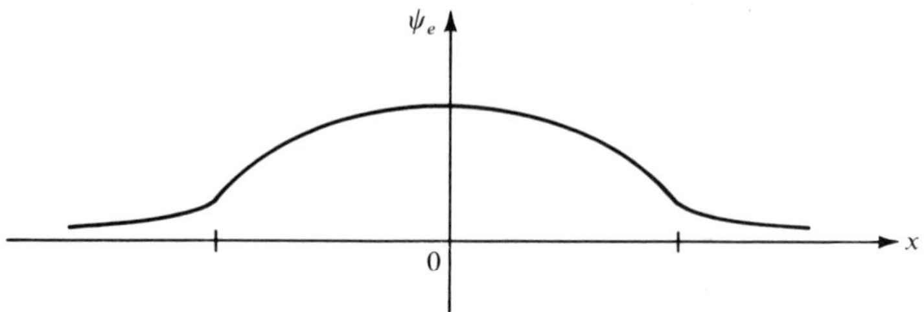

If the energy is just right, the derivative is continuous:

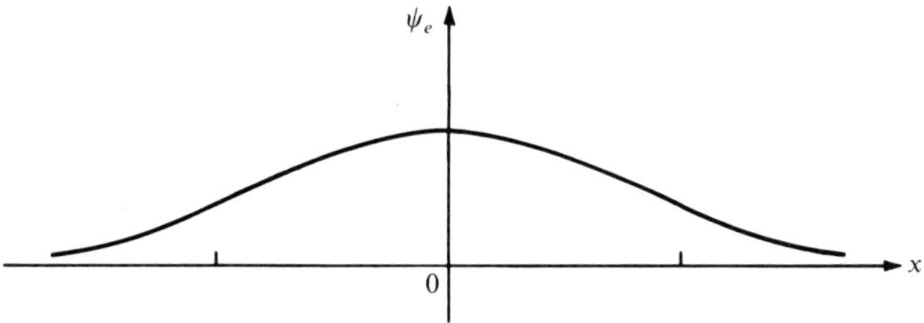

If the potential well is deep enough, then, substantially above the energy that produces this satisfactory picture, there is a second even bound state:

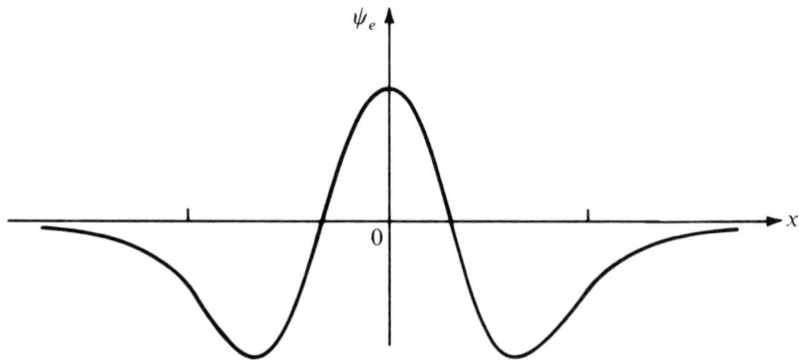

Still deeper wells have higher even states ψ_e that cross the axis 4, 6, 8, . . . times.

Now look at the *odd* solutions. These are described in $-a/2 \leq x \leq a/2$ by $I \sin Kx$; it is now J that must be zero in (3). Again A and D must be zero to keep ψ finite. Because ψ is odd, $C = -B$. The set of equations (3) becomes, with $\psi = \psi_o$ for *odd*,

$$\begin{aligned}\psi_o &= Be^{\alpha x}, & x &< \frac{-a}{2},\\ &= I\sin Kx, & \frac{-a}{2} &\le x \le \frac{+a}{2},\\ &= -Be^{-\alpha x}, & \frac{+a}{2} &< x.\end{aligned} \tag{2.5-8}$$

Continuity of ψ_o at $a/2$ yields

$$I\sin\left(\frac{Ka}{2}\right) = -Be^{-\alpha a/2}.$$

Continuity of $d\psi_o/dx$ at $a/2$ yields

$$KI\cos\left(\frac{Ka}{2}\right) = \alpha Be^{-\alpha a/2}.$$

Divide the second of these equations by the first:

$$K\cot\left(\frac{Ka}{2}\right) = -\alpha. \tag{2.5-9}$$

Like (7), (9) is a transcendental equation that can be solved numerically for E. The two lowest odd solutions look like this:

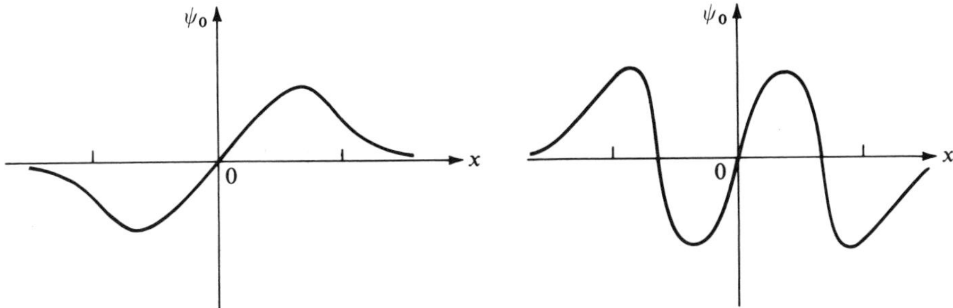

The next job is to obtain the actual values that the energy can equal, that is, to obtain the energy eigenvalues. With

$$u \equiv \frac{Ka}{2}, \qquad v \equiv \frac{\alpha a}{2},$$

(7), for the even solutions, becomes

$$v = u \tan u, \tag{2.5-10}$$

and (9), for the odd solutions, becomes

$$v = -u \cot u. \tag{2.5-11}$$

Also,

$$u^2 + v^2 = \frac{(K^2 + \alpha^2)a^2}{4} = \frac{MV_0 a^2}{2\hbar^2} \equiv w^2. \tag{2.5-12}$$

The dimensionless quantity w^2 describes the essential characteristics of the well; for example, it determines the number of possible bound states.

Look at the situation in the u, v plane. Equation (12) requires that any point (u, v) that describes a solution must lie on a circle of radius w. For an even solution u and v must also satisfy (10). Therefore each intersection of a graph of (10) with a circle of radius w represents a possible solution. Similarly, each intersection of a graph of (11) with a circle of radius w represents a possible odd solution. Since u and v are defined so that they are both positive, only intersections in the first quadrant are of interest.

Figure 2.5-5 summarizes these observations and is a device for solving the transcendental equations (10) and (11). Here are some examples that show how to extract information from this graph. Suppose that the parameter $w \equiv a\sqrt{2MV_0}/2\hbar$ is equal to 1.0. The circle of radius 1.0 intersects the curve $v = u \tan u$ once, at $u = 0.74$, $v = 0.68$. Since $v \equiv \alpha a/2 = a\sqrt{2M(-E)}/2\hbar$, it follows that the energy of this state is given by

$$E = -\frac{2\hbar^2}{Ma^2}v^2 = -\frac{2\hbar^2}{Ma^2}\cdot 0.46.$$

Since there are no other intersections with the $v = 1.0$ circle, there are no other bound states. If V_0, or a, or both are decreased so that w decreases, then the magnitude of E decreases, but there is always one bound state. If w is increased so that it lies between $\pi/2$ and π, then there are two intersections, one with the $u \tan u$ curve that gives the lowest even solution, and one with the $-u \cot u$ curve that gives the lowest odd solution. If w is increased to 6.5, there are five bound states. The lowest state is always even, and the parity of the states then alternates as the energy is increased. Figure 2.5-5 shows at a glance the number and approximate energies of the bound states. More accurate values of the energies are easily obtained by one or two numerical iterations beginning with the value of u read from the graph.

Figure 2.5-5 shows that there is one bound state if $0 < w < \pi/2$, that there are two bound states if $\pi/2 < w < \pi$, three bound states if $\pi < w < 3\pi/2, \ldots, n$ bound states if $(n - 1)\pi/2 < w < n\pi/2$. This result also follows from a little

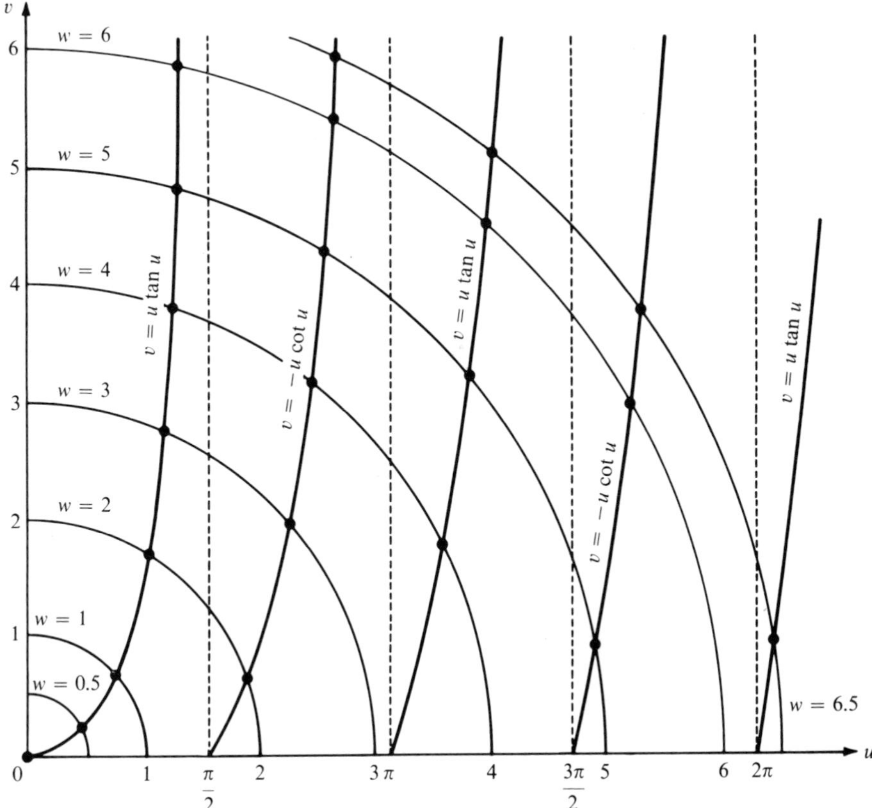

Fig. 2.5-5. Graphs for solving (2.5-10) and (2.5-11) to obtain the energy eigenvalues of a one-dimensional rectangular well.

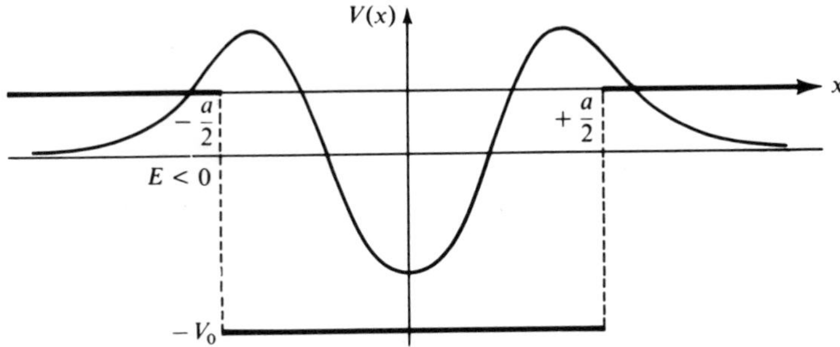

Fig. 2.5-6. The third state superposed on a graph of the potential energy for w slightly greater than π. (Plotting wave functions over their energy levels is a convenient device. It is, however, important to recognize that, while the same x axis applies to both, the zeros and the units of the vertical scales of V and ψ are different.)

thought about the behavior of the wave functions. Suppose, for example, that w is slightly larger than π, so that $\sqrt{2MV_0}$ is slightly larger than h/a. Then the electron can be bound in the well, with energy less than zero, and still have enough kinetic energy for more than a whole wavelength to fit into the distance a. As shown in Figure 2.5-6, the wave function turns over just before the edges of the well, and therefore permits decreasing exponentials to be hooked on smoothly. If the well were narrower or more shallow so that less than one wavelength would fit, the wave function would be heading away from the axis at $\pm a/2$, and the decreasing exponentials could not be hooked on properly. The two lower-lying states,

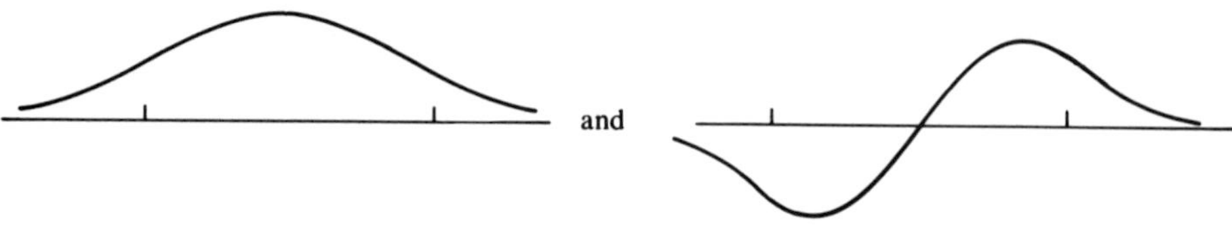

obviously exist if the state

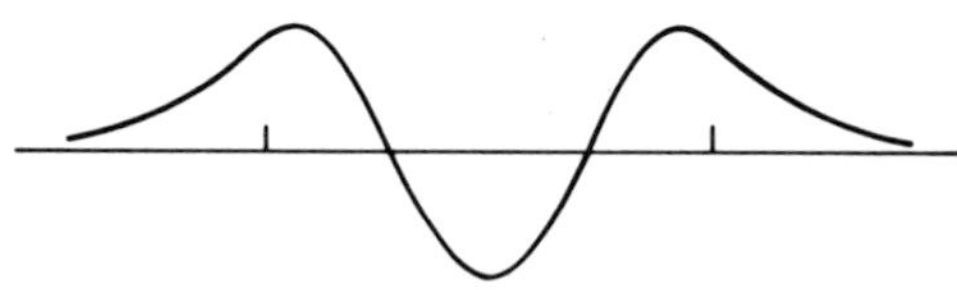

exists, and so there are three bound states.

PROBLEM 2.10

An electron is in a one-dimensional rectangular well with width $a = 4.00 \times 10^{-8}$ cm, and a depth $V_0 = 24.0$ eV. Use Figure 2.5-5 to find roughly the energies, in eV, of all possible bound states. For the lowest-lying state, the ground state, find the energy to three significant figures. Sketch qualitatively the wave functions of all the bound states. Sketch qualitatively the wave function of a typical $E > 0$ state.

PROBLEM 2.11

Here is a problem you have already solved, whether you know it or not: $V = \infty$, $x < 0$. $V = -V_0$, $0 \leq x \leq b$. $V = 0$, $x > b$. See Figure 2.5-7. In what sense have you already solved it? What condition is there on V_0 and b if no bound states exist? If three and only three bound states exist? Sketch these three bound states. (*Hint*: Insert an impenetrable wall in the middle of the symmetric rectangular well treated above.)

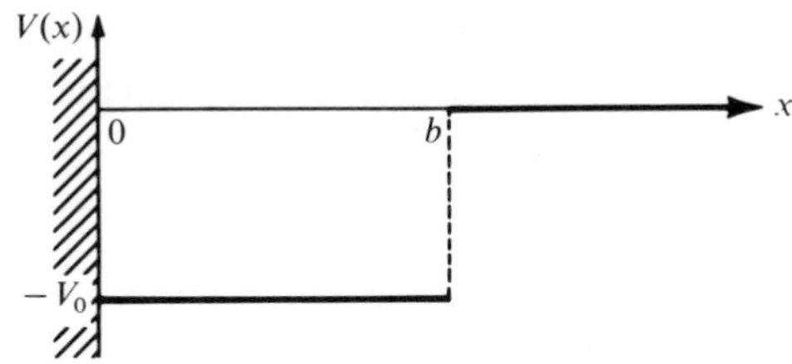

Fig. 2.5-7. A rectangular well with an impenetrable region at one edge.

The preceding deals with the question, "Given the potential energy function $V(\vec{r})$, what are E and $\psi(\vec{r})$?" This question is important, but its inverse should not be forgotten. The Schrödinger equation can be written

$$V(\vec{r}) = \frac{\hbar^2}{2M}\frac{\nabla^2\psi(\vec{r})}{\psi(\vec{r})} + E \tag{2.5-13}$$

to emphasize that if E is known and $\psi(\vec{r})$ is known within a multiplying constant, then $V(\vec{r})$ is determined.

EXAMPLE

Suppose that the $x < 0$ region is excluded because $V(x) = \infty$ for $x < 0$, and that the (unnormalized) wave function in $x \geq 0$ is

$$\psi(x) = (1 - e^{-\beta x})e^{-\alpha x}.$$

Obtain E and $V(x)$.

SOLUTION

At large x, $\psi \sim e^{-\alpha x}$. If $V(x) \to 0$ at large x, then

$$\alpha = \frac{\sqrt{2M(-E)}}{\hbar}, \qquad E = -\frac{\hbar^2\alpha^2}{2M}.$$

$$\frac{d^2\psi}{dx^2} = \alpha^2 e^{-\alpha x} - (\alpha + \beta)^2 e^{-(\alpha+\beta)x}.$$

Substitute into the one-dimensional form of (13).

$$V(x) = \frac{\hbar^2}{2M}\frac{\alpha^2 e^{-\alpha x} - (\alpha + \beta)^2 e^{-(\alpha+\beta)x}}{(1 - e^{-\beta x})e^{-\alpha x}} - \frac{\hbar^2\alpha^2}{2M}$$

$$= -\frac{\hbar^2}{2M}\frac{2\alpha\beta + \beta^2}{e^{\beta x} - 1}.$$

With the coordinate x replaced by the radius r, this form of V is called the Hulthén potential, and is of some use in describing the interaction between neutrons and protons. It is instructive to sketch ψ and V, and to examine their behavior near the origin and at large distances.

The nuclear force problem is a relevant example. As pointed out at the end of Section 1.4, there is no classical guidance for the choice of a nuclear potential, and one must deduce the potential from measurements that give energy levels and various pieces of information about nuclear wave functions. A common variation of this approach is to choose a plausible form for $V(\vec{r})$ and to adjust parameters in this form to fit the data.

PROBLEM 2.12

A neutron is in the ground state of a well approximated by (1). The lowest possible energy of a gamma ray that can remove the neutron from the well is 8.0 MeV. The width a is 3.0×10^{-13} cm. Obtain the depth V_0 in MeV.

A singular extreme of the rectangular well model is the attractive delta function potential,

$$V(x) = -U\delta(x),$$

where U is a positive constant.

$$-\frac{\hbar^2}{2M}\frac{d^2\psi}{dx^2} - U\delta(x)\psi = E\psi$$

must be solved for a bound state, $E < 0$. If $x \neq 0$, the potential is zero and

$$\frac{d^2\psi}{dx^2} = \alpha^2\psi.$$

For $x > 0$ the only acceptable solution is $Ae^{-\alpha x}$. For $x < 0$ it is $Ae^{+\alpha x}$. Continuity at $x = 0$ demands that the same constant A appear in front of both exponentials. The constant α, and therefore the energy E, are determined by the derivative condition (2.3-6). There is only a single bound state. The complete set of energy eigenfunctions consists of this single bound state plus the positive energy states which exist for any $E > 0$.

PROBLEM 2.13

Complete the calculation for the energy of the bound state; it is $-MU^2/2\hbar^2$. Obtain and sketch the normalized wave function.

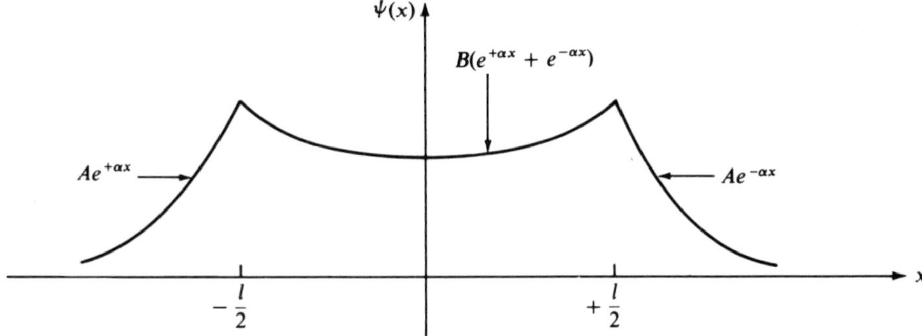

Fig. 2.5-8. The even bound state for two attractive delta function wells separated by a distance l.

We are still some distance from calculations that can give quantitative agreement with measurements. There are no rectangular and delta function barriers in nature, and the world is three-dimensional. We can, however, begin to understand some things qualitatively. The rectangular well calculation indicates that all bound, $E < 0$, states have wave functions with an interior oscillatory part, an exterior exponential, and a quantization of energy traceable to some matching or fitting condition. Unbound, $E > 0$, states exist for any energy. Even for these states, however, there will be resonances, marked by a relatively large probability of finding the particle in the region of the well, whenever the wave fits nicely into the well.

More complicated things can also be understood on the basis of simple models involving only rectangular and delta function wells. Here is a crude picture that yields some insight into the reason why the singly ionized hydrogen molecule is a bound system. The electron sees the sum of the attractive Coulomb potentials of the two separated protons. Replace this actual situation by a one-dimensional problem in which the electron sees two narrow attractive wells,

$$V(x) = -U\left[\delta\left(x + \frac{l}{2}\right) + \delta\left(x - \frac{l}{2}\right)\right], \tag{2.5-14}$$

as illustrated in Figure 2.5-8. The parameter U might reasonably be chosen so that $MU^2/2\hbar^2 = 13.6$ eV, or $U = 0.53$ Å $\cdot$ 27.2 eV, to give the right binding for an isolated well. Now solve the Schrödinger equation in the three regions separated by the two delta functions for the *even* solution. For $x < -l/2$, $\psi = Ae^{+\alpha x}$. For $x > +l/2$, $\psi = Ae^{-\alpha x}$; the coefficient A is the same because ψ has even parity. Between $-l/2$ and $+l/2$ the general solution is $Be^{+\alpha x} + Ce^{-\alpha x}$, with $B = C$ because ψ is even. Continuity at $+l/2$ implies

$$B(e^{+\alpha l/2} + e^{-\alpha l/2}) = Ae^{-\alpha l/2}.$$

Condition (2.3-6) on the derivative becomes

$$(-\alpha A e^{-\alpha l/2}) - (\alpha e^{\alpha l/2} - \alpha e^{-\alpha l/2})B = -\frac{2MU}{\hbar^2} A e^{-\alpha l/2}.$$

Eliminate B and rearrange:

$$\frac{\hbar^2}{MU}\alpha = 1 + e^{-\alpha l}.$$

For very large l, where $\alpha l \gg 1$, $\alpha = MU/\hbar^2 \equiv \alpha_\infty$, just as in Problem 2.13, because the two wells are so far apart that they cannot influence each other. For very small l, where $\alpha l \ll 1$, $\alpha = 2\alpha_\infty$, because the two wells sitting on top of each other simply become one well. For intermediate values of l, $\alpha_\infty < \alpha < 2\alpha_\infty$. For example, if $l = 1/\alpha_\infty$, then $\alpha = 1.28\alpha_\infty$. The energy, given by $E = -\alpha^2\hbar^2/2M$, therefore decreases as the protons are brought closer together, and the system becomes better bound. In the actual H_2^-, the effect indicated here is counterbalanced by the Coulomb repulsion between the protons to give an equilibrium for l near 1 Å.

PROBLEM 2.14

Figure 2.5-8 shows the *even* parity bound state for the $V(x)$ given in (14). There is also an *odd* parity state. Find its wave function and its energy as a function of l. Can you understand qualitatively why the energy of the odd state is higher than that of the even state?

2.6 Complex Potential Energies

Look again at Section 1.5. With the definition of the current (1.5-6), (1.5-5) can be written

$$\frac{\partial \rho}{\partial t} + \vec{\nabla} \cdot \vec{j} = \frac{V - V^*}{i\hbar} \rho. \tag{2.6-1}$$

If V is not real, particles are not in general conserved. The usual meaning of potential energy does not suggest having a complex V, but its introduction is a useful device for constructing phenomenological models of the absorption and decay of particles.

As an example, consider the consequence of choosing

$$\begin{aligned} V &= 0, && x < 0, \\ &= -V_0 - iW, && x \geq 0, \end{aligned} \tag{2.6-2}$$

where V_0 and W are real constants. In the $x < 0$ region, the solution is, for $k \equiv \sqrt{2ME}/\hbar$,

$$Ae^{ikx} + Be^{-ikx}.$$

In $x \geq 0$, the Schrödinger equation is

$$\frac{d^2\psi}{dx^2} + \Lambda^2\psi = 0,$$

where

$$\Lambda^2 = \frac{2M}{\hbar^2}(E + V_0) + \frac{2iMW}{\hbar^2}.$$

The quantity Λ must be complex if $W \neq 0$. Let

$$\Lambda \equiv K + i\zeta,$$

where K and ζ are real. Then

$$\Lambda^2 = K^2 - \zeta^2 + 2iK\zeta = \frac{2M}{\hbar^2}(E + V_0) + \frac{2iMW}{\hbar^2}.$$

Since the real and imaginary terms must separately equal each other,

$$K^2 = \frac{M}{\hbar^2}[(E + V_0) + \sqrt{(E + V_0)^2 + W^2}], \qquad \zeta = \frac{MW}{\hbar^2 K}.$$

In $x \geq 0$,

$$\begin{aligned}\psi &= Ce^{+i\Lambda x} + De^{-i\Lambda x}\\ &= Ce^{+iKx}e^{-\zeta x} + De^{-iKx}e^{+\zeta x}.\end{aligned}$$

For a wave traveling toward $+x$, choose $D = 0$. Then it is necessary to choose $\zeta > 0$, and therefore $W > 0$, so that ψ does not diverge as x gets large. Now

$$\rho = |\psi|^2 = |C|^2 e^{-2\zeta x}; \tag{2.6-3}$$

the probability density decreases as x increases from zero, and we have produced a model of an absorbing medium. Look at this result from the point of view of (1). Since the state has a definite energy, the wave function is $\psi(x)e^{-iEt/\hbar}$, and ρ is independent of the time. With $\partial\rho/\partial t = 0$, (1) becomes in one dimension

$$\frac{dj}{dx} = \frac{V - V^*}{i\hbar}\rho. \tag{2.6-4}$$

Here

$$\begin{aligned} j &= \frac{\hbar}{2Mi}[C^* e^{-iKx-\zeta x}(iK - \zeta)Ce^{+iKx-\zeta x} - \text{complex conjugate}]\\ &= \frac{\hbar K}{M}|C|^2 e^{-2\zeta x} = \frac{\hbar K}{M}\rho. \end{aligned} \tag{2.6-5}$$

The term on the right of (4) becomes

$$-\frac{2W\rho}{\hbar}.$$

The quantity $-(dj/dx)\Delta x$ is the number per second crossing a plane at x, minus the number per second crossing the plane at $x + \Delta x$. This quantity is positive and equals $(2W\rho/\hbar)\Delta x$, which therefore measures the number absorbed per second in Δx.

PROBLEM 2.15

Very light nuclei are rather transparent to 100-MeV neutrons, while heavy nuclei are not. That is, if a 100-MeV neutron has a head-on encounter with a He^3 nucleus, it has a good chance of passing through without anything happening, but if its encounter is with a U^{238} nucleus, it will probably be absorbed. Use $V_0 = 30$ MeV, your general ideas about magnitudes, and (3) to estimate a numerical value for W in (2) that gives a reasonable picture of nuclear matter as seen by a 100-MeV neutron.

Next, look at the effect of a complex potential on a bound state. To avoid labor, consider a delta function potential

$$V(x) = (-U - iY)\delta(x).$$

Just as in Problem 2.13, for $x \neq 0$,

$$\frac{d^2\psi}{dx^2} = \alpha^2\psi, \qquad \alpha = \frac{\sqrt{2M(-E)}}{\hbar},$$

$$\begin{aligned} \psi &= Ae^{+\alpha x}, \qquad x < 0, \\ &= Ae^{-\alpha x}, \qquad x > 0. \end{aligned}$$

Condition (2.3-6) on the derivative at the origin is now

$$\left(\frac{d\psi}{dx}\right)_{+\varepsilon} - \left(\frac{d\psi}{dx}\right)_{-\varepsilon} = \frac{2M}{\hbar^2}(-U - iY)\psi(0),$$

or

$$\alpha = \frac{M(U + iY)}{\hbar^2},$$

so that E acquires an imaginary part:

$$E = E_0 - \frac{i\Gamma}{2},$$

where

$$E_0 \equiv \frac{-M(U^2 - Y^2)}{2\hbar^2},$$
$$\Gamma \equiv \frac{2MUY}{\hbar^2}. \qquad (2.6\text{-}5)$$

The wave function that includes position and time dependence is

$$\Psi(x, t) = \psi(x)e^{-iEt/\hbar} = \psi(x)e^{-iE_0t/\hbar}e^{-\Gamma t/2\hbar}.$$

The probability of finding the particle somewhwere is

$$\int_{-\infty}^{+\infty} |\Psi(x, t)|^2\, dx = \left[\int_{-\infty}^{+\infty} |\psi(x)|^2\, dx\right] e^{-\Gamma t/\hbar} = P_0 e^{-t/\tau}, \qquad (2.6\text{-}6)$$

where the mean life

$$\tau \equiv \frac{\hbar}{\Gamma}, \qquad (2.6\text{-}7)$$

and P_0 is the probability of finding the particle somewhere at $t = 0$.

We have then produced a model of a decaying state. Think, for example, about a muon which decays with a mean life of 2.2×10^{-6} seconds into an electron and two neutrinos:

$$\mu^- \to e^- + \bar{\nu}_e + \nu_\mu.$$

Suppose that the muon was trapped in some state in the attractive force field of a nucleus at $t = 0$. Then with $P_0 = 1$ and $\tau = 2.2 \times 10^{-6}$ sec, we have a useful model of the situation. The muon does not, of course, vanish into nothingness, but if the apparatus is not sensitive to electrons and neutrinos, decay is equivalent to disappearance. Other examples should come to mind, because $e^{-t/\tau}$ approximates quite generally the behavior of unstable particles, radioactive nuclei, and decaying atomic states.

PROBLEM 2.16

Suppose that $V = -iW$, a constant, for $0 \leq x \leq a$, and $V = \infty$ otherwise. Obtain the wave function for the lowest state. Normalize it at $t = 0$, and calculate the probability of still finding the particle at time t.

At the beginning of this section, the introduction of complex potentials was called a device for constructing a *phenomenological* model. This unlovely bit of jargon is used in physics in a rather special way. It is used to describe models that are arranged to fit the observations, but that are some distance away from the most fundamental theories that are available. The description of neutrons absorbed in nuclear matter could be pursued to the deeper level of interactions of the incident neutron with the nucleons in the nucleus. The disappearance of the muon could be described in terms of the interactions that lead to the decay. In both cases we chose not to go more deeply, and instead constructed phenomenological models, that is, models with parameters adjusted to describe some of the aspects of the phenomena.

2.7 *The Harmonic Oscillator*

For most systems small excursions in the neighborhood of a position of stable equilibrium can be approximated by harmonic oscillations. In classical mechanics the same calculation treats small vibrations of a mass at the end of a spring, a pendulum, a rocking chair, and the balance wheel of a watch. Review the classical harmonic oscillator and make sure you understand:

a. The equation of motion and the various ways of writing the solution.

b. The potential energy as a function of displacement, the kinetic energy as a function of displacement, and the fact that the average value of potential energy equals the average value of kinetic energy.

c. That small oscillations around an equilibrium position are usually, but not always, harmonic.

d. That it is a poor approximation for large displacements.

In quantum physics the applicability of the harmonic oscillator model to things like atoms vibrating around equilibrium positions in molecules and crystals is obvious. In addition, because of the relative simplicity of harmonic oscillator wave functions, these functions are used as a first approximation in a great variety of problems, such as those dealing with nucleons moving in the potential energy well of a nucleus.

We want then to look at the consequences of choosing the potential energy

$$V(x, y, z) = \frac{K_x x^2}{2} + \frac{K_y y^2}{2} + \frac{K_z z^2}{2}. \tag{2.7-1}$$

This function rises parabolically everywhere and grows indefinitely at large distances. It has only bound states, and since the potential is zero at the bottom of the well, these bound states all have positive energy. The energy bookkeeping is somewhat like that used for the problem of a particle in a box, described in Section 2.4.

The three spring constants K_x, K_y, and K_z may or may not be the same. For example, an atom in a crystal may feel less resistance to displacement in one direction than in another. If the three spring constants are the same, then

$V = Kr^2/2$, and the problem can be solved in spherical polar coordinates as well as in Cartesian coordinates; see Section 7.2.

Since (1) is of the form

$$V(x, y, z) = V_x(x) + V_y(y) + V_z(z),$$

separation in Cartesians works as described in Section 2.1, the three equations (2.1-7) result, and we need to look at only one of these equations,

$$-\frac{\hbar^2}{2M}\frac{d^2\psi_x(x)}{dx^2} + \frac{K_x x^2}{2}\psi_x(x) = E_x\psi_x(x). \tag{2.7-2}$$

To save writing, let

$$\begin{aligned} \beta &\equiv \sqrt{\frac{M\omega_x}{\hbar}}, \\ \xi &\equiv \beta x, \\ \eta(\xi) &\equiv \psi_x(x), \\ \varepsilon &\equiv \frac{2E_x}{\hbar\omega_x}, \end{aligned} \tag{2.7-3}$$

where $\omega_x \equiv \sqrt{K_x/M}$ is the *classical* angular frequency of the oscillator. Equation (2) then becomes

$$-\frac{d^2\eta(\xi)}{d\xi^2} + \xi^2\eta(\xi) = \varepsilon\eta(\xi). \tag{2.7-4}$$

This equation is called *Weber's equation*, and it has been treated in detail in the mathematical literature. Look first at its behavior in the *asymptotic* regions $\xi \to \pm\infty$. There $\xi^2\eta \gg \varepsilon\eta$ for any finite ε, so that (4) is approximately

$$\frac{d^2\eta}{d\xi^2} \simeq \xi^2\eta.$$

This equation is solved to leading powers of ξ by

$$\eta \simeq \xi^n e^{-\xi^2/2} \qquad \text{and} \qquad \xi^n e^{+\xi^2/2}$$

for any n. The second possibility must be rejected because it diverges. We therefore look for a solution of the form

$$\eta(\xi) = H(\xi)e^{-\xi^2/2}, \tag{2.7-5}$$

where $H(\xi)$ remains to be determined. The fact that $\xi^n e^{-\xi^2/2}$ is an approximate solution might inspire the correct guess that $H(\xi)$ will turn out to be a polynomial in ξ.

Substitution of (5) into (4) yields

$$\frac{d^2H(\xi)}{d\xi^2} - 2\xi\frac{dH(\xi)}{d\xi} + (\varepsilon - 1)H(\xi) = 0. \tag{2.7-6}$$

Now look for a series solution

$$H(\xi) = \xi^s(a_0 + a_1\xi + a_2\xi^2 + a_3\xi^3 + \cdots), \tag{2.7-7}$$

with $a_0 \neq 0$ so that ξ^s is the lowest power of ξ that occurs. Substitute (7) into (6).

$$\begin{aligned}&[s(s-1)a_0\xi^{s-2} + (s+1)sa_1\xi^{s-1} + \cdots + (s+\nu)(s+\nu-1)a_\nu\xi^{s+\nu-2} + \cdots]\\ &\quad - 2[sa_0\xi^s + (s+1)a_1\xi^{s+1} + \cdots + (s+\nu)a_\nu\xi^{s+\nu} + \cdots]\\ &\quad + (\varepsilon - 1)[a_0\xi^s + a_1\xi^{s+1} + \cdots + a_\nu\xi^{s+\nu} + \cdots] = 0.\end{aligned}$$

Since this equation is to be valid for all ξ, the coefficient of each power of ξ must separately be zero. Since $a_0 \neq 0$, the requirement that the coefficient of ξ^{s-2} be zero demands

$$s(s-1) = 0. \tag{2.7-8}$$

This equation gives the two possibilities, $s = 0$ and $s = 1$, for the lowest power of ξ that can occur in the solution. It is called the *indicial* equation; the occurrence of an equation of this kind is common in power series solutions to differential equations.

The requirement that the coefficient of ξ^{s-1} be zero yields

$$(s+1)sa_1 = 0. \tag{2.7-9}$$

For the higher powers of ξ, the equations are the following:

$$\begin{aligned}\xi^s&: (s+2)(s+1)a_2 - (2s+1-\varepsilon)a_0 = 0.\\ \xi^{s+1}&: (s+3)(s+2)a_3 - (2s+3-\varepsilon)a_1 = 0.\\ &\vdots\\ \xi^{s+\nu}&: (s+\nu+2)(s+\nu+1)a_{\nu+2} - (2s+2\nu+1-\varepsilon)a_\nu = 0.\\ &\vdots\end{aligned} \tag{2.7-10}$$

Suppose we choose some value for $a_0 \neq 0$. The ξ^s equation then determines a_2, the ξ^{s+2} equation determines a_4, the ξ^{s+4} equation determines a_6, and so forth. Similarly, if $a_1 \neq 0$, then the ξ^{s+1} equation determines a_3, the ξ^{s+3} equation determines a_5, and so forth. These *recursion relations* develop all of the series once the lowest terms have been chosen.

Now look at the ratio of successive terms in these series. For large ν, in both the series that builds on a_0 and the one that builds on a_1,

$$\frac{a_{\nu+2}}{a_\nu} = \frac{2s+2\nu+1-\varepsilon}{(s+\nu+2)(s+\nu+1)} \to \frac{2}{\nu}.$$

This ratio is the same as that for a power series expansion of $\xi^n e^{+\xi^2}$ for any n.

$$\xi^n e^{\xi^2} = \xi^n \left[1 + \xi^2 + \frac{\xi^4}{2!} + \cdots + \frac{\xi^\nu}{(\nu/2)!} + \frac{\xi^{\nu+2}}{(1 + \nu/2)!} + \cdots \right],$$

$$\frac{(\nu/2)!}{(1 + \nu/2)!} = \frac{2}{\nu + 2} \to \frac{2}{\nu}.$$

If the function $H(\xi)$ is indeed an infinite series of such behavior, then the wave function $H(\xi)e^{-\xi^2/2}$ would diverge as $\xi \to \pm\infty$. The series must be cut off, and (10) shows how to do it: The energy must be chosen so that, for some ν,

$$\varepsilon = 2s + 2\nu + 1. \tag{2.7-11}$$

Since $a_0 \neq 0$, ε must be selected to cut off the series that is proportional to a_0. This value of ε will not cut off the series proportional to a_1, and to avoid divergence as $\xi \to \pm\infty$, a_1 must be set equal to zero.

That $a_1 = 0$ can also be seen another way. The potential energy $K_x x^2/2$ is symmetric, so that, as shown in Section 2.5, the parity must be definitely even or definitely odd. Since $a_0 \neq 0$, the parity is $(-1)^s$ for the leading term $a_0 \xi^s e^{-\xi^2/2}$. If $a_1 \neq 0$, there would also be the term $a_1 \xi^{s+1} e^{-\xi^2/2}$, which would have parity $(-1)^{s+1}$, and the solution would be neither even nor odd.

Equation (9) is then always satisfied, and the two possible solutions of the indicial equation (8) can be explored without further restrictions. If $s = 0$, (10) becomes

$$a_{\nu+2} = \frac{2\nu + 1 - \varepsilon}{(\nu + 2)(\nu + 1)} a_\nu.$$

To cut off the series, $\varepsilon = 1$ or 5 or 9 or . . . because ν is even. Suppose for example that $\varepsilon = 5$. Then

$$a_2 = \frac{2 \cdot 0 + 1 - 5}{(0 + 2)(0 + 1)} a_0 = -2a_0,$$

$$a_4 = \frac{2 \cdot 2 + 1 - 5}{(2 + 2)(2 + 1)} a_2 = 0,$$

and the polynomial

$$H_2(\xi) = a_0(1 - 2\xi^2).$$

It is conventional and convenient to label the polynomials with a subscript that equals the highest power of the variable that occurs. If $s = 1$, (10) becomes

$$a_{\nu+2} = \frac{2\nu + 3 - \varepsilon}{(\nu + 3)(\nu + 2)} a_\nu.$$

Now ε must be 3 or 7 or 11 or .... Suppose for example $\varepsilon = 11$. Then

$$a_2 = \frac{2\cdot 0 + 3 - 11}{(0+3)(0+2)} a_0 = -\tfrac{4}{3}a_0,$$

$$a_4 = \frac{2\cdot 2 + 3 - 11}{(2+3)(2+2)} a_2 = -\tfrac{1}{5}a_2 = \tfrac{4}{15}a_0,$$

$$a_6 = \frac{2\cdot 4 + 3 - 11}{(4+3)(4+2)} a_4 = 0,$$

and

$$H_5(\xi) = a_0(\xi - \tfrac{4}{3}\xi^3 + \tfrac{4}{15}\xi^5).$$

The $H_n(\xi)$ are called *Hermite polynomials.*

The requirement that the wave function must remain finite at large distances has again led to the appearance of a discrete spectrum of allowed energies. The $s = 0$ and $s = 1$ possibilities together permit

$$\varepsilon = 1, 3, 5, 7, \ldots.$$

From (3) the x contribution to the energy is

$$E_{xn} = (n + \tfrac{1}{2})\hbar\omega_x, \qquad n = 0, 1, 2, \ldots. \tag{2.7-12}$$

The energy E_x therefore has a lowest possible value, $\hbar\omega_x/2$, and a set of evenly spaced higher states that differ from each other by $\hbar\omega_x$.

The wave functions can now be assembled. Each energy eigenstate is, according to (5), a Gaussian $e^{-\xi^2/2}$ times a Hermite polynomial, which can be obtained by the technique used for H_2 and H_5 in the sample calculations above. The first few H_n are given below. The arbitrary multiplying constants a_0 have been chosen to match the conventional definitions of the H_n.

$$\begin{aligned} H_0 &= 1 \\ H_1 &= 2\xi \\ H_2 &= 4\xi^2 - 2 \\ H_3 &= 8\xi^3 - 12\xi \\ H_4 &= 16\xi^4 - 48\xi^2 + 12 \\ H_5 &= 32\xi^5 - 160\xi^3 + 120\xi \\ H_6 &= 64\xi^6 - 480\xi^4 + 720\xi^2 - 120. \end{aligned} \tag{2.7-13}$$

Our approach is adequate here, but there are various useful and interesting relations involving the H_n which you should learn when you get involved in more detailed calculations.[8]

[8] Two such results are $e^{-t^2+2t\xi} = \sum_{n=0}^{\infty} H_n(\xi)t^n/n!$, where $e^{-t^2+2t\xi}$ is called a *generating function*, and the Rodrigues formula, $H_n = (-1)^n e^{\xi^2}\,\partial^n/\partial\xi^n e^{-\xi^2}$. The form of the H_n given above is that which matches these expressions. Verify that the second of these expressions is a convenient way of calculating the H_n. For these and other properties of the H_n, see L. I. Schiff, *Quantum Mechanics*, 3d ed. (New York: McGraw-Hill Book Company, 1968), Sec. 13.

The wave function is now given in terms of the dimensionless variable ξ by (5), or, in terms of x, by

$$\psi_{xn}(x) = A_n H_n(\beta x) e^{-\beta^2 x^2/2}, \tag{2.7-14}$$

where A_n is a normalization constant, determined by the requirement that

$$\int_{-\infty}^{+\infty} |\psi_{xn}(x)|^2 \, dx = A_n^2 \int_{-\infty}^{+\infty} [H_n(\beta x)]^2 e^{-\beta^2 x^2} \, dx = 1.$$

Any desired normalization constant can be found directly, or results of the kind mentioned in footnote 8 can be used to obtain the general form

$$A_n = \sqrt{\frac{\beta}{2^n n! \sqrt{\pi}}}. \tag{2.7-15}$$

Since the Gaussian $e^{-\beta^2 x^2/2}$ is even, the parity of the wave functions is that of the H_n, which is $(-1)^n$. The alternation of even and odd states is like that found in the rectangular well problem.

As in Section 2.4, the behavior of the ground state can be viewed as an illustration of the uncertainty principle. The probability density, $A_0^2 e^{-\beta^2 x^2}$, is $1/e$ times its $x = 0$ value at $x = \pm 1/\beta$, so that $\Delta x \simeq 1/\beta$. At the origin $E - V(x) = \hbar\omega_x/2 - 0$, so that an estimate of the kinetic energy there is $p^2/2M = \hbar\omega_x/2$. With $p \simeq \Delta p$, $\Delta p \simeq \sqrt{\hbar M \omega_x} = \hbar\beta$, and $\Delta x \, \Delta p \simeq \hbar$. For a particle in any potential well, the energy, measured from the bottom of the well, cannot be zero. If the particle is confined in some Δx, it must jiggle at least enough so that its Δp satisfies $\Delta x \, \Delta p \simeq \hbar$. This lowest possible energy, $\hbar\omega_x/2$ for the oscillator, is called the *zero point energy*; the energy per atom approaches this value as the absolute temperature approaches zero. The very low temperature behavior of helium is a good illustration. Helium atoms attract each other very slightly. Classically one would expect to be able to solidify helium at any pressure; the kinetic energy could be made negligible, and the attractive forces would have to win at sufficiently low temperatures. Actually there is a lowest possible nonzero amount of kinetic energy that wins over the weak attractive forces, and helium does not solidify at any temperature unless the pressure is raised to about 26 atmospheres.

PROBLEM 2.17

In the rectangular well problem, the wave function dies out in the classically forbidden region like $e^{-\alpha|x|}$. For the harmonic oscillator, however, it dies out more rapidly, like $e^{-\beta^2 x^2/2}$. Give a qualitative explanation of the difference.

PROBLEM 2.18

Calculate H_0 and H_2. Obtain $\Psi_{x0}(x, t)$ and $\Psi_{x2}(x, t)$ with the normalization constants derived by you, not lifted from (15). Sketch the five lowest wave functions $\psi_{x0,1,2,3,4}(x)$.

PROBLEM 2.19

Transitions between vibrational levels of diatomic molecules give rise to infrared radiation. Use this fact to estimate the spring constant that approximately describes the force between two atoms in a molecule.

PROBLEM 2.20

Suppose that

$$\begin{aligned} V &= \infty, \qquad & x < 0, \\ &= \frac{K_x x^2}{2}, \qquad & x \geq 0. \end{aligned}$$

Use the ideas of Problem 2.11 to obtain the wave functions $\psi_x(x)$ and energy levels E_x.

The solution in three dimensions can now be assembled by multiplying together three solutions of the form developed above. With similar notation for the x, y, and z contributions,

$$\begin{aligned} \Psi_{n_x n_y n_z}(x, y, z, t) &= \Psi_{n_x}(x, t)\Psi_{n_y}(y, t)\Psi_{n_z}(z, t) \\ &= \psi_{n_x}(x)\psi_{n_y}(y)\psi_{n_z}(z)e^{-iEt/\hbar}, \end{aligned} \tag{2.7-16}$$

where

$$E = E_x + E_y + E_z = (n_x + \tfrac{1}{2})\hbar\omega_x + (n_y + \tfrac{1}{2})\hbar\omega_y + (n_z + \tfrac{1}{2})\hbar\omega_z.$$

If each of the one-dimensional wave functions is separately normalized, then the three-dimensional wave function is also normalized;

$$\begin{aligned} &\int_{-\infty}^{+\infty}\int_{-\infty}^{+\infty}\int_{-\infty}^{+\infty} |\Psi(x, y, z, t)|^2\,dx\,dy\,dz \\ &\quad = \int_{-\infty}^{+\infty} |\psi_x|^2\,dx \int_{-\infty}^{+\infty} |\psi_y|^2\,dy \int_{-\infty}^{+\infty} |\psi_z|^2\,dz = 1 \cdot 1 \cdot 1 = 1. \end{aligned}$$

If the oscillator is isotropic, so that $K_x = K_y = K_z \equiv K$, then with $\omega \equiv \sqrt{K/M}$,

$$E = (n_x + n_y + n_z + \tfrac{3}{2})\hbar\omega. \tag{2.7-17}$$

The ground state, with $E = 3\hbar\omega/2$, is not degenerate, but all higher states are. The even spacing of the levels produces an unusually high degree of degeneracy. The isotropic harmonic oscillator can be separated in Cartesian, spherical polar, and cylindrical coordinates. It is generally true that separability in several different coordinate systems is associated with high degeneracy.

PROBLEM 2.21

a. Show that, for an isotropic harmonic oscillator, the degree of degeneracy of a level with energy $(n + \frac{3}{2})\hbar\omega$ is

$$\frac{(n+1)(n+2)}{2}.$$

b. Suppose we have cylindrical symmetry, $K_x = K_y \neq K_z$. Discuss the degree of degeneracy.

2.8 *Some Comments about Wave Functions*

Sections 2.1 through 2.7 show how ψ acts in the presence of various mathematically tractable potential energy functions. This section uses the experience gained to discuss general features of the way $\psi(x)$ responds to $V(x)$, and also to deal more honestly with beams of particles.

Suppose that the Schrödinger equation has been separated so that, for definite energy E, the x behavior satisfies

$$\frac{d^2\psi}{dx^2} = \frac{2M}{\hbar^2}[V(x) - E]\psi.$$

Then $d^2\psi/dx^2$ has the same sign as ψ if $V > E$, and opposite sign if $V < E$. If $V(x)$ is *everywhere* greater than E, the wave function curves away from the axis everywhere, and diverges as $x \to +\infty$, or as $x \to -\infty$, or both. If an acceptable ψ exists, there must be somewhere a region in which $E > V$. In this region ψ curves toward the axis and therefore has oscillatory behavior. Of course, the oscillations will be sinusoidal only in intervals where V is constant. The magnitude of the ratio of curvature to value is large if the kinetic energy $E - V(x)$ is large, and small if $E - V(x)$ is small. The amplitude tends to be large where $E - V(x)$ is small because the particle is most likely to be found where it moves slowly; see Problem 1.9.

An inflection point occurs wherever $V(x) = E$ because $d^2\psi/dx^2$ is zero there. If $V(x) = E$ in an extended interval, ψ is a straight line; see Problem 2.3.

PROBLEM 2.22

Verify that the ground state and the first excited state wave functions of the one-dimensional harmonic oscillator have inflection points wherever $V(x) = E$.

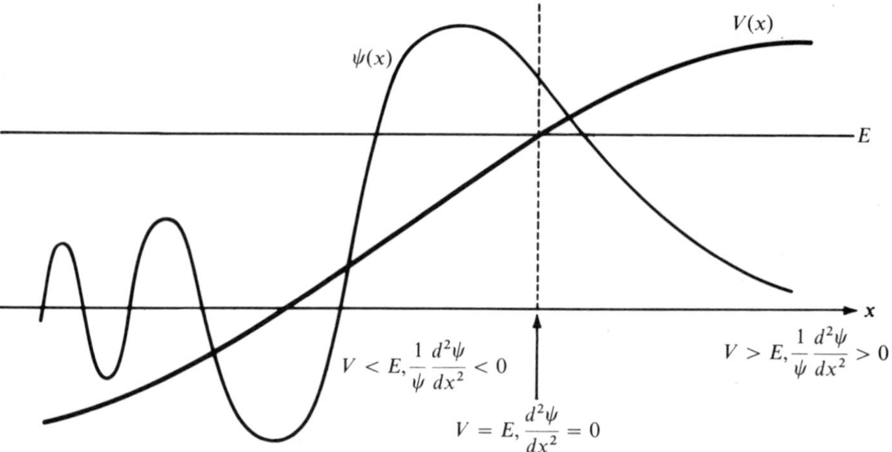

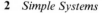

Fig. 2.8-1. The general behavior of ψ. For $V < E$, ψ oscillates with short wavelength and small amplitude where $E - V$ is large and with long wavelength and large amplitude where $E - V$ is small. There is an inflection point where $V = E$. For $V > E$, the form of ψ is roughly exponential.

In a region where $V(x) > E$, $d^2\psi/dx^2$ and ψ have the same sign, ψ curves away from the axis, and its behavior is roughly exponential. Of course, it is actually exponential only in intervals where V is constant.

If $V(-\infty)$ and $V(+\infty)$ both exceed E, the particle is trapped in a *bound state*. Such a state can exist only for certain discrete energies because ψ must approach zero as $x \to -\infty$ and also as $x \to +\infty$. One can choose, for any energy, a solution that goes to zero at one extreme, say as $x \to -\infty$, but it is a very restrictive requirement that the solution obtained by integrating through the potential well also goes to zero as $x \to +\infty$.

PROBLEM 2.23

Review your solution to Problem 2.10. $E = -15$ eV was *not* one of the possible energies. Nevertheless, try to construct a wave function with $E = -15$ eV for the parameters of Problem 2.10 as follows. Arrange that ψ behaves decently in $x < -a/2$ by giving it the form $e^{+\alpha x}$ there. Require continuity and continuity of derivative at $-a/2$ and at $+a/2$ to obtain the rest of ψ. Is the ψ so obtained acceptable? Discuss what happens if E is changed continuously from -15 eV until it passes through one of the energy eigenvalues found in Problem 2.10.

To normalize a bound state, one requires that

$$\int_{-\infty}^{+\infty} |\psi(x)|^2 \, dx = 1.$$

This integral will typically have contributions from a classically accessible region where $E > V(x)$, and from regions where $E < V(x)$. If ψ describes a low-lying state in a wide and deep well, most of the contribution comes from the classically accessible region. Otherwise there is a substantial probability of finding the particle where $V(x) > E$. For the extreme case of the delta function well, the probability of finding the particle outside the well is 100%. For a well with parameters chosen so as to describe the deuteron, the probability of finding the neutron and proton separated by more than the range of the force is about 60%. One can easily stumble into complaining, "There is something wrong here. After all, once the neutron gets out of the proton's force range, it will just drift off."

PROBLEM 2.24

Answer this complaint.

If $E > V$ as $x \to +\infty$, or as $x \to -\infty$, or both, then ψ describes an *unbound* or *scattering* state. The particle comes from and goes to very large distances, and ψ oscillates in these distant regions with no restriction on its phase. There is nothing that forces ψ to fit into some interval, and the allowed values of the energy form a continuum. For such circumstances we have repeatedly written $Ae^{ipx/\hbar}$, or, in three dimensions, with time dependence included again,

$$Ae^{i(\vec{p}\cdot\vec{r}-Et)/\hbar} \equiv Ae^{i(\vec{k}\cdot\vec{r}-\omega t)} \tag{2.8-1}$$

to describe the incident beam. The amplitude A was selected so that $|A|^2\vec{p}/M$ equals the incident flux density.

In expression (1) the momentum is definitely $\vec{p}$, so that the uncertainty principle forces the position to be completely indefinite. The energy is definitely E, so that the time of arrival of the particle is completely indefinite also. Only a wave of infinite extent that has been going on forever is described by (1). Most real sources, however, emit finite wave packets with random relative phases. The emitter of the particles would typically be a large group of atoms with some lifetime τ, so that the energy of the particles would have an uncertainty $\Delta E \simeq \hbar/\tau$. The use of (1) is therefore an approximation, perhaps even a swindle, and the rest of this section deals with this point.

The computations of this chapter could be done with wave packets rather than infinite plane waves to describe the incident particles. As one should

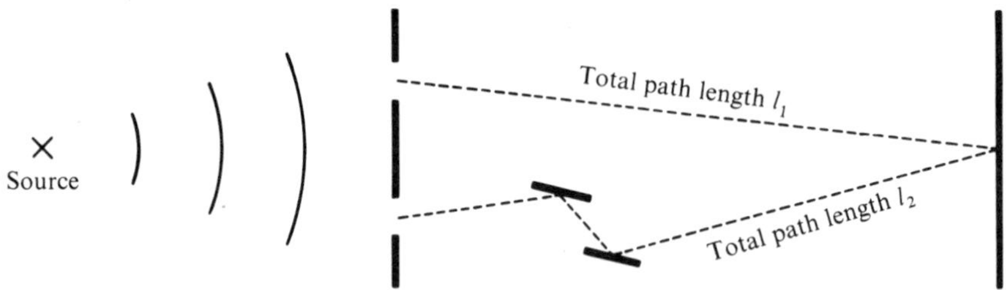

Fig. 2.8-2. An interference experiment with unequal path lengths.

expect, the results for quantities such as transmission and reflection coefficients change very little if $\Delta E \ll E$. However, even if $\Delta E \ll E$, one can show that expression (1) is really wrong. Consider a two-slit interference experiment, for either photons or electrons, in which the two interfering beams can be made to travel two different distances, l_1 and l_2 (Figure 2.8-2). If $l_2 > l_1$, will we see a diffraction pattern? If the incident wave is described by (1), we will. As $l_2 - l_1$ is varied, the pattern would shift but not disappear. Suppose, however, that the source is an ordinary discharge tube that supplies a typical line in the visible spectrum. The emission is governed by some lifetime τ, and the source provides randomly phased wave packets of duration τ and spatial length $v\tau$, where v is the speed. At any instant, if $\Delta E \ll E$, expression (1) is a good approximation, but the phase would drift about randomly and would change drastically in times of the order of τ. Therefore if $l_2 - l_1 > v\tau$, the light from the lower slit would have some random phase with respect to that from the upper slit; the results would be the same as if two separate and incoherent sources were illuminating the two slits. There would, of course, be an interference pattern at each instant, but this pattern would dance about in times of the order of τ and give the impression of uniform illumination to the usual slow observer.

Optical transitions have $\tau \simeq 10^{-8}$ sec, but collision and Doppler broadening usually give an effective τ of about 10^{-11} sec; see Problem 1.6. Therefore $v\tau \simeq 0.3$ cm. If $l_2 - l_1 \simeq 0.3$ cm, there would be substantial blurring of the pattern, and if $l_2 - l_1$ is increased much beyond 0.3 cm, the pattern would disappear completely. The quantity $v\tau$ is of great importance in optics, and is called the *coherence length* or the *extinction length*.[9] One is likely to meet these concepts if one plays with the Michelson interferometer; if the arms are unequal, the interference pattern is poor or even totally invisible. For electrons the same ideas are valid. In low-energy electron diffraction experiments, an energy spread ΔE of around 0.2 eV is typical, so that $\tau = \hbar/\Delta E$ is about 3×10^{-15} sec. Speeds of a few times 10^8 cm/sec then give coherence lengths of a few hundred Å,

[9] M. Born and E. Wolf, *Principles of Optics*, 4th ed. (New York: Pergamon Press, 1970), Sec. 7.5.8.

and one cannot study without additional considerations larger regions with electron diffraction.[10]

Suppose that an incident beam consists of N individual packets, that $\Delta E \ll E$, and that each packet is well approximated by

$$ae^{i(\vec{k}\cdot\vec{r}-\omega t+\delta_n)}.$$

What should one use for the wave function if there is no information about the individual phases δ_n?

There is no wave function that will give the right answer for all possible experiments because there is a random *mixture* rather than a coherent *superposition* of the contributions. Suppose, however, that we agree to do nothing as subtle as unequal-path-length interferometry. Can one fake a wave function that gives right answers for equal-path-length interferometry, transmission and reflection coefficients, and similar things? One can, and we have been doing it throughout this chapter, by writing expression (1) with

$$A = a\sqrt{N}, \tag{2.8-2}$$

which gives a probability density proportional to

$$|A|^2 = |a|^2 N,$$

the density for each particle multiplied by the number of particles. That this is right can be seen as follows. The wave function is at any instant

$$\begin{aligned}\Psi &= ae^{i(\vec{k}\cdot\vec{r}-\omega t+\delta_1)} + ae^{i(\vec{k}\cdot\vec{r}-\omega t+\delta_2)} + \cdots + ae^{i(\vec{k}\cdot\vec{r}-\omega t+\delta_N)} \\ &= ae^{i(\vec{k}\cdot\vec{r}-\omega t)}[e^{i\delta_1} + e^{i\delta_2} + \cdots + e^{i\delta_N}],\end{aligned} \tag{2.8-3}$$

where the phases $\delta_1, \delta_2, \ldots, \delta_N$ are unknown and changing erratically in time. The probability density is then at any instant

$$\begin{aligned}|\Psi|^2 &= |a|^2[e^{-i\delta_1} + e^{-i\delta_2} + \cdots + e^{-i\delta_N}][e^{i\delta_1} + e^{i\delta_2} + \cdots + e^{i\delta_N}] \\ &= |a|^2[N + e^{i(\delta_1-\delta_2)} + e^{i(\delta_2-\delta_1)} + \cdots + e^{i(\delta_{N-1}-\delta_N)} + e^{i(\delta_N-\delta_{N-1})}] \\ &= |a|^2[N + 2\cos(\delta_1 - \delta_2) + \cdots + 2\cos(\delta_{N-1} - \delta_N)].\end{aligned} \tag{2.8-4}$$

Things like the blackening on a photographic plate and the count rate in a particle detector yield the time average over the random phases. The terms $\cos(\delta_i - \delta_j)$ are as likely to be positive as negative, and the average result is simply $|a|^2 N$.

[10] J. B. Pendry, *Low Energy Electron Diffraction* (New York: Academic Press, Inc., 1975), Sec. 1B.

PROBLEM 2.25

a. Suppose that there are two beams, each consisting of many incoherent packets that are rather well defined in energy. The fake wave functions, in the sense of the preceding paragraph, are $Ae^{i(\vec{p}\cdot\vec{r}-Et)/\hbar}$ for the first beam and $Be^{i(\vec{p}\cdot\vec{r}-Et)/\hbar}$ for the second beam. Write and defend a fake wave function for the situation in which the two beams are superposed in the same region.

b. Are ten fiddles ten times as loud as one fiddle? Why?

2.9 *The Constant Force Problem*

The potential energy function gx gives a constant force $-g\hat{e}_x$ if g is a constant. Familiar examples include uniform electric and gravitational fields. A linear potential might describe some of the interaction between *quarks*, the particles that are, according to rather successful speculations, constituents of nucleons and mesons. There are approximation methods that use a linear potential to represent a more complicated potential in a small region. It is therefore interesting to look at the separated equation for the x behavior with $V(x) = gx$,

$$-\frac{\hbar^2}{2M}\frac{d^2\psi(x)}{dx^2} + gx\psi(x) = E\psi(x). \tag{2.9-1}$$

To save writing, let

$$\xi \equiv \left(x - \frac{E}{g}\right)\left(\frac{2Mg}{\hbar^2}\right)^{1/3}, \tag{2.9-2}$$

$$\eta(\xi) = \psi(x).$$

Then (1) becomes

$$\frac{d^2\eta}{d\xi^2} - \xi\eta = 0. \tag{2.9-3}$$

This equation is not as harmless as it looks, because its solution cannot be written in terms of a finite number of elementary functions. It is this complication that has postponed the discussion of the constant force problem. The difficulty is not unique to linear potentials. In regions of constant potential the solutions of the Schrödinger equation are sines, cosines, or exponentials. The harmonic oscillator is described by Gaussians multiplied by polynomials. Chapter 7 shows that the solution for the Coulomb potential can be expressed in terms of exponentials and polynomials. However, nearly all potentials, including many with beguilingly simple functional forms, are not so tractable.

An unfamiliar $V(\vec{r})$ should always be studied first in terms of the qualitative ideas of Section 2.8. With $V = gx$, ψ has an inflection point where $gx = E$, that is, where $\xi = 0$. For $V = gx < E$, or $\xi < 0$, the wave function must oscillate. As ξ becomes more negative and V falls further below E, the kinetic energy becomes larger and the wavelength and amplitude become smaller. In the classically inaccessible region where $gx > E$, or $\xi > 0$, the wave function must die away monotonically. It should decrease faster than the $e^{-\alpha x}$ found for constant V in Section 2.2, and more slowly than the $e^{-\beta^2 x^2/2}$ found for $V \sim x^2$ in Section 2.7. Since gx rises indefinitely as $x \to +\infty$, the large x region presents an impenetrable barrier, there can be no net current of particles, and therefore, as argued in Section 2.2, ψ can be written as a purely real function. The qualitative behavior is much like that displayed in Figure 2.8-1, although there $V(x)$ is not quite a straight line; see the exact $\eta(\xi)$ graphed in Figure 2.9-1 below.

Whenever the form of $V(\vec{r})$ makes it impossible to find the solution in terms of an elementary function, numerical integration or approximation methods of the kind described in Chapter 10 can be used. In addition, some forms of $V(\vec{r})$ give differential equations that have been studied with care, and for which tabulated solutions are available. Such is indeed the case for $V(x) = gx$.

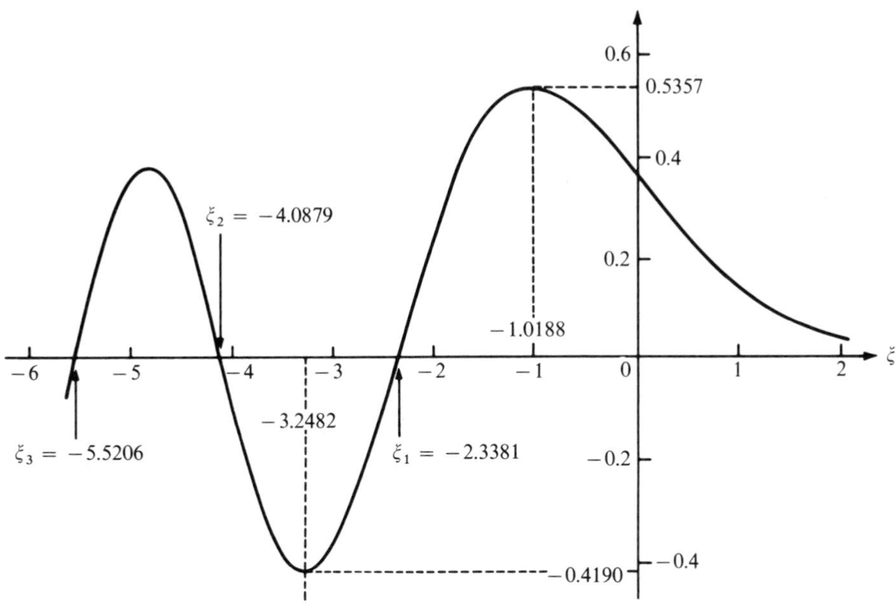

Fig. 2.9-1. The function Ai(ξ). The three zeros of Ai closest to the origin occur at ξ_1, ξ_2, and ξ_3. The values and positions of the first maximum and the first minimum are also shown.

The solution of (3) that vanishes as $\xi \to \pm\infty$ is any constant times the *Airy integral*[11]

$$\mathrm{Ai}(\xi) \equiv \frac{1}{\pi} \lim_{\varepsilon \to 0} \int_0^\infty e^{-\varepsilon u} \cos\left(\frac{u^3}{3} + \xi u\right) du. \tag{2.9-4}$$

A proper study of this function would require too long a detour, but it is not hard to understand the essential results. To see that Ai satisfies (3), substitute and exchange the order of differentiation with respect to ξ and integration over u.

$$\begin{aligned}
&\frac{d^2}{d\xi^2} \pi\mathrm{Ai}(\xi) - \xi\pi\mathrm{Ai}(\xi) \\
&\quad = \lim_{\varepsilon \to 0} \int_0^\infty e^{-\varepsilon u}\left[\frac{\partial^2}{\partial \xi^2} \cos\left(\frac{u^3}{3} + \xi u\right) - \xi \cos\left(\frac{u^3}{3} + \xi u\right)\right] du \\
&\quad = -\lim_{\varepsilon \to 0} \int_0^\infty e^{-\varepsilon u}(u^2 + \xi)\cos\left(\frac{u^3}{3} + \xi u\right) du \\
&\quad = -\lim_{\varepsilon \to 0} \int_0^\infty e^{-\varepsilon u} \frac{\partial}{\partial u} \sin\left(\frac{u^3}{3} + \xi u\right) du \\
&\quad = -\lim_{\varepsilon \to 0}\left[e^{-\varepsilon u} \sin\left(\frac{u^3}{3} + \xi u\right)\right]_0^\infty - \lim_{\varepsilon \to 0} \varepsilon \int_0^\infty e^{-\varepsilon u} \sin\left(\frac{u^3}{3} + \xi u\right) du.
\end{aligned} \tag{2.9-5}$$

In the last step an integration by parts shifts $\partial/\partial u$ from the sine onto the $e^{-\varepsilon u}$. Now all contributions are zero, and it follows that Ai is a solution. The sole purpose of the $e^{-\varepsilon u}$ is to give convergence at the upper limit, and (4) is often written in the form

$$\frac{1}{\pi} \int_0^\infty \cos\left(\frac{u^3}{3} + \xi u\right) du$$

with the understanding that contributions from $u \to +\infty$ are to be suppressed.

For large $|\xi|$ the integrand in (4) oscillates very rapidly, the integral is the sum of nearly equal positive and negative contributions, $\mathrm{Ai}(\xi)$ is very small and does go to zero for $\xi \to \pm\infty$ as claimed. The asymptotic forms of Ai are

$$\xi \to +\infty\colon\ \mathrm{Ai}(\xi) \sim \frac{1}{2\sqrt{\pi}\,\xi^{1/4}} e^{-(2/3)\xi^{\frac{3}{2}}}, \tag{2.9-6}$$

$$\xi \to -\infty\colon\ \mathrm{Ai}(\xi) \sim \frac{1}{\sqrt{\pi}(-\xi)^{1/4}} \sin\left[\frac{2}{3}(-\xi)^{3/2} + \frac{\pi}{4}\right]. \tag{2.9-7}$$

[11] See M. Abramowitz and I. A. Stegun, eds., *Handbook of Mathematical Functions*, N.B.S. Applied Mathematics Series, No. 55 (Washington: U.S. Government Printing Office, 1964). The properties of Ai and references to detailed treatments are given in Secs. 10.4 and 10.5. Numerical values are given in Tables 10.11, 10.12, and 10.13. G. B. Airy introduced the integral in his 1838 study of the rainbow; see H. M. Nussenzweig, "The Theory of the Rainbow," *Scientific American* **236** (4), 116 (1977).

As anticipated, (6) shows that in the classically inaccessible region Ai decreases faster than the $e^{-\alpha x}$ found for constant potentials and more slowly than the $e^{-\beta^2 x^2/2}$ found for harmonic oscillators. The general behavior of Ai in the classically accessible $\xi < 0$ is also easy to understand. There $(-\xi)^{1/4}$ is proportional to $(E + g|x|)^{1/4}$, or (kinetic energy)$^{1/4}$, so that the amplitude is proportional to (speed)$^{-1/2}$. The probability density $\rho = |\psi|^2$, averaged over the oscillations of the sine, is therefore inversely proportional to the speed, just as in Problem 1.9. The faster a particle moves through a neighborhood, the less likely it is to be caught there. The phase Φ of the sine is given by

$$\Phi = \frac{2}{3}(-\xi)^{3/2} + \frac{\pi}{4} = \frac{2}{3}\left(\frac{E}{g} - x\right)^{3/2} \frac{\sqrt{2Mg}}{\hbar} + \frac{\pi}{4},$$

so that

$$-\frac{d\Phi}{dx} = \frac{\sqrt{2M(E - gx)}}{\hbar} \equiv \frac{p(x)}{\hbar}, \tag{2.9-8}$$

where $p(x)$ is the momentum that a classical particle would have at x. If the potential energy varies slowly enough to permit the approximate determination of a wavelength $\lambda(x)$, then this $\lambda(x)$ is the distance required for the phase to change by 2π,

$$\lambda(x)\left|\frac{d\Phi}{dx}\right| = 2\pi. \tag{2.9-9}$$

Therefore (8) becomes

$$p(x) = \frac{h}{\lambda(x)}, \tag{2.9-10}$$

and the connection between momentum and wavelength in each neighborhood matches the familiar $p = h/\lambda$. Since $2i \sin \Phi = e^{+i\Phi} - e^{-i\Phi}$, Ai in the $\xi \to -\infty$ region can be viewed as a superposition of equal amounts of incident and reflected waves in which the momentum changes rather slowly with position.

Numerical values of Ai are available in various tables; see footnote 11. Figure 2.9-1 shows the results in the vicinity of the inflection point, and also gives the locations of the three zeros of Ai that are closest to the origin. Other zeros can be calculated from the asymptotic form (7) to high accuracy. Even for ξ_3, (7) gives -5.5172 rather than the exact -5.5206, and the agreement improves rapidly with increasing $-\xi$.

PROBLEM 2.26

Electrons with 3.0-eV kinetic energy enter a uniform electric field through a small hole in a grounded plate as shown in Figure 2.9-2. The field is 6.0 volts/20.0 cm. The incident current is 1.0×10^{-3} amp. Take the x axis

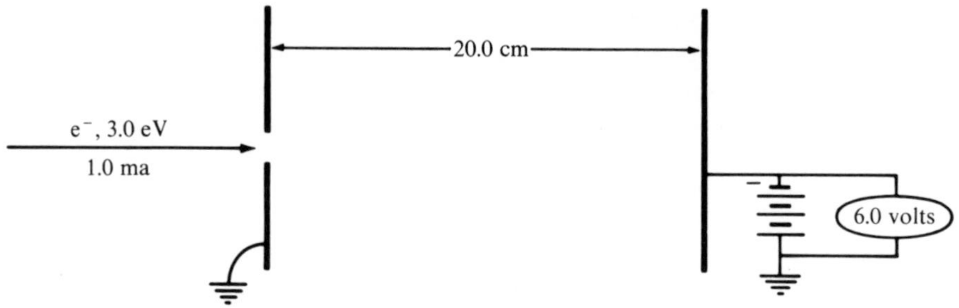

Fig. 2.9-2. Electrons in a constant electric field.

parallel to the electron beam, with origin halfway between the plates. Plot from $-5. \times 10^{-5}$ cm to $+2. \times 10^{-5}$ cm the probability per cm that an electron is found at x.

a. according to classical mechanics.
b. according to the quantum calculation.

If

$$\begin{aligned} V(x) &= gx, \qquad x \geq 0, \\ &= \infty, \qquad x < 0, \end{aligned} \tag{2.9-11}$$

then the particle cannot escape in either the positive or the negative direction, all states are bound, and there is a discrete spectrum of energy levels. These energy levels can be found from the zeros of Ai. The function Ai vanishes as $x \to +\infty$ for any energy. Now the wave function must also vanish at $x = 0$, and therefore one of the zeros of Ai must occur at $x = 0$. The ground state is the one for which the first zero, at $\xi_1 = -2.3381$, corresponds to $x = 0$, and (2) gives

$$-2.3381 = \left(0 - \frac{E_1}{g}\right)\left(\frac{2Mg}{\hbar^2}\right)^{1/3}.$$

The nth state has the energy

$$E_n = |\xi_n| \left(\frac{g^2\hbar^2}{2M}\right)^{1/3}, \tag{2.9-12}$$

where $\mathrm{Ai}(\xi_n) = 0$, with $\xi_2 = -4.0879$, $\xi_3 = -5.5206$, and so forth.

PROBLEM 2.27

a. A quark with twice the mass of a proton is in a $V(x)$ given by

$$\begin{aligned} V(x) &= -V_0 + gx, \qquad x \geq 0, \\ &= \infty, \qquad x < 0, \end{aligned}$$

with $g = 1.30 \times 10^{16}$ MeV/cm $= 1.30$ GeV/10^{-13} cm, and $V_0 = 1.27$ GeV. Find its five lowest energy eigenvalues, and sketch the corresponding wave functions. (The parameters are similar to those used in current quark speculations. Since the energies turn out to be a substantial fraction of the rest energy, the use of nonrelativistic quantum mechanics is only a fair approximation.)

b. Compare the relative energy level spacings $(E_3 - E_2)/(E_2 - E_1)$, $(E_4 - E_3)/(E_2 - E_1)$, $(E_5 - E_4)/(E_2 - E_1)$ found in (**a**) with those for the rigid-walled box and those for the harmonic oscillator. Give a qualitative explanation of your findings.

2.10 *Additional Chapter 2 Problems*

PROBLEM 2.28

Suppose that $V(\vec{r}) = -V_0$ inside the cube defined by $0 \le x \le a$ *and* $0 \le y \le a$ *and* $0 \le z \le a$, and $V(\vec{r}) = 0$ outside the cube. Can the Schrödinger equation be separated into the three equations (2.1-7)?

PROBLEM 2.29

Suppose that $V = 0$ for $x < 0$ and $V = -V_0$ for $x \ge 0$. Neutrons with energy E come from $-\infty$. The incident flux density, number/sec m^2, is f. Obtain the wave function and the transmission coefficient. Evaluate all constants in terms of E, V_0, f, M, and $\hbar$. Obtain and plot the probability density $\rho = |\psi|^2$ as a function of x.

PROBLEM 2.30

$V(x) = 0$, $x < 0$, and $V(x) = U\delta(x) + V_0$, $x \ge 0$. That is, $V(x)$ has a sharp spike *and* a finite step at the origin.

a. Obtain the condition, analogous to (2.3-6), satisfied by the derivative of ψ at the origin.

b. Electrons with energy $E > V_0$ are incident from the right, from $+\infty$. Obtain the transmission coefficient, that is, the ratio of the current in $x < 0$ to the incident current.

PROBLEM 2.31

$V = 0$, $x < 0$, and $V = V_0$, $x \ge 0$. In $x < 0$ there is a wave, Ae^{ikx}, heading toward the potential step, and a wave heading away from it. In $x \ge 0$ there is also a wave, Be^{-iKx}, heading toward the potential step, and a wave heading away from it. The total energy is the same for all four waves.

a. Obtain the wave function in terms of A, B, k, V_0, M, and $\hbar$.

b. Describe an apparatus, with electron guns, collimators, electric and magnetic devices, and so forth, that produces this wave function. You will have to think a little about coherence.

PROBLEM 2.32

$V = 0$, $x < 0$, and $V = V_0$, $x \geq 0$. Particles come in with momentum $\vec{p}_i$ that makes an angle θ_i with the x axis. Show that the angle of incidence equals the angle of reflection, just as in optics.

PROBLEM 2.33

A proton is in the ground state of an unknown potential well $V(x)$. From high-energy scattering experiments, one learns that the probability density is proportional to e^{-x^2/b^2}, where $b = 2 \times 10^{-13}$ cm. What is $V(x)$?

PROBLEM 2.34

$V(x) = 0$, $x < 0$, and $V(x)$ is a real constant V_0 for $x \geq 0$. Neutrons with energy E come from negative x, and half of them are reflected. Deduce what you can about V_0.

PROBLEM 2.35

Find the first excited state wave function and energy for $V = Kx^2/2$ by guessing that $\psi_1 = Axe^{-x^2/c^2}$, substituting and evaluating constants. Argue that this ψ_1 is indeed the first excited state rather than some other state.

PROBLEM 2.36

Consider the vee-shaped potential $V(x) = g|x|$ for all x. Sketch the three lowest energy eigenfunctions and find the corresponding eigenvalues.

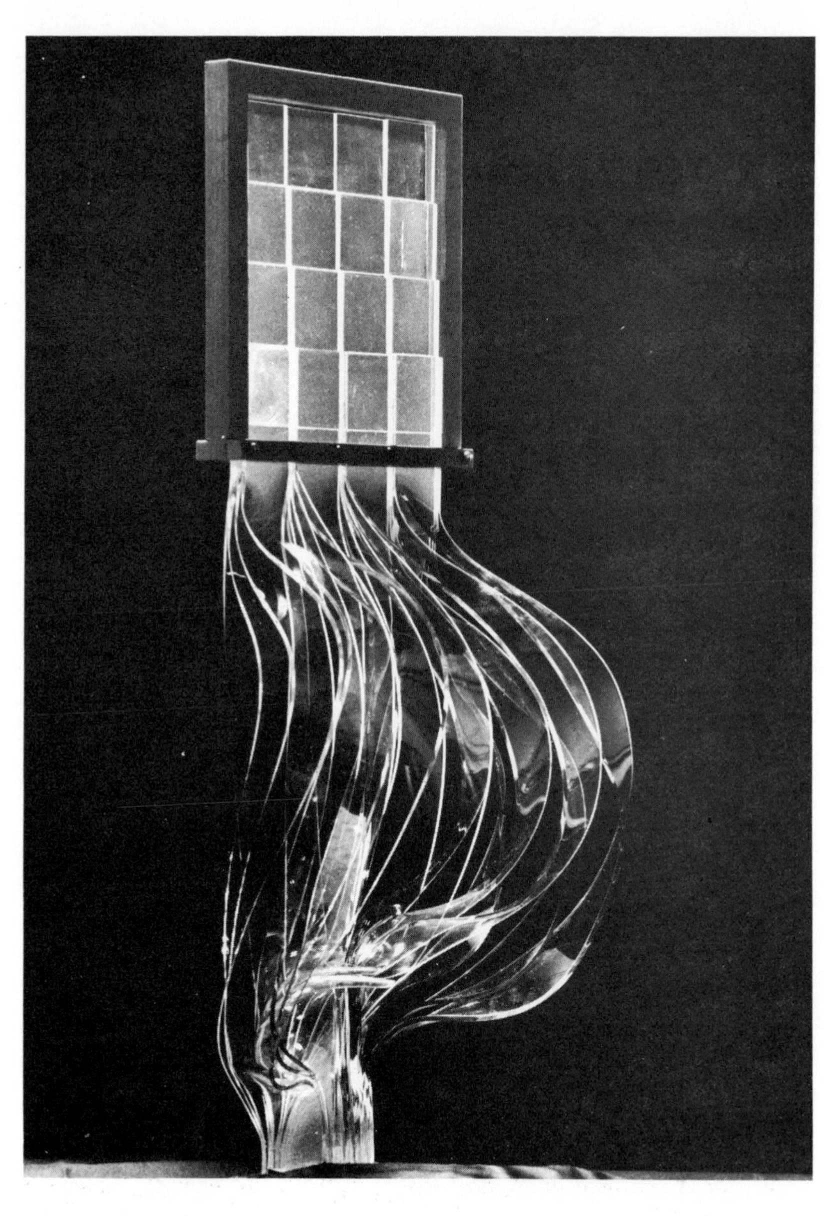

A particle detector that permits accurate timing. Each of the 5 cm × 5 cm squares in the frame is a piece of scintillant, that is, a piece of material that emits light when a charged particle shoots through it. The light is conveyed to a photomultiplier tube by the bent plastic strips. These strips all have the same length, so that the light takes nearly the same time to travel from any part of the detector to the photomultiplier tube. The speed of light in the scintillant is 2×10^{10} cm/sec. With the scintillant divided into 5 cm squares, the arrival of a charged particle can therefore be determined to about 5 cm/(2×10^{10} cm/sec) = 3×10^{-10} sec. Such time intervals are short in terms of human perception, but they are long in the sense that $\hbar/(3 \times 10^{-10}$ sec) is only 2×10^{-6} eV. (Courtesy of S. G. Hummel, Physics Department, College of William and Mary)

3

Tools

Chapters 1 and 2 show how to calculate the wave functions of various simple systems and how to get the probability density ρ and the probability current density $\vec{j}$ from these $\Psi(\vec{r}, t)$. The qualitative idea that information about momentum resides in the distribution of wavelengths in Ψ has come up repeatedly. There is however much more that can be extracted once Ψ is known, and the concept of *expectation values* is the right tool for the job. Also, wave functions and operators have general properties that are useful both in specific calculations and in understanding the structure of the theory. You may, in working with the examples of Chapter 2, have caught preliminary glimpses of such properties, and the next step is to take a more careful look.

3.1 *Expectation Values of $F(\vec{r})$*

Suppose that one wants to compute from $\Psi(\vec{r}, t)$ the *mean value*, or *expectation value* of x, denoted by $\langle x \rangle$, that is, a prediction for the result of the following procedure. Prepare a large number of identical systems, each described by Ψ. Measure the x coordinate of the particle in each system, add the results, and divide by the number of systems. This quantity $\langle x \rangle$ will be x, multiplied by the normalized probability of finding the value between x and $x + dx$, regardless of what y and z are, added over all x:

$$\langle x \rangle = \int_{-\infty}^{+\infty} x \left[\int_{-\infty}^{+\infty} \int_{-\infty}^{+\infty} |\Psi(\vec{r}, t)|^2 \, dy \, dz \right] dx = \int_{-\infty}^{+\infty} \int_{-\infty}^{+\infty} \int_{-\infty}^{+\infty} x |\Psi(\vec{r}, t)|^2 \, d\tau. \tag{3.1-1}$$

What can one predict regarding measurements of a function of position $F(\vec{r})$? The expectation value of F is this function evaluated at $\vec{r}$, multiplied by the probability of finding the particle in the volume element $d\tau$ at $\vec{r}$, added over all space:

$$\langle F \rangle = \int_{-\infty}^{+\infty} \int_{-\infty}^{+\infty} \int_{-\infty}^{+\infty} F(\vec{r}) |\Psi(\vec{r}, t)|^2 \, d\tau. \tag{3.1-2}$$

It is useful to know the width of the distribution about $\langle F \rangle$. Do all measurements produce results very close to the mean, or are the results spread over a large range? A good measure of the width is the *mean squared deviation,* which is also called the *variance* or the *square of the standard deviation.* Take the deviation from the mean, $F - \langle F \rangle$. Square it, and then find the mean of the square.

$$(\Delta F)^2 = \langle (F - \langle F \rangle)^2 \rangle, \tag{3.1-3}$$

or

$$\begin{aligned} (\Delta F)^2 &= \langle (F^2 - 2F\langle F \rangle + \langle F \rangle^2) \rangle \\ &= \langle F^2 \rangle - \langle F \rangle^2. \end{aligned} \tag{3.1-4}$$

The mean squared deviation is the mean of the square minus the square of the mean. You may have met the root mean squared deviation, ΔF, in laboratory data analysis. Verify that what is meant here by $\langle F \rangle$ and ΔF matches your previous experience.

EXAMPLE

Let s_z be the z component of the spin angular momentum of an electron. A measurement of s_z yields either $+\hbar/2$ or $-\hbar/2$. Suppose that P_+, the probability of finding $+\hbar/2$, is 0.30. Obtain $\langle s_z \rangle$ and Δs_z.

SOLUTION

$P_+ + P_- = 1. \quad \therefore P_- = 0.70.$

$$\langle s_z \rangle = \left(\frac{+\hbar}{2} \right) P_+ + \left(\frac{-\hbar}{2} \right) P_- = -0.40 \frac{\hbar}{2}.$$

Note that while this is the *mean* of a large number of measurements, no measurement ever yields $-0.40\hbar/2$. The result is always either $+\hbar/2$ or $-\hbar/2$.

From (3),

$$
\begin{aligned}
(\Delta s_z)^2 &= \langle (s_z - \langle s_z \rangle)^2 \rangle \\
&= 0.30\left[\left(\frac{+\hbar}{2}\right) - \left(-0.40\frac{\hbar}{2}\right)\right]^2 + 0.70\left[\left(\frac{-\hbar}{2}\right) - \left(-0.40\frac{\hbar}{2}\right)\right]^2 \\
&= 0.84\left(\frac{\hbar}{2}\right)^2.
\end{aligned}
$$

Alternatively, from (4),

$$
(\Delta s_z)^2 = 0.30\left(\frac{+\hbar}{2}\right)^2 + 0.70\left(\frac{-\hbar}{2}\right)^2 - \left(-0.40\frac{\hbar}{2}\right)^2 = 0.84\left(\frac{\hbar}{2}\right)^2.
$$

Either way,

$$
\Delta s_z = 0.92\frac{\hbar}{2}.
$$

PROBLEM 3.1

A crooked die has equal probability of turning up $N = 1, 2, 3, 4$, or 5, but the probability of turning up $N = 6$ is 2.0 times as great as that of turning up 1. Obtain the normalized probability function $\rho(N)$ as a function of N. Obtain $\langle N \rangle$ and ΔN.

EXAMPLE

Suppose, as in Problem 2.11 and Figure 2.5-7, that

$$
\begin{aligned}
V(x) &= \infty, & x &< 0, \\
&= -V_0, & 0 &\le x \le b, \\
&= 0, & b &< x.
\end{aligned}
$$

Obtain $\langle x \rangle$ for the lowest bound state in this $V(x)$.

SOLUTION

$$
0 \le x \le b\colon\ \psi = A \sin Kx, \qquad K = \frac{\sqrt{2M(E + V_0)}}{\hbar}
$$

$$
b < x\colon\ \psi = Be^{-\alpha x}, \qquad \alpha = \frac{\sqrt{2M(-E)}}{\hbar}
$$

Continuity and continuity of the derivative at b require

$$A \sin Kb = Be^{-\alpha b}$$
$$KA \cos Kb = -\alpha Be^{-\alpha b}.$$

Divide the second equation by the first.

$$K \cot Kb = -\alpha.$$

If M, V_0, and b are specified, this transcendental equation can be solved for E by the methods of Section 2.5. Then K and α are determined.

Normalize ψ.

$$\int_0^\infty |\psi|^2\,dx = \int_0^b (A \sin Kx)^2\,dx + \int_b^\infty (Be^{-\alpha x})^2\,dx = 1.$$

From continuity at b,

$$B = Ae^{\alpha b} \sin Kb.$$

$$\therefore \quad A^2\left[\int_0^b \sin^2 Kx\,dx + e^{2\alpha b} \sin^2 Kb \int_b^\infty e^{-2\alpha x}\,dx\right] = 1,$$

$$A^2 = \left[\frac{Kb - \sin Kb \cos Kb}{2K} + \frac{\sin^2 Kb}{2\alpha}\right]^{-1}.$$

$$\langle x \rangle = \int_0^\infty x|\psi|^2\,dx = \int_0^b x(A \sin Kx)^2\,dx + \int_b^\infty x(Be^{-\alpha x})^2\,dx$$

$$= A^2\left[\int_0^b x \sin^2 Kx\,dx + e^{2\alpha b} \sin^2 Kb \int_b^\infty xe^{-2\alpha x}\,dx\right]$$

$$= \frac{b^2 - \left(\dfrac{b}{K}\right)\sin 2Kb + \left(\dfrac{1}{K^2} + \dfrac{1}{\alpha^2}\right)\sin^2 Kb}{\dfrac{2(Kb - \sin Kb \cos Kb)}{K} + \dfrac{2(\sin^2 Kb)}{\alpha}}$$

This rather messy expression has two understandable limits. If the well is just barely deep enough to bind one state, then Kb is slightly larger than $\pi/2$, and α is very small compared to $1/b$ and to K. Then $\langle x \rangle \simeq 1/2\alpha$, a large distance. Very little of the probability density is in $0 \leq x \leq b$; most of it is in the long exponential tail, and it is no surprise that

$$\langle x \rangle \simeq \frac{\int_0^\infty xe^{-2\alpha x}\,dx}{\int_0^\infty e^{-2\alpha x}\,dx} = \frac{1}{2\alpha}.$$

If on the other hand the well is very deep, then Kb is only slightly smaller than π, and α is very large. Then $\langle x \rangle \simeq b/2$. This result makes sense because the exponential tail disappears within a very short distance and almost all of the probability density is in $0 \le x \le b$.

PROBLEM 3.2

Consider $V(x) = 0$, $0 \le x \le a$; $V = \infty$ otherwise, with energy eigenstates $\psi_n(x) = \sqrt{2/a}\,\sin(\pi n x/a)$, $n = 1, 2, 3, \ldots$. Compute $\langle x \rangle$ and Δx for general ψ_n.

PROBLEM 3.3

For the ground state and first excited state of the harmonic oscillator, obtain $\langle V(x) \rangle$. Express $\langle V(x) \rangle$ as a fraction of the total energy. (The answer is that $\langle V(x) \rangle = E/2$ for both states. This is, in fact, the answer for *any* energy eigenstate of the harmonic oscillator, and also for the classical harmonic oscillator.) Write $\langle x \rangle$ and Δx for these two states.

3.2 Fourier Integrals and Expectation Values of $F(\vec{p})$

To obtain the momentum content of a given $\Psi(\vec{r}, t)$, one needs to analyze the distribution in Ψ of wavelengths λ, or of values of $k \equiv 2\pi/\lambda$. Fourier series and Fourier integrals[1] provide the means to do this analysis. The ideas will be developed first for a one-dimensional $\psi(x)$, and then extended to include time and three-dimensional $\vec{r}$ dependence.

If $\psi(x)$ is piece-wise continuous and periodic with period L,

$$\psi(x + L) = \psi(x),$$

then it can be represented in $-L/2 < x < +L/2$ by the complex Fourier series

$$\psi(x) = \sum_{n=-\infty}^{+\infty} a_n e^{2\pi i n x/L}. \tag{3.2-1}$$

The series converges to $\psi(x)$ where $\psi(x)$ is continuous, and to the average value of $\psi(x)$ at discontinuities. To evaluate the coefficients a_n, multiply by $(1/L)e^{-2\pi i m x/L}$ and integrate from $-L/2$ to $+L/2$.

$$\int_{-L/2}^{+L/2} \psi(x) e^{-2\pi i m x/L} \frac{dx}{L} = \sum_n a_n \int_{-L/2}^{+L/2} e^{2\pi i (n-m)x/L} \frac{dx}{L}.$$

[1] See F. S. Crawford, Jr., *Waves, Berkeley Physics Course*, Vol. 3 (New York: McGraw-Hill Book Company, 1968), Secs. 2.3, 6.3–6.5; J. Matthews and R. L. Walker, *Mathematical Methods of Physics*, 2d ed. (New York: W. A. Benjamin, Inc., 1970), Chap. 4.

The integral on the right gives one if $n = m$ and zero if $n \neq m$. The *Kronecker delta* δ_{nm} is standard and convenient shorthand for this behavior;

$$\begin{aligned} \delta_{nm} &= 1 \qquad \text{if } n = m, \\ &= 0 \qquad \text{if } n \neq m. \end{aligned} \tag{3.2-2}$$

It plays much the same role for discrete quantities n, m as the Dirac delta function $\delta(x - u)$ does for the continuous quantities x, u. With this notation,

$$\int_{-L/2}^{+L/2} \psi(x) e^{-2\pi i m x/L} \frac{dx}{L} = \sum_n a_n \delta_{nm} = a_m \tag{3.2-3}$$

because the summation gives nothing unless $n = m$.

PROBLEM 3.4

If $\psi(x)$ is real, it can be described by the usual real Fourier series,

$$\psi(x) = b_0 + \sum_{n=1}^{\infty} \left(b_n \cos \frac{2\pi n x}{L} + c_n \sin \frac{2\pi n x}{L} \right).$$

Obtain the b_n and the c_n in terms of the a_n. Note that, if $\psi(x)$ is real, $a_n^* = a_{-n}$.

Substitute the result (3) for the a_n into (1).

$$\psi(x) = \sum_{n=-\infty}^{+\infty} \left(\int_{-L/2}^{+L/2} \psi(u) e^{-2\pi i n u/L} \frac{du}{L} \right) e^{2\pi i n x/L}.$$

With $2\pi n/L \equiv k_n$, $\qquad k_{n+1} - k_n = 2\pi/L \equiv \Delta k$,

$$\psi(x) = \sum_{n=-\infty}^{+\infty} \frac{\Delta k}{2\pi} \int_{-L/2}^{+L/2} \psi(u) e^{-ik_n(u-x)}\, du. \tag{3.2-4}$$

Let $L \to \infty$. Then k_n becomes a continuous variable k, $\Delta k \to dk$, and

$$\sum_{n=-\infty}^{+\infty} \Delta k \to \int_{-\infty}^{+\infty} dk.$$

Equation (4) becomes

$$\psi(x) = \int_{-\infty}^{+\infty} \frac{dk}{2\pi} \int_{-\infty}^{+\infty} du\, \psi(u) e^{-ik(u-x)}. \tag{3.2-5}$$

If this expression is written in the form

$$\psi(x) = \int_{-\infty}^{+\infty} du\, \psi(u) \left[\int_{-\infty}^{+\infty} dk \frac{e^{ik(x-u)}}{2\pi} \right], \tag{3.2-6}$$

comparison with (2.3-4) shows that the integral over k is a representation of the Dirac delta function,

$$\int_{-\infty}^{+\infty} dk \frac{e^{ik(x-u)}}{2\pi} = \delta(x - u). \tag{3.2-7}$$

Now view (5) as an expression that gives $\psi(x)$ as an integral over k:

$$\psi(x) = \frac{1}{\sqrt{2\pi}} \int_{-\infty}^{+\infty} g(k)e^{ikx}\, dk, \tag{3.2-8}$$

where

$$g(k) = \frac{1}{\sqrt{2\pi}} \int_{-\infty}^{+\infty} \psi(x)e^{-ikx}\, dx. \tag{3.2-9}$$

Equation (8) expresses $\psi(x)$ as a *Fourier integral*, and (9) gives $g(k)$, the *Fourier transform* of $\psi(x)$. The Fourier integral expresses $\psi(x)$ as a superposition of plane waves e^{ikx} with coefficients $g(k)/\sqrt{2\pi}$; it has the same form as the wave packets discussed in Sections 1.4 and 2.1.

A single component $g(k)e^{ikx}$ describes a particle of definite momentum $\hbar k$. The wave function (8) is a superposition of such waves, and since the probability varies as the square of the magnitude of the amplitude, $|g(k)|^2\, dk$ is proportional to the probability of finding the momentum between $\hbar k$ and $\hbar(k + dk)$. From (8) and (9),

$$\begin{aligned}
\int_{-\infty}^{+\infty} g(k)^* g(k) dk &= \int_{-\infty}^{+\infty} g(k)^* \left[\frac{1}{\sqrt{2\pi}} \int_{-\infty}^{+\infty} \psi(x)e^{-ikx}\, dx\right] dk \\
&= \int_{-\infty}^{+\infty} \left[\frac{1}{\sqrt{2\pi}} \int_{-\infty}^{+\infty} g(k)^* e^{-ikx}\, dk\right] \psi(x) dx \qquad (3.2\text{-}10) \\
&= \int_{-\infty}^{+\infty} \psi^*(x)\psi(x) dx.
\end{aligned}$$

If $\psi(x)$ is normalized, then so is $g(k)$. Defining $g(k)$ with the factor $1/\sqrt{2\pi}$ not only made (8) and (9) look symmetric, it also avoided the need for an additional normalization constant. The quantity $|g(k)|^2\, dk$ is not just proportional, it is *equal* to the probability of finding the momentum between $\hbar k$ and $\hbar(k + dk)$.

PROBLEM 3.5

Consider the ground state wave function for the harmonic oscillator

$$\psi(x) = \sqrt{\frac{\beta}{\sqrt{\pi}}}\, e^{-\beta^2 x^2/2}.$$

Use (9) to calculate the Fourier transform of this wave function. Obtain $P(p)$, where $P(p)dp$ is the probability of finding the momentum between p and $p + dp$. Sketch a graph of the result. Calculate Δp, the root mean squared deviation in momentum. Use the result of Problem 3.3 to obtain $\Delta x\, \Delta p$.

PROBLEM 3.6

Consider $V(x) = 0$, $0 \leq x \leq a$; $V = \infty$ otherwise, with energy eigenstates $\psi_n(x) = \sqrt{2/a}\,\sin(\pi n x/a)$, $n = 1, 2, 3, \ldots$. Use (9) to calculate the Fourier transform of this wave function. Obtain $P(p)$, where $P(p)dp$ is the probability of finding the momentum between p and $p + dp$. Sketch a graph of the result.

So far, time dependence has been omitted. Actually each component plane wave, as described in Section 1.4, is

$$g(k)e^{i(kx - \omega t)},$$

with $\omega(k) = \hbar k^2/2M$. Introduction of the time dependence means that $g(k)e^{-i\omega t}$ now plays the role of $g(k)$, so that

$$\Psi(x, t) = \frac{1}{\sqrt{2\pi}} \int_{-\infty}^{+\infty} g(k)e^{i(kx - \omega t)}\, dk, \tag{3.2-11}$$

where $e^{-i\omega t}$ cannot be taken out of the integral because $\omega = \omega(k)$, and

$$g(k) = \frac{1}{\sqrt{2\pi}} \int_{-\infty}^{+\infty} \Psi(x, t)e^{-i(kx - \omega t)}\, dx, \tag{3.2-12}$$

where $e^{i\omega t}$ *can* be taken out of the integral because $\omega \neq \omega(x)$. If the specification $\Psi(x, t)$ of the state in *position space* is known, then $g(k)$, the specification of the state in *momentum space*, can be computed, and *vice versa*.

Now we can turn to the problem of computing $\langle p \rangle$. Since $|g(k)|^2\, dk$ is the probability of finding the momentum between $\hbar k$ and $\hbar(k + dk)$,

$$\langle p \rangle = \int_{-\infty}^{+\infty} \hbar k |g(k)|^2\, dk. \tag{3.2-13}$$

This expression is useful if the momentum space description is available, but one often wants $\langle p \rangle$ in terms of $\Psi(x, t)$. Use (12) to replace $g(k)$.

$$\begin{aligned} \langle p \rangle &= \int_{-\infty}^{+\infty} \int_{-\infty}^{+\infty} \hbar k g^*(k) \frac{\Psi(x, t)}{\sqrt{2\pi}} e^{-i(kx - \omega t)}\, dx\, dk \\ &= \frac{\hbar}{\sqrt{2\pi}} \int_{-\infty}^{+\infty} \int_{-\infty}^{+\infty} g^*(k)e^{i\omega t}\Psi(x, t)\left(i \frac{\partial}{\partial x} e^{-ikx}\right) dx\, dk. \end{aligned} \tag{3.2-14}$$

Do the x integration by parts to shift $\partial/\partial x$ from e^{-ikx} onto $\Psi(x,t)$. Use the physical requirement that $\Psi \to 0$ when $x \to \pm\infty$ because the particle is known always to be in some finite region.

$$\langle p \rangle = \frac{-i\hbar}{\sqrt{2\pi}} \int_{-\infty}^{+\infty} \int_{-\infty}^{+\infty} g^*(k) e^{i\omega t} e^{-ikx} \left(\frac{\partial}{\partial x} \Psi(x,t) \right) dx\, dk.$$

Finally, recognize the k integral as the complex conjugate of (11).

$$\langle p \rangle = \int_{-\infty}^{+\infty} \Psi^* \left(\frac{\hbar}{i} \frac{\partial}{\partial x} \Psi \right) dx. \tag{3.2-15}$$

A simple generalization, involving n partial integrations, gives

$$\langle p^n \rangle = \int_{-\infty}^{+\infty} \Psi^* \left(\frac{\hbar}{i} \frac{\partial}{\partial x} \right)^n \Psi\, dx. \tag{3.2-16}$$

With this equation, $\langle F(p) \rangle$ can be calculated for any $F(p)$ that can be written as a power series in p. These results are not surprising; we have already used the observation that the operator $(\hbar/i)(\partial/\partial x)$ pulls out the x component of momentum.

PROBLEM 3.7

Begin with the momentum space statement $\langle p^2 \rangle = \int_{-\infty}^{+\infty} (\hbar k)^2 |g(k)|^2\, dk$ and obtain (16) for $n = 2$.

PROBLEM 3.8

For the ground state and the first excited state of the harmonic oscillator, obtain the expectation value of the kinetic energy by using (16). Compare what you did here with the method you used in Problem 3.5.

PROBLEM 3.9

It is easy to get $\langle x \rangle$ in position space and $\langle p \rangle$ in momentum space. Now derive $\langle x \rangle$ in momentum space. The result is, with $g(k,t) = g(k)e^{-i\omega t}$,

$$\langle x \rangle = \int_{-\infty}^{+\infty} g^*(k,t) i \frac{\partial}{\partial k} g(k,t) dk = \int_{-\infty}^{+\infty} g^*\left(\frac{p}{\hbar}, t\right) i \frac{\partial}{\partial p} g\left(\frac{p}{\hbar}, t\right) dp.$$

Except for a minus sign, the structure looks like (15). The minus sign is related to the one that makes (12) look different from (11).

So far, everything has been done in one dimension, but the generalization is easy:

$$\langle p_x\rangle = \int_{-\infty}^{+\infty}\int_{-\infty}^{+\infty}\int_{-\infty}^{+\infty}\Psi^*\frac{\hbar}{i}\frac{\partial}{\partial x}\Psi\,d\tau, \qquad \langle p_y\rangle = \int_{-\infty}^{+\infty}\int_{-\infty}^{+\infty}\int_{-\infty}^{+\infty}\Psi^*\frac{\hbar}{i}\frac{\partial}{\partial y}\Psi\,d\tau,$$

$$\langle \vec{p}\rangle = \langle \hat{e}_x p_x + \hat{e}_y p_y + \hat{e}_z p_z\rangle = \int_{-\infty}^{+\infty}\int_{-\infty}^{+\infty}\int_{-\infty}^{+\infty}\Psi^*\frac{\hbar}{i}\vec{\nabla}\Psi\,d\tau. \tag{3.2-17}$$

$$\Psi(\vec{r}, t) = \left(\frac{1}{2\pi}\right)^{3/2}\int_{-\infty}^{+\infty}\int_{-\infty}^{+\infty}\int_{-\infty}^{+\infty} g(\vec{k})e^{i(\vec{k}\cdot\vec{r}-\omega t)}\,dk_x\,dk_y\,dk_z. \tag{3.2-18}$$

$$g(\vec{k}) = \left(\frac{1}{2\pi}\right)^{3/2}\int_{-\infty}^{+\infty}\int_{-\infty}^{+\infty}\int_{-\infty}^{+\infty}\Psi(\vec{r}, t)e^{-i(\vec{k}\cdot\vec{r}-\omega t)}\,dx\,dy\,dz. \tag{3.2-19}$$

The results of this and the previous section can be made to look similar and summarized. Each dynamical quantity is represented by an operator Ω, where an *operator* is anything, such as a derivative or a multiplying function, that changes Ψ into something else. We assume that the expectation value of the quantity is obtained by calculating $\Omega\Psi$, multiplying by Ψ^*, and integrating;

$$\langle\Omega\rangle = \int_{-\infty}^{+\infty}\int_{-\infty}^{+\infty}\int_{-\infty}^{+\infty}\Psi^*\Omega\Psi\,d\tau. \tag{3.2-20}$$

Equations (3.1-2) and (3.2-16) then become special cases of this statement, with the operator Ω identified as the multiplying function $F(\vec{r})$ in the first case, and $((\hbar/i)\,\partial/\partial x)^n$ in the second case.

Here the wave function and the operators have been written in position space, but (20) is also valid in momentum space if the integration is over $\vec{k}$ rather than $\vec{r}$, as can be seen from (13) and from Problem 3.9.

3.3 *Hermitian Operators*

The manipulations of quantum theory involve two kinds of things, *wave functions* and *operators*. The wave function $\Psi(\vec{r}, t)$ of a system is the repository of information about that system. Operators do things to the wave function, perhaps by differentiating it and multiplying it by functions; they form equations that Ψ must satisfy and they are the means by which information is extracted from Ψ in expectation value calculations. For example, the Schrödinger equation asks that the Hamiltonian operator

$$H = \frac{p^2}{2M} + V = \frac{-\hbar^2}{2M}\nabla^2 + V$$

operating on Ψ gives the same thing as $i\hbar\,\partial/\partial t$ operating on Ψ. The expectation value of the energy is obtained by inserting the operator H between Ψ^* and Ψ and integrating:

$$\langle H\rangle = \int_{-\infty}^{+\infty}\int_{-\infty}^{+\infty}\int_{-\infty}^{+\infty}\Psi^*\left(\frac{-\hbar^2}{2M}\nabla^2 + V\right)\Psi\,d\tau.$$

The expectation value of position is obtained from Ψ and the operator $\vec{r}$:

$$\langle\vec{r}\rangle = \int_{-\infty}^{+\infty}\int_{-\infty}^{+\infty}\int_{-\infty}^{+\infty}\Psi^*\vec{r}\Psi\,d\tau = \int_{-\infty}^{+\infty}\int_{-\infty}^{+\infty}\int_{-\infty}^{+\infty}\vec{r}\,|\Psi|^2\,d\tau.$$

We shall now study some properties of operators that can represent measurable quantities. The operators H, $\vec{r}$, derivatives of any order with respect to coordinates, and multiplying functions are *linear* as defined by (1.4-5) and (1.4-6). We assume that all operators of interest here are linear.

Look again at (3.2-15). It is a prescription for predicting the mean of measurements of momentum and should therefore produce a real number. To explore whether it does, compute $\langle p\rangle^*$ and compare with $\langle p\rangle$.

$$\langle p\rangle^* = \int_{-\infty}^{+\infty}\Psi\left(\frac{\hbar}{-i}\frac{\partial}{\partial x}\Psi^*\right)dx.$$

Integrate by parts.

$$\langle p\rangle^* = -\int_{-\infty}^{+\infty}\left(\frac{\hbar}{-i}\frac{\partial\Psi}{\partial x}\right)\Psi^*\,dx - \frac{\hbar}{i}[\Psi^*\Psi]_{-\infty}^{+\infty}.$$

If the particle is definitely in some finite region, then $\Psi^*\Psi$ must be zero at great distances, and

$$\langle p\rangle^* = \int_{-\infty}^{+\infty}\Psi^*\frac{\hbar}{i}\frac{\partial}{\partial x}\Psi\,dx = \langle p\rangle.$$

The result could be viewed as a minor miracle. An operator with an i in it is sandwiched between an almost arbitrary complex Ψ and Ψ^*, the result is integrated, and somehow the number produced must be real.

The restriction on an operator Ω that demands

$$\langle\Omega\rangle = \langle\Omega\rangle^* \tag{3.3-1}$$

for any acceptable wave function is a stringent one, and most operators that one might casually construct fail to satisfy it. Those that do are called *Hermitian operators*[2] and only they can represent measurable quantities.

Equation (1) states that[3]

$$\int\Psi^*\Omega\Psi = \left(\int\Psi^*\Omega\Psi\right)^* = \int\Psi\Omega^*\Psi^* = \int(\Omega\Psi)^*\Psi. \tag{3.3-2}$$

[2] In honor of Charles Hermite, 1822–1901, French mathematician. The two different topics, Hermitian operators and Hermite polynomials, are only a small part of his work.

[3] In this and similar expressions, $\int$ stands for integration over the space, that is, usually, $\int_{-\infty}^{+\infty}\int_{-\infty}^{+\infty}\int_{-\infty}^{+\infty}d\tau$.

One can describe this result by observing that, if Ω is Hermitian, it can be shifted from Ψ onto Ψ^* if it is replaced by Ω^*. Suppose $\Psi = \Psi_1 + c\Psi_2$, where Ψ_1 and Ψ_2 are acceptable wave functions and c is any constant. Then

$$\int (\Psi_1^* + c^*\Psi_2^*)\Omega(\Psi_1 + c\Psi_2) = \int \Psi_1^*\Omega\Psi_1 + c^*c \int \Psi_2^*\Omega\Psi_2 + c \int \Psi_1^*(\Omega\Psi_2) + c^* \int \Psi_2^*(\Omega\Psi_1)$$

must be real for arbitrary c because a linear combination of two acceptable wave functions is an acceptable wave function. The first two terms are real because of (2). For any c the sum of the last two terms must then be real and equal to its complex conjugate.

$$c \int \Psi_1^*(\Omega\Psi_2) + c^* \int \Psi_2^*(\Omega\Psi_1) = c^* \int \Psi_1(\Omega^*\Psi_2^*) + c \int \Psi_2(\Omega^*\Psi_1^*).$$

Write this equation first for $c = b$, a real constant, then for $c = ib$, and add:

$$\int \Psi_1^*\Omega\Psi_2 = \int (\Omega\Psi_1)^*\Psi_2. \tag{3.3-3}$$

The shifting of the operator seen in (2) works even if the functions to the left and to the right of the operator are different. The result (3) follows from (2), and (2) follows from (3) if $\Psi_1 = \Psi_2$. The two statements are equivalent, and either can be used as the definition of Hermitian operators.

Real constants, well-behaved real functions of position, well-behaved real functions of momentum, and sums of Hermitian operators are Hermitian operators.

PROBLEM 3.10

Prove the preceding statements. Why are the words "well-behaved" necessary?

The product of two Hermitian operators is generally not Hermitian.

$$\begin{aligned}\langle\Lambda\Omega\rangle^* &= \left(\int \Psi^*\Lambda\Omega\Psi\right)^* = \int (\Lambda\Omega\Psi)^*\Psi \\ &= \int (\Omega\Psi)^*\Lambda\Psi = \int \Psi^*\Omega\Lambda\Psi = \langle\Omega\Lambda\rangle,\end{aligned} \tag{3.3-4}$$

where (3) has been used twice. If Ω and Λ do not commute, that is, if $\Lambda\Omega \neq \Omega\Lambda$, the product will not be Hermitian. For example,

$$xp_x\Psi = x\frac{\hbar}{i}\frac{\partial}{\partial x}\Psi,$$

$$p_x x\Psi = \frac{\hbar}{i}\frac{\partial}{\partial x}(x\Psi) = x\frac{\hbar}{i}\frac{\partial\Psi}{\partial x} + \frac{\hbar}{i}\Psi\frac{\partial x}{\partial x}$$

$$= x\frac{\hbar}{i}\frac{\partial\Psi}{\partial x} + \frac{\hbar}{i}\Psi = xp_x\Psi + \frac{\hbar}{i}\Psi,$$

xp_x and $p_x x$ are different operators, neither will, in general, produce real expectation values, and neither will do as the quantal operator *position · momentum.* However, $\frac{1}{2}(xp_x + p_x x)$ will do; it is Hermitian and has the correct classical limit. Quite generally, (4) guarantees that $(\Lambda\Omega + \Omega\Lambda)$ is Hermitian because

$$\begin{aligned}\langle\Lambda\Omega + \Omega\Lambda\rangle^* &= \langle\Lambda\Omega\rangle^* + \langle\Omega\Lambda\rangle^* = \langle\Omega\Lambda\rangle + \langle\Lambda\Omega\rangle \\ &= \langle\Lambda\Omega + \Omega\Lambda\rangle.\end{aligned}$$

Now consider the *commutator* of Λ and Ω, defined by

$$\Lambda\Omega - \Omega\Lambda \equiv [\Lambda, \Omega]. \tag{3.3-5}$$

Another Hermitian operator is $i[\Lambda, \Omega]$, because

$$\begin{aligned}\langle i(\Lambda\Omega - \Omega\Lambda)\rangle^* &= (i)^*(\langle\Lambda\Omega\rangle^* - \langle\Omega\Lambda\rangle^*) \\ &= \langle i(\Lambda\Omega - \Omega\Lambda)\rangle.\end{aligned}$$

Is $i[p_x, x]$ a reasonable thing to try as the operator for *position · momentum*? It turns out that it is not.

$$i(p_x x - xp_x)\Psi = \hbar\frac{\partial}{\partial x}(x\Psi) - x\hbar\frac{\partial}{\partial x}\Psi = \hbar\Psi \tag{3.3-6}$$

for arbitrary Ψ, so that

$$\langle i(p_x x - xp_x)\rangle = \int \Psi^*\hbar\Psi = \hbar.$$

This is a useful and interesting result, but it contains no information about position and momentum.

PROBLEM 3.11

What is $i[p_x, x]$ in momentum space?

PROBLEM 3.12

Prove:

a. If Ω is Hermitian, then Ω^n is Hermitian for positive integers n.

b. If n is even, then, fortunately, $\langle \Omega^n \rangle$ can never be negative. (If it were otherwise, the expectation value of the square of a length might turn out to be $-4.0\ \text{cm}^2$.)

Operators that are to represent dynamical quantities must give the correct classical limit, and they must be Hermitian. These guides are very useful, but they are not in general sufficient to determine operators uniquely.

PROBLEM 3.13

Consider the classical dynamical quantity x^2p^2. Write two different quantum mechanical operators that might represent this quantity. Use (6) to show that the difference between these two possibilities is not detectable in the classical limit.

Section 2.6 introduced complex potentials as a device for the description of the absorption and decay of particles. It was clear there that these complex potentials were invented for a particular purpose and that they could not claim to be legitimate descendents of the usual potential energy functions of classical mechanics. Here is further evidence of their disreputableness.

$$\langle V(\vec{r}) + iW(\vec{r}) \rangle = \int \Psi^* V \Psi + i \int \Psi^* W \Psi,$$

which is a complex number. Complex potentials are therefore not Hermitian, and their expectation values do not give the mean of any direct measurements.

3.4 *The Precise Uncertainty Principle*

The qualitative arguments in Section 1.2 show that it is impossible to measure some pairs of quantities simultaneously with unlimited accuracy. Among such pairs are x, p_x and y, p_y, but not x, p_y. This section is devoted to a general theorem that contains these results.

Let P and Q be Hermitian operators.[4] Let the system be described by a state function Ψ. Let the root mean squared deviations be ΔP and ΔQ. Then, as is proved below,

$$(\Delta P)(\Delta Q) \geq \tfrac{1}{2} |\langle i[P, Q] \rangle|, \tag{3.4-1}$$

[4] In classical mechanics it has long been the fashion to call canonically conjugate pairs of dynamical variables P and Q. One way to make the transition from classical to quantum mechanics is to replace the *Poisson brackets* by the commutators of the corresponding quantities. The canonically conjugate pairs are those that lead to simple uncertainty principles. This approach has charm and utility, and you will want to learn about it eventually. See for example L. I. Schiff, *Quantum Mechanics*, 3d ed. (New York: McGraw-Hill Book Company, 1968), pp. 173–176.

where $[P, Q] \equiv PQ - QP$ is the commutator of P and Q as defined by (3.3-5). The expectation value $\langle i[P, Q]\rangle$, and those contained in ΔP and ΔQ according to (3.1-3), must be obtained from the same Ψ, evaluated at the same time.

To prove this statement, we need the *Schwarz inequality*: For any two functions $F(x)$ and $G(x)$, and any interval a to b,

$$\left(\int_a^b |F|^2\,dx\right)\left(\int_a^b |G|^2\,dx\right) \geq \left|\int_a^b F^*G\,dx\right|^2. \qquad (3.4\text{-}2)$$

To obtain this result, note that

$$\int\left|F(x)\left(\int G^*(x')G(x')dx'\right) - G(x)\left(\int G^*(x')F(x')dx'\right)\right|^2 dx \geq 0 \quad (3.4\text{-}3)$$

because the integrand has been designed to be the square of a magnitude, which cannot be negative. The equality holds only if the integrand is zero everywhere, that is, if for some constant c,

$$G(x) = cF(x). \qquad (3.4\text{-}4)$$

The inequality (3) can be written as

$$\int\left[F^*\left(\int G^*G\,dx'\right) - G^*\left(\int F^*G\,dx'\right)\right]\left[F\left(\int G^*G\,dx'\right)\right.$$
$$\left. - G\left(\int G^*F\,dx'\right)\right]dx \geq 0,$$

or

$$\left(\int F^*F\,dx\right)\left(\int G^*G\,dx\right)^2 - \left(\int F^*G\,dx\right)\left(\int G^*F\,dx\right)\left(\int G^*G\,dx\right)$$
$$- \left(\int G^*F\,dx\right)\left(\int F^*G\,dx\right)\left(\int G^*G\,dx\right)$$
$$+ \left(\int G^*G\,dx\right)\left(\int F^*G\,dx\right)\left(\int G^*F\,dx\right) \geq 0.$$

Cancelling $\int G^*G\,dx$ and rearranging yields (2).

Now define the two Hermitian operators

$$\Omega \equiv P - \langle P\rangle$$
$$\Lambda \equiv Q - \langle Q\rangle$$

and apply the Schwarz inequality by letting

$$F \equiv \Omega\Psi, \qquad G \equiv \Lambda\Psi.$$

$$\int |F|^2\,dx = \int (\Omega^*\Psi^*)(\Omega\Psi)dx = \int \Psi^*\Omega^2\Psi\,dx$$

$$= \int \Psi^*(P - \langle P\rangle)^2\Psi\,dx \equiv (\Delta P)^2,$$

where (3.3-3) has been used to shift Ω^* from Ψ^* to $\Omega\Psi$. Similarly,

$$\int |G|^2\,dx = (\Delta Q)^2.$$

$$\therefore \quad (\Delta P)^2(\Delta Q)^2 \geq \left|\int (\Lambda^*\Psi^*)(\Omega\Psi)dx\right|^2 = \left|\int \Psi^*\Lambda\Omega\Psi\,dx\right|^2, \tag{3.4-5}$$

where (3.3-3) has been used to shift Λ. The right side of (5) is

$$|\langle\Lambda\Omega\rangle|^2 = \left|\tfrac{1}{2}\langle\Lambda\Omega + \Omega\Lambda\rangle + \frac{1}{2i}\langle i(\Lambda\Omega - \Omega\Lambda)\rangle\right|^2.$$

The two operators

$$\Lambda\Omega + \Omega\Lambda \qquad \text{and} \qquad i(\Lambda\Omega - \Omega\Lambda) \equiv i[\Lambda, \Omega]$$

are both Hermitian, and their expectation values are both real.

$$\therefore \quad (\Delta P)^2(\Delta Q)^2 \geq \tfrac{1}{4}|\langle\Lambda\Omega + \Omega\Lambda\rangle|^2 + \tfrac{1}{4}|\langle i[\Lambda, \Omega]\rangle|^2. \tag{3.4-6}$$

From the definitions of Λ and Ω,

$$[\Lambda, \Omega] = (Q - \langle Q\rangle)(P - \langle P\rangle) - (P - \langle P\rangle)(Q - \langle Q\rangle) = [Q, P].$$

Since $|\langle\Lambda\Omega + \Omega\Lambda\rangle|^2$ is never negative, the theorem (1) follows. Notice that the equal sign in (1) obtains only if first, from (4),

$$(Q - \langle Q\rangle)\Psi = c(P - \langle P\rangle)\Psi \tag{3.4-7}$$

everywhere, and second,

$$\langle\Lambda\Omega + \Omega\Lambda\rangle = 0. \tag{3.4-8}$$

Suppose that the operators P and Q are p_x and x. As observed in (3.3-6),

$$[p_x, x] = \frac{\hbar}{i} \tag{3.4-9}$$

is an identity. The inequality (1) therefore says that, for any Ψ,

$$(\Delta p_x)(\Delta x) \geq \frac{\hbar}{2}. \tag{3.4-10}$$

The uncertainty principle has become precise: For any Ψ, if Δp_x and Δx are the root mean squared deviations of p_x and x, then their product must be at least $\hbar/2$. Look again at the solution to Problem 3.5. For the Gaussian wave function of the form $e^{-\beta^2 x^2/2}$, $\Delta p_x \, \Delta x = \hbar/2$. Such a wave function is sometimes called a *minimum packet*; regardless of the value of β, the product $\Delta p_x \cdot \Delta x$ is as small as it can be. It is not hard to show that if (7) and (8) are satisfied, then $|\Psi|^2$ must be a Gaussian.[5]

Suppose that P and Q are p_x and y. Then

$$[p_x, y]\Psi = \frac{\hbar}{i}\frac{\partial}{\partial x}(y\Psi) - y\frac{\hbar}{i}\frac{\partial}{\partial x}\Psi = 0.$$

There is no limitation on the simultaneous measurability of p_x and y.

PROBLEM 3.14

a. Obtain the uncertainty principle for p_z and z.

b. Classically the orbital angular momentum $\vec{L}$ is $\vec{r} \times \vec{p}$. The quantal operator for $\vec{L}$ is also $\vec{r} \times \vec{p}$, with $\vec{r}$ and $\vec{p}$ equal to the position and momentum operators; this choice is examined more carefully in Chapter 6. Use

$$L_z = (\vec{r} \times \vec{p})_z = xp_y - yp_x = x\frac{\hbar}{i}\frac{\partial}{\partial y} - y\frac{\hbar}{i}\frac{\partial}{\partial x}.$$

Let $\phi = \tan^{-1}(y/x)$ be the angle of rotation around the z axis. Use the theorem (1) to deduce that

$$\Delta L_z \, \Delta\phi \geq \frac{\hbar}{2}. \tag{3.4-11}$$

This result is qualitatively valid; it is impossible to measure ϕ and L_z simultaneously and precisely. However, the fact that ϕ is restricted to a finite range, 0 to 2π, limits the validity of (11).[6]

c. Explore the possibility of measuring simultaneously the energy $H = p^2/2M + V(\vec{r})$ and x. Show that

$$\Delta H \, \Delta x \geq \frac{\hbar}{2M}\langle p_x \rangle. \tag{3.4-12}$$

This result depends on the form of Ψ and is not a simple uncertainty principle of the form (10) or (11).

---

[5] D. Bohm, *Quantum Theory* (Englewood Cliffs, New Jersey: Prentice-Hall, Inc., 1951), Sec. 10.9; D. S. Saxon, *Elementary Quantum Mechanics* (San Francisco: Holden-Day, Inc., 1968), Chap. V. Sec. 8.

[6] P. Carruthers and W. N. Nieto, *Rev. Mod. Phys.* **40**, 411 (1968), *Sec. 4.A.*

Look at the result (12) in the problem above. Suppose that Ψ describes a bound state in a potential well, so that $\langle p_x \rangle = 0$. Then (12) says that $\Delta H \, \Delta x \geq 0$. Does it follow that the energy and position can then be known simultaneously to arbitrary accuracy? It does not. A measurement of position will generally mess up one's knowledge of the energy, and this example is a good reminder of what the theorem does *not* say. If for some state $\langle [P, Q] \rangle = 0$, it does *not* follow that in that state $\Delta P = 0$ and $\Delta Q = 0$. For a bound definite energy state, $\Delta H = 0$, but Δx is of the order of the width of the well. In this particular example $\Delta H \, \Delta x = 0$, but Δx is greater than zero. In other examples it is important to recognize that violation of (7) or (8) may also raise the floor under the uncertainty product.

The energy-time uncertainty principle requires special consideration. Section 1.2 shows that, if the time at which a particle passes a point is uncertain by Δt, then its energy uncertainty satisfies

$$\Delta E \, \Delta t \gtrsim \hbar. \tag{3.4-13}$$

Section 1.3 mentions, and Section 3.11 quantifies, the observation that a state with mean life τ has an energy uncertainty $\Delta E \simeq \hbar/\tau$. The uncertainty principle (13) is true and useful, but it does not follow from the theorem (1). The reason is that the time t is not an operator, it is a parameter, and there is no general meaning that can be attached to its expectation value. The expression $\int \Psi^* t \Psi$ is of no interest; it gives t, no matter what Ψ is used. The $\Delta E \, \Delta t$ uncertainty principle is examined further in Section 3.10.

3.5 *Orthogonality*

Suppose that Ψ_n and Ψ_m are eigenfunctions of the Hermitian operator Ω with different eigenvalues,

$$\Omega \Psi_n = \omega_n \Psi_n, \qquad \Omega \Psi_m = \omega_m \Psi_m, \qquad \omega_n \neq \omega_m.$$

If a state is described by an eigenstate of an operator Ω, the expectation value is the eigenvalue:

$$\langle \Omega \rangle = \int \Psi_n^* \Omega \Psi_n = \int \Psi_n^* \omega_n \Psi_n = \omega_n.$$

Since Ω is Hermitian, all eigenvalues must be real. Consider

$$\int \Psi_n^* \Omega \Psi_m = \int (\Omega \Psi_n)^* \Psi_m.$$

The left side is

$$\int \Psi_n^* \omega_m \Psi_m = \omega_m \int \Psi_n^* \Psi_m,$$

and the right side is

$$\int (\omega_n \Psi_n)^* \Psi_m = \omega_n^* \int \Psi_n^* \Psi_m = \omega_n \int \Psi_n^* \Psi_m.$$

$$\therefore \quad (\omega_m - \omega_n) \int \Psi_n^* \Psi_m = 0, \tag{3.5-1}$$

and, since $\omega_m \neq \omega_n$,

$$\int \Psi_n^* \Psi_m = 0. \tag{3.5-2}$$

The overlap integral of Ψ_n^* and Ψ_m is zero. For a particle in a box, (2) merely states again that

$$\int_0^a \left(\sqrt{\frac{2}{a}} \sin \frac{n\pi x}{a}\right)\left(\sqrt{\frac{2}{a}} \sin \frac{m\pi x}{a}\right) dx = \delta_{nm}. \tag{3.5-3}$$

For the harmonic oscillator wave functions it is worthwhile to sketch a few of the lowest states and their products, and to see how the overlap integrals conspire to give zero through the cancellation of positive and negative contributions.

If (2) is satisfied, then Ψ_n and Ψ_m are said to be *orthogonal.* The use of the word *orthogonal* comes from an analogy with a many-dimensional complex vector space. If two vectors $\vec{a}$ and $\vec{b}$ in such a space have components $(a_1, a_2, \ldots, a_\nu, \ldots)$ and $(b_1, b_2, \ldots, b_\nu, \ldots)$, the scalar product is defined by

$$\vec{a} \cdot \vec{b} \equiv a_1^* b_1 + a_2^* b_2 + \cdots + a_\nu^* b_\nu + \cdots = (\vec{b} \cdot \vec{a})^*. \tag{3.5-4}$$

Unless the scalar product is real, $\vec{a} \cdot \vec{b} \neq \vec{b} \cdot \vec{a}$. The complex conjugate of the first vector's components is used so that the scalar product of a vector with itself is interpretable as the square of its magnitude.

$$|\vec{a}|^2 \equiv \vec{a} \cdot \vec{a} = a_1^* a_1 + a_2^* a_2 + \cdots = \sum_\nu |a_\nu|^2, \tag{3.5-5}$$

a real quantity that is never negative. If each $\Psi(x)$ is viewed as a vector, with its value in each small neighborhood as one of its components, a rough analogue between expression (4) and $\int \Psi_n^* \Psi_m$ begins to emerge:[7]

$$\int \Psi_n^*(v) \Psi_m(v) dv \leftrightarrow \sum_\nu a_\nu^* b_\nu, \tag{3.5-6}$$

and one can understand why Ψ_n and Ψ_m are said to be orthogonal if (2) is satisfied. Section 3.8 examines these ideas in more detail.

Suppose that there is degeneracy, so that $\omega_n = \omega_m$ even though Ψ_n and Ψ_m are different. Equation (1) still obtains, but (2) no longer follows, and Ψ_n and

[7] This observation also permits the Schwarz inequality (3.4-2) to be viewed as an analogue of the vector inequality $(\vec{a} \cdot \vec{a})(\vec{b} \cdot \vec{b}) \geq (\vec{a} \cdot \vec{b})^2$, where the equality holds if $\vec{a}$ and $\vec{b}$ are parallel.

Ψ_m may or may not be orthogonal. The particle in an $a \times a \times a$ cube is a good example. Look at the states $\Psi_n \equiv (1, 2, 1)$ and $\Psi_m \equiv (2, 1, 1)$, in the notation of Section 2.4. For them, $\int \Psi_n^* \Psi_m = 0$, but this conclusion does not follow from the observation that Ψ_n and Ψ_m are different energy eigenfunctions; they are degenerate.[8] There are, in fact, infinitely many wave functions that all have the same energy; any

$$\Psi = \alpha\Psi_n + \beta\Psi_m, \qquad |\alpha|^2 + |\beta^2| = 1,$$

will do. For the degenerate pair Ψ_n and $\alpha\Psi_n + \beta\Psi_m$,

$$\int \Psi_n^*(\alpha\Psi_n + \beta\Psi_m) = \alpha.$$

In general, if there is degeneracy, (i) the wave functions may or may not be orthogonal, (ii) those choices that fit the symmetry of the situation often turn out to be orthogonal, and (iii) we can always find pairs that are orthogonal. Statement (iii) is best verified by showing explicitly how to do it with a construction known as the *Schmidt orthogonalization method*. If Ψ_n and Ψ_m are degenerate and not orthogonal, replace them by the pair Ψ_n and

$$\Psi_m' = \frac{\Psi_m - \Psi_n \int \Psi_n^* \Psi_m}{\sqrt{1 - |\int \Psi_n^* \Psi_m|^2}}. \tag{3.5-7}$$

This replacement clearly does the job, because

$$\int \Psi_n^* \Psi_m' = \frac{\int \Psi_n^* \Psi_m - (\int \Psi_n^* \Psi_n) \int \Psi_n^* \Psi_m}{\sqrt{1 - |\int \Psi_n^* \Psi_m|^2}} = 0.$$

The square root in the denominator normalizes Ψ_m'. If three or more degenerate eigenfunctions need to be made orthogonal, similar constructions can be devised.

PROBLEM 3.15

The three eigenfunctions Ψ_a, Ψ_b, and Ψ_c are linearly independent and degenerate, and are not mutually orthogonal. Devise a scheme for producing linear combinations of Ψ_a, Ψ_b, and Ψ_c that are orthogonal; that is, produce a prescription that does for this case what (7) does for two functions.

---

[8] One could of course observe that Ψ_n and Ψ_m are nondegenerate eigenfunctions of p_x^2, or of p_y^2, and that they must therefore be orthogonal.

We can then produce a set of mutually orthogonal eigenfunctions, automatically if there is no degeneracy, by construction if there is degeneracy. Since these functions can always be normalized, it is always possible to arrange the eigenfunctions of any Hermitian operator so that

$$\int \Psi_n^* \Psi_m = \delta_{nm} \tag{3.5-8}$$

if the eigenvalues and eigenfunctions are discrete, as in (3).

If the eigenvalues are continuously distributed, a change in language is needed. Suppose for example that the wave functions describe free particles and are the eigenfunctions of the x component of momentum, with eigenvalues that can be any value p, or k,

$$\psi_k(x) = Ae^{ipx/\hbar} = Ae^{ikx}.$$

The overlap integral $\int \psi_{k'}^* \psi_k$ should be zero if $k' \neq k$, and be something useful if $k' = k$. Since here

$$\int \psi_{k'}^* \psi_k = A^*A \int_{-\infty}^{+\infty} e^{i(k-k')x}\, dx, \tag{3.5-9}$$

an integral which is infinite if $k = k'$, plane waves cannot be normalized according to the prescription used for bound states. Until now, we have sometimes chosen the amplitude A to make the incident flux come out right, and at other times observed that the infinite plane wave is an unrealizable idealization because the electron is always confined in some region. Still, there is a formal difficulty here that can now be resolved. Instead of the kind of normalization implied by (8), *Dirac delta function normalization* is used:

$$\int \psi_{k'}^* \psi_k = \delta(k - k'). \tag{3.5-10}$$

Look back at (3.2-7) and change the variables there to match the present situation.

$$\int_{-\infty}^{+\infty} \frac{e^{i(k-k')x}}{2\pi}\, dx = \delta(k - k'). \tag{3.5-11}$$

With the choice $A = 1/\sqrt{2\pi}$ in (9), (10) is now satisfied and the wave functions

$$\psi_k(x) = \frac{e^{ikx}}{\sqrt{2\pi}}, \tag{3.5-12}$$

where k can take on any positive or negative value, are orthogonal and normalized, but in the Dirac delta function sense.

The eigenfunctions of a Hermitian operator can always be arranged so that they form a set of functions that are both orthogonal and normalized, or *orthonormal*, in the sense of (8) where the eigenvalues are discrete, and in the sense of (10) where they are continuously distributed.

3.6 *Eigenfunction Expansions*

Section 3.5 shows that the eigenfunctions of a Hermitian operator can be arranged as an orthonormal set, in the sense of (3.5-8) or (3.5-10). Such sets of functions are often useful for constructing series expansions; Fourier series are a familiar and relevant example. However, the question of *completeness* must be faced before one can proceed.

Suppose that one wants to expand $\psi(x)$ in the interval $-a/2 < x < +a/2$ by writing

$$\psi(x) = \sum_{n=0}^{\infty} b_n \cos \frac{2\pi n x}{a}.$$

Generally, the attempt will not succeed, because the set of functions $\cos 2\pi n x/a$ is not *complete*. Unless $\psi(x)$ happens to be an even function of x, the functions $\sin 2\pi n x/a$ also are needed in the expansion. What about the eigenfunctions of Hermitian operators: Do they form complete sets?

The answer is *yes*, according to the remarkable *expansion postulate*: If then Ψ_n are the eigenfunctions of a Hermitian operator Ω,

$$\Omega\Psi_n = \omega_n \Psi_n,$$

then any acceptable wave function Ψ, that satisfies the same boundary conditions as the Ψ_n, can be represented by

$$\Psi = \sum_n C_n \Psi_n, \tag{3.6-1}$$

where the C_n are constants and the sum may need to include all of the eigenfunctions of Ω. In this equation $\sum$ is indeed a discrete sum where the eigenstates are discrete, but has to be replaced by an integral where the eigenstates are continuously distributed. It may happen that both a sum and an integral occur in the same expansion. For the energy eigenstates of a potential well with a few bound states, one needs to sum over these and to integrate over the continuously distributed positive energy states.

Let us check the expansion postulate for some examples. If the Hermitian operator is the energy for a particle in the box considered in Section 2.4, a superposition of the eigenfunctions (2.4-3) can be arranged to converge to any Ψ that is zero on and outside the walls of the box. This observation is nothing but the statement that Ψ can be represented by a Fourier series. Expansion in the momentum eigenfunctions of a free particle, $e^{i\vec{p}\cdot\vec{r}/\hbar}$, means making a Fourier integral representation (3.2-18), where the $\vec{k}$ play the part of the labels n and the $g(\vec{k})$ correspond to the C_n. It can be shown that any normalizable Ψ can be expressed as a linear combination of harmonic oscillator wave functions.[9] The parity operator $\mathscr{P}$, defined by

$$\mathscr{P}\Psi(\vec{r}, t) = \Psi(-\vec{r}, t),$$

is Hermitian, and its eigenfunctions are all even functions, with eigenvalue $+1$, and all odd functions, with eigenvalue -1. Since any Ψ can be written as a superposition of an even function and an odd function, the eigenfunctions of $\mathscr{P}$ do form a complete set. Chapter 8 shows that s_z, the z component of the spin of an electron, can be either $+\hbar/2$ or $-\hbar/2$, and that the two corresponding eigenfunctions form a complete set in their space: Any arbitrary spin $\frac{1}{2}$ state can be written as a linear combination of a state in which s_z is definitely $+\hbar/2$ and a state in which s_z is definitely $-\hbar/2$.

Now look at two examples of non-Hermitian operators. For the operator $ip_x = \hbar\, \partial/\partial x$, since

$$\hbar \frac{\partial}{\partial x} Ae^{\alpha x/\hbar} = \alpha A e^{\alpha x/\hbar},$$

the real exponentials $Ae^{\alpha x/\hbar}$ are the eigenfunctions. Since all of these exponentials diverge, either as $x \to +\infty$ or as $x \to -\infty$, there is no linear combination of these functions that can converge to an acceptable Ψ; they are *not* complete. However, the eigenfunctions of the complex conjugation operator K,

$$K\Psi \equiv \Psi^*,$$

are complete; the eigenfunctions are all real functions and all purely imaginary functions.

For non-Hermitian operators, the eigenfunctions may or may not form a complete set. For every Hermitian operator that has been examined, the eigenfunctions have turned out to be complete, and the expansion postulate is the distillation of this experience. This way of approaching the completeness question looks like a peculiar kind of experimental mathematics, and will disturb intermediate-level students, who tend toward an excessive faith in the orderliness of their discipline. There are in fact theorems that deal with the question for important classes of operators,[10] but no theorem is known that deals with all of the classes of operators that one might encounter.

[9] D. Park, *Introduction to the Quantum Theory*, 2d ed. (New York: McGraw-Hill Book Company, 1974), Appendix 2.

[10] R. Courant and D. Hilbert, *Methods of Mathematical Physics*, Vol. I (New York: Interscience Publishers, Inc., 1953), Sec. II.§1.3.

Because the Ψ_n are arranged to form an orthonormal set, it is easy to evaluate the coefficients C_n in (1). Multiply (1) by Ψ_m^* and integrate over the space of the variables involved:

$$\int \Psi_m^* \Psi = \sum_n C_n \int \Psi_m^* \Psi_n = \sum_n C_n \delta_{mn} = C_m. \tag{3.6-2}$$

Equation (3.2-3) is an example of this procedure. Remember that $\sum$ becomes an integral and that δ_{mn} becomes a Dirac delta function if there is a continuum of states.

PROBLEM 3.16

The position x is a Hermitian operator. Its eigenfunctions ψ_b satisfy

$$x\psi_b = b\psi_b,$$

where b is a constant. Find the ψ_b (do not worry about normalization). Show how any $\psi(x)$ can be written as a linear combination of the $\psi_b(x)$.

Suppose that in (1) the Ψ_n are the energy eigenfunctions of the Hamiltonian H that governs the development of Ψ; that is, suppose that, with x standing for all the coordinates,

$$H\Psi(x, t) = i\hbar \frac{\partial}{\partial t} \Psi(x, t). \tag{3.6-3}$$

Assume that H does not depend explicitly on t. Expand as in (1),

$$\Psi(x, t) = \sum_E C_E \Psi_E(x, t), \tag{3.6-4}$$

$$H\Psi_E(x, t) = E\Psi_E(x, t). \tag{3.6-5}$$

Here the label E, rather than the general label n, is used to emphasize that the expansion is in terms of the energy eigenfunctions. From Section 2.1,

$$\Psi_E(x, t) = \psi_E(x)e^{-iEt/\hbar}.$$

Now expand, as in (4), at $t = 0$, and evaluate C_E as in (2):

$$\Psi(x, 0) = \sum_E C_E \psi_E(x), \tag{3.6-6}$$

$$C_E = \int \psi_E^*(x)\Psi(x, 0)dx. \tag{3.6-7}$$

At any later time each of the energy eigenfunctions has developed into $\psi_E(x)e^{-iEt/\hbar}$, and $\Psi(x, t)$ is obtained by assembling these functions with the constants C_E evaluated at $t = 0$:

$$\Psi(x, t) = \sum_E C_E \psi_E(x) e^{-iEt/\hbar}. \tag{3.6-8}$$

It is easy to check that this expression does solve (3). Operate on the sum with H:

$$\sum_E C_E H\psi_E(x)e^{-iEt/\hbar} = \sum_E C_E E\psi_E(x)e^{-iEt/\hbar}.$$

Operate on the sum with $i\hbar\,\partial/\partial t$:

$$\sum_E C_E \psi_E(x) i\hbar \frac{\partial}{\partial t} e^{-iEt/\hbar} = \sum_E C_E \psi_E(x) E e^{-iEt/\hbar},$$

the same thing.

PROBLEM 3.17

$V = 0, 0 \leq x \leq a$, and $V = \infty$ otherwise. At $t = 0$ the particle is definitely in the left half of the well because, as shown in Figure 3.6-1,

$$\Psi(x, 0) = \frac{2}{\sqrt{a}} \sin \frac{2\pi x}{a}, \qquad 0 \leq x \leq \frac{a}{2},$$
$$= 0 \qquad \text{otherwise.}$$

Expand $\Psi(x, 0)$ in terms of the energy eigenfunctions, as in (6). Evaluate enough terms to that $\Psi(x, 0)$ is given to 5% accuracy at $x = a/4$. Then write $\Psi(x, t)$ in the form (8).

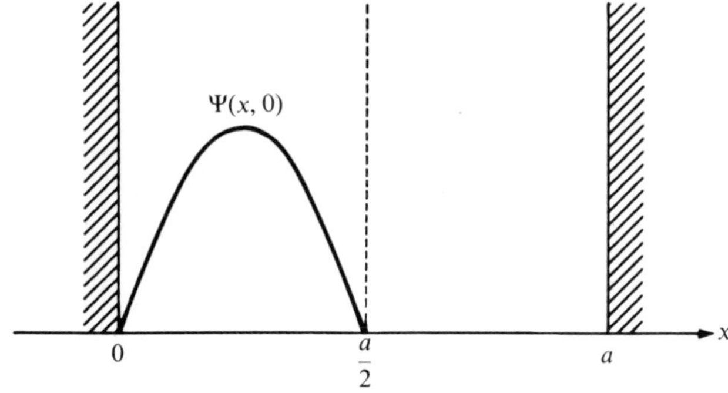

Fig. 3.6-1. The initial conditions for Problem 3.17.

PROBLEM 3.18

$V = 0$, $-a/2 \leq x \leq +a/2$, and $V = \infty$ otherwise. At $t = 0$,

$$\begin{aligned}\Psi(x, 0) &= C_1\psi_1(x) + C_2\psi_2(x) \\ &= \frac{1}{\sqrt{2}}\left(\sqrt{\frac{2}{a}}\cos\frac{\pi x}{a}\right) + \frac{1}{\sqrt{2}}\left(\sqrt{\frac{2}{a}}\sin\frac{2\pi x}{a}\right);\end{aligned}$$

that is, (6) happens to have only two nonzero terms because $\Psi(x, 0)$ is a superposition of equal amounts of the lowest and next-to-the lowest energy eigenstates. Obtain $\Psi(x, t)$. Use this $\Psi(x, t)$ to calculate $\langle H \rangle$, ΔH, $\langle x \rangle$, and $\langle p \rangle$. You will find that both $\langle x \rangle$ and $\langle p \rangle$ oscillate in time. Sketch $|\Psi(x, t)|^2$ at different times to show how these oscillations come about.

One can present the content of (8) in a different form. Write (8) for the time t' at x'.

$$\Psi(x', t') = \sum_E C_E \psi_E(x') e^{-iEt'/\hbar}.$$

Multiply by $\psi^*_{E'}(x')$ and integrate over x'.

$$\begin{aligned}\int \psi^*_{E'}(x')\Psi(x', t')dx' &= \sum_E C_E \int \psi^*_{E'}(x')\psi_E(x')e^{-iEt'/\hbar}\,dx' \\ &= \sum_E C_E \delta_{E'E} e^{-iEt'/\hbar} \\ &= C_{E'} e^{-iE't'/\hbar}.\end{aligned} \tag{3.6-9}$$

Now write (8) with this form for the C_E.

$$\Psi(x, t) = \sum_E \left[e^{iEt'/\hbar} \int \psi^*_E(x')\Psi(x', t')dx' \right] \psi_E e^{-iE/\hbar},$$

or

$$\Psi(x, t) = \int G(x, t; x', t')\Psi(x', t')dx', \tag{3.6-10}$$

where

$$G(x, t; x', t') \equiv \sum_E \psi^*_E(x')\psi_E(x)e^{-iE(t-t')/\hbar}. \tag{3.6-11}$$

The function $G(x, t; x', t')$ is called the *propagator* because it determines how Ψ propagates in space and time. It depends on the Hamiltonian because the Hamiltonian determines the ψ_E, but it is the same for any initial Ψ that one might specify. Its time dependence involves only the difference $t - t'$, and (10) shows that, when $t = t'$, the propagator becomes $\delta(x - x')$, so that (11) becomes

$$\delta(x - x') = \sum_E \psi_E^*(x')\psi_E(x). \tag{3.6-12}$$

Equation (12) is called the *closure* property. A review of the steps leading to (12) shows that it is valid for any complete orthonormal set of eigenfunctions.

PROBLEM 3.19

Obtain the closure property for the general case. That is, show that if the $\Psi_n(x, t)$ are any complete orthonormal set, so that one can write

$$\Psi(x, t) = \sum_n C_n \Psi_n(x, t)$$

for any $\Psi(x, t)$ and evaluate the C_n as in (2), then

$$\sum_n \Psi_n^*(x', t)\Psi_n(x, t) = \delta(x' - x). \tag{3.6-13}$$

Equation (10) looks like the answer to the quantum mechanic's dream: If $\Psi(x', t')$ is known at one time t' for all x', then an integral gives the wave function $\Psi(x, t)$ at all other times for all values of the coordinates. However, the job of actually constructing the propagator is painful, and doing the integration in (10) exactly is impossible, for all but the most trivial situations.[11] The real use of this construction occurs in approximation methods.

3.7 Operators, Eigenvalues, and Measurements

The results of the previous sections can be used to obtain some of the key theorems of quantum theory. *If Ψ is an eigenstate of Ω with eigenvalue ω_n, then a measurement of Ω must give the result ω_n.* Since

$$\Omega\Psi_n = \omega_n \Psi_n,$$

$$\langle\Omega\rangle = \int \Psi_n^* \Omega \Psi_n = \int \Psi_n^* \omega_n \Psi_n = \omega_n,$$

[11] See D. S. Saxon, *Elementary Quantum Mechanics* (San Francisco: Holden-Day, Inc., 1968). The propagator for a free particle is developed and discussed on pp. 60–61, and the general case is discussed on pp. 110–111.

as was already pointed out in Section 3.5.

$$\langle\Omega^2\rangle = \int \Psi_n^*\Omega^2\Psi_n = \omega_n \int \Psi_n^*\Omega\Psi_n = \omega_n^2.$$

From (3.1-4)

$$(\Delta\Omega)^2 = \langle\Omega^2\rangle - \langle\Omega\rangle^2 = \omega_n^2 - \omega_n^2 = 0.$$

If there is any chance that a measurement of Ω gives a result other than $\langle\Omega\rangle$, then $(\Delta\Omega)^2 > 0$. Therefore the expectation value is the eigenvalue, and the deviations from this value are zero. Conversely, *for a Hermitian operator* Ω, *if* $\Delta\Omega = 0$, *then the state* Ψ *must be an eigenstate of* Ω *with eigenvalue* $\langle\Omega\rangle$.

$$(\Delta\Omega)^2 \equiv \int \Psi^*(\Omega - \langle\Omega\rangle)^2\Psi = 0.$$

Since $\Omega - \langle\Omega\rangle$ is Hermitian,

$$\int (\Omega - \langle\Omega\rangle)^*\Psi^*(\Omega - \langle\Omega\rangle)\Psi = \int |(\Omega - \langle\Omega\rangle)\Psi|^2 = 0.$$

The integrand is never negative, so that the integral can be zero only if the integrand is zero everywhere, that is, if Ψ satisfies the eigenvalue equation

$$\Omega\Psi = \langle\Omega\rangle\Psi.$$

Therefore Ψ *being an eigenfunction of* Ω *is a necessary and a sufficient condition for the precise predictability of a measurement of* Ω. Like several of the important results of this chapter, this conclusion is not an entirely new idea. Throughout Chapter 2 $H\Psi = E\Psi$ is used as the condition for the energy to be definitely E, and $-i\hbar\vec{\nabla}\Psi = \vec{p}\Psi$ is used as the condition for the momentum to be definitely $\vec{p}$.

Suppose one begins with any state Ψ, not necessarily an eigenstate of Ω, measures Ω, and obtains some number ω. At that moment Ω is known as accurately as the measuring instruments permit; it is ω and nothing else. Since Ω is known precisely, the particle must be described by an eigenstate, and the known value of Ω must be the corresponding eigenvalue ω_n. It follows that a measurement of Ω changes the state of the system from the initial state Ψ into one of the eigenstates Ψ_n of Ω, and that ω, the result of the measurement, must be the corresponding eigenvalue ω_n.

The list of eigenvalues ω_n is then the list of possible results of a measurement of Ω, regardless of the state Ψ. If for example $V = 0$, $0 \le x \le a$, and $V = \infty$ otherwise, then a measurement of E_x must yield $\hbar^2\pi^2/2Ma^2$, or $4\hbar^2\pi^2/2Ma^2$, or $9\hbar^2\pi^2/2Ma^2$, or .... Now use the expansion (3.6-1) of Ψ in terms of the eigenfunctions of Ω,

$$\Psi = \sum_n C_n\Psi_n,$$

to evaluate the expectation value of Ω.

$$\langle\Omega\rangle = \int \Psi^*\Omega\Psi = \int\left(\sum_m C_m^*\Psi_m^*\right)\Omega\left(\sum_n C_n\Psi_n\right)$$

$$= \sum_{m,n} C_m^* C_n \int \Psi_m^* \omega_n \Psi_n \tag{3.7-1}$$

$$= \sum_{m,n} C_m^* C_n \omega_n \delta_{mn} = \sum_n |C_n|^2 \omega_n.$$

Since the ω_n are the possible values and $\langle\Omega\rangle$ is the mean value, $|C_n|^2$ must be the probability of finding ω_n, and therefore the probability of the state being Ψ_n after the measurement. One might guess this result without writing (1). In the expansion of Ψ in terms of the eigenfunctions, C_n is the amplitude with which Ψ_n contributes, and probabilities are proportional to the square of the magnitude of amplitudes. This idea was used to write (3.2-13).

PROBLEM 3.20

Show that

$$\sum_n |C_n|^2 = 1 \tag{3.7-2}$$

if the state $\Psi = \sum_n C_n\Psi_n$ is normalized. Without (2), $|C_n|^2$ could not be interpreted as the probability of finding ω_n. Equation (3.2-10) is an illustration of this result.

EXAMPLE

Suppose that $V = Kx^2/2$, and that

$$\Psi(x, 0) = \sqrt{\frac{\beta}{\sqrt{\pi}}}\, e^{-\beta^2(x-b)^2/2}, \qquad \beta = \sqrt{\frac{M\omega}{\hbar}}, \qquad \omega = \sqrt{\frac{K}{M}}.$$

That is, the wave function at $t = 0$ is just like that for the harmonic oscillator ground state, except that it is centered at $x = b$ rather than $x = 0$. If the energy is measured, what is the probability of finding $\hbar\omega/2$? $3\hbar\omega/2$? More than $3\hbar\omega/2$? What is $\Psi(x, t)$?

SOLUTION

Expand as in (3.6-6).

$$\Psi(x, 0) = C_0\psi_0(x) + C_1\psi_1(x) + \cdots,$$

where

$$\psi_0 = \sqrt{\frac{\beta}{\sqrt{\pi}}}\, e^{-\beta^2x^2/2}, \qquad \psi_1 = \sqrt{\frac{\beta}{2\sqrt{\pi}}}\,(2\beta x)e^{-\beta^2x^2/2}, \ldots.$$

Evaluate the coefficients as in (3.6-7).

$$C_0 = \int_{-\infty}^{+\infty} \psi_0^*(x)\Psi(x, 0)dx$$

$$= \frac{\beta}{\sqrt{\pi}} e^{-\beta^2 b^2/4} \int_{-\infty}^{+\infty} e^{-\beta^2(x-b/2)^2}\, dx = e^{-\beta^2 b^2/4}.$$

$$C_1 = \int_{-\infty}^{+\infty} \psi_1^*(x)\Psi(x, 0)dx$$

$$= \beta^2 \sqrt{\frac{2}{\pi}} e^{-\beta^2 b^2/4} \int_{-\infty}^{+\infty} \left[\left(x - \frac{b}{2}\right)e^{-\beta^2(x-b/2)^2} + \frac{b}{2} e^{-\beta^2(x-b/2)^2}\right] d\left(x - \frac{b}{2}\right)$$

$$= \frac{\beta b}{\sqrt{2}} e^{-\beta^2 b^2/4}.$$

The probability of finding $\hbar\omega/2$ is

$$|C_0|^2 = e^{-\beta^2 b^2/2}.$$

The probability of finding $3\hbar\omega/2$ is

$$|C_1|^2 = \frac{\beta^2 b^2}{2} e^{-\beta^2 b^2/2}.$$

The probability of finding more than $3\hbar\omega/2$ is

$$\sum_{n=2}^{\infty} |C_n|^2.$$

Since

$$\sum_{n=0}^{\infty} |C_n|^2 = 1,$$

$$\sum_{n=2}^{\infty} |C_n|^2 = 1 - |C_0|^2 - |C_1|^2 = 1 - e^{-\beta^2 b^2/2}\left(1 + \frac{\beta^2 b^2}{2}\right).$$

Notice how the chance of finding a high energy increases as βb increases.

$$\Psi(x, t) = \sum_{n=0}^{\infty} C_n \psi_n(x) e^{-iE_n t/\hbar}$$

$$= C_0 \psi_0(x) e^{-i\omega t/2} + C_1 \psi_1(x) e^{-3i\omega t/2} + \cdots,$$

where C_0, ψ_0, C_1, ψ_1 are given above, and the higher terms can be found in the same way. (By using the equations given in footnote 8 of Chapter 2, one can show that

$$C_n = \frac{\beta^n b^n e^{-\beta^2 b^2/4}}{\sqrt{2^n n!}}.$$

The series for $\Psi(x, t)$ can be summed, and the result describes a wave packet that oscillates back and forth with the classical frequency ω.)

We now have the concepts and language needed for a general summary: The state function Ψ contains all information that can be known about the system. The behavior of Ψ is determined by the Schrödinger equation,

$$H\Psi = i\hbar \frac{\partial \Psi}{\partial t}.$$

Each measurable dynamical quantity is represented by a Hermitian operator Ω with a complete set of orthonormal eigenfunctions Ψ_n,

$$\Omega \Psi_n = \omega_n \Psi_n.$$

Since the Ψ_n are complete, the state function can be represented by the expansion

$$\Psi = \sum_n C_n \Psi_n, \qquad C_n = \int \Psi_n^* \Psi.$$

The possible results of a measurement of Ω are the eigenvalues, and the probability of obtaining ω_n is $|C_n|^2$.

In our approach these statements were developed through discussions of what happens and what is reasonable, but they are really the essential axioms of quantum mechanics. They contain what we have done and will do. If you are formally inclined and mathematically skillful, you might consider clipping the previous paragraph and throwing the rest of the book away.

PROBLEM 3.21

Suppose that $\Psi(x, t)$ is a solution of $H\Psi = i\hbar\, \partial\Psi/\partial t$, where H does not depend explicitly on t, so that (3.6-3) through (3.6-8) are valid. Show that $\langle H \rangle$ and ΔH are independent of the time.

PROBLEM 3.22

Review your solution to Problem 3.17. For the $\Psi(x, t)$ obtained there, calculate $\langle H \rangle$. Keep a few more terms than you did in Problem 3.17, and try to guess the exact result.

PROBLEM 3.23

Consider again the three-dimensional isotropic harmonic oscillator. Suppose that Ψ is a superposition only of states with $E = \frac{3}{2}\hbar\sqrt{K/M}$ and with $E = \frac{5}{2}\hbar\sqrt{K/M}$. Suppose also that $\langle H\rangle = 2.20\hbar\sqrt{K/M}$. Describe all possible wave functions that satisfy these conditions.

If some Ψ_n is an eigenfunction of *two* operators Ω and Λ,

$$\Omega\Psi_n = \omega_n\Psi_n, \qquad \Lambda\Psi_n = \lambda_n\Psi_n,$$

then both Ω and Λ are known precisely. The general uncertainty principle (3.4-1) says that $\langle[\Lambda, \Omega]\rangle = 0$ for the state Ψ_n, and it is obvious that $[\Lambda, \Omega]\Psi_n = 0$.

$$\begin{aligned}(\Lambda\Omega - \Omega\Lambda)\Psi_n &= \Lambda\omega_n\Psi_n - \Omega\lambda_n\Psi_n \\ &= \lambda_n\omega_n\Psi_n - \omega_n\lambda_n\Psi_n = 0.\end{aligned} \tag{3.7-3}$$

If, further, *all* eigenstates of Λ are also eigenstates of Ω, then, since any Ψ can be expanded in terms of the Ψ_n,

$$[\Lambda, \Omega]\Psi = [\Lambda, \Omega]\sum_n C_n\Psi_n = \sum_n C_n[\Lambda, \Omega]\Psi_n = 0.$$

An operator that gives zero when applied to any Ψ is the zero operator, and $[\Lambda, \Omega] = 0$. If all eigenfunctions of one operator are also eigenfunctions of another operator, then the two operators *commute*.

In what sense can this statement be turned around? That is, what follows if $[\Lambda, \Omega] = 0$? Suppose that one of the operators, say Ω, has a nondegenerate set of eigenfunctions Ψ_n; $\Omega\Psi_n = \omega_n\Psi_n$. If $[\Lambda, \Omega] = 0$,

$$\Omega(\Lambda\Psi_n) = \Lambda\Omega\Psi_n = \Lambda\omega_n\Psi_n = \omega_n(\Lambda\Psi_n), \tag{3.7-4}$$

so that $\Lambda\Psi_n$ is also an eigenfunction of Ω with the same eigenvalue ω_n. Since there is no degeneracy, $\Lambda\Psi_n$ must be a constant times Ψ_n, $\Lambda\Psi_n = \lambda_n\Psi_n$, and therefore Ψ_n is also an eigenfunction of Λ. If Λ and Ω both have degenerate eigenstates, the argument given here does not work. However, one can show that then simultaneous eigenfunctions can always be obtained. As in most complications involving degeneracy, it is also possible to argue as follows: Change the Hamiltonian very slightly so as to remove the degeneracy. For example, for an isotropic harmonic oscillator, change the spring constants a very small amount so that they are all very slightly different. Then prove your theorems for the nondegenerate case, and observe that the change in the Hamiltonian, if sufficiently small, cannot make any difference.

The example of the energy and momentum of a free particle illustrates some of these comments. The two operators commute;

$$\left[-\frac{\hbar^2}{2M}\nabla^2, \frac{\hbar}{i}\vec{\nabla}\right] = 0. \tag{3.7-5}$$

The momentum eigenfunctions are not degenerate. Only one state, $Ae^{i\vec{p}\cdot\vec{r}/\hbar}$, satisfies the momentum eigenvalue equation with eigenvalue $\vec{p}$;

$$\frac{\hbar}{i}\vec{\nabla}(Ae^{i\vec{p}\cdot\vec{r}/\hbar}) = \vec{p}(Ae^{i\vec{p}\cdot\vec{r}/\hbar}).$$

The energy eigenfunctions are highly degenerate; any linear combination of momentum eigenfunctions that have the same magnitude of momentum has the same energy E,

$$-\frac{\hbar^2}{2M}\nabla^2\sum_n A_n e^{i\vec{p}_n\cdot\vec{r}/\hbar} = E\sum_n A_n e^{i\vec{p}_n\cdot\vec{r}/\hbar}$$

if each of the $\vec{p}_n$ satisfies $\vec{p}_n\cdot\vec{p}_n/2M = E$. As the general arguments show, the eigenfunctions of the nondegenerate momentum operator are simultaneously eigenfunctions of the energy operator because of (5). Not all energy eigenfunctions are momentum eigenfunctions, but many of them are. These simultaneous eigenfunctions are the $Ae^{i\vec{p}\cdot\vec{r}/\hbar}$, and they form a complete set.

Suppose that all pairs of several operators $\Lambda, \Omega, \mathrm{X}, \ldots,$ commute:

$$[\Lambda, \Omega] = 0, \qquad [\Lambda, \mathrm{X}] = 0, \qquad [\Omega, \mathrm{X}] = 0, \ldots.$$

Then there is a set of wave functions that are simultaneous eigenfunctions of all these operators,

$$\Lambda\Psi_n = \lambda_n\Psi_n, \qquad \Omega\Psi_n = \omega_n\Psi_n, \qquad \mathrm{X}\Psi_n = \chi_n\Psi_n, \ldots,$$

and all of the dynamical quantities represented by these operators can be known precisely at the same time. Chapter 6 shows that in a spherically symmetric potential, the energy, the total angular momentum, and any component, say the z component, of the angular momentum all commute, and there exist simultaneous eigenfunctions of these three operators. For this example the three eigenvalues specify the state completely, so that energy, total angular momentum, and the z component of angular momentum are called a *complete set of commuting Hermitian operators.*

3.8 *Vectors, Matrices, and Dirac Bracket Notation*

Section 3.5 provided some murky hints of an analogy between wave functions and vectors, and much of this section is devoted to exploring this topic further. Consider a space with Cartesian axes labeled $1, 2, \ldots, n, \ldots,$ mutually orthogonal unit vectors $\hat{e}_1, \hat{e}_2, \ldots, \hat{e}_n, \ldots$, along these axes, and general vectors $\vec{\Psi}$. This space differs from the familiar x, y, z space because the number of dimensions may be infinite, and the vectors may be complex. Define the scalar product of two vectors as in (3.5-4), and the square of the magnitude of a vector as in (3.5-5). The unit vectors satisfy

$$\hat{e}_m \cdot \hat{e}_n = \delta_{mn}. \tag{3.8-1}$$

Any vector $\vec{\Psi}$ can be expanded in terms of the $\hat{e}_n$,

$$\vec{\Psi} = \sum_n C_n \hat{e}_n. \tag{3.8-2}$$

To evaluate the C_n, operate on both sides of (2) with $\hat{e}_m\cdot$.

$$\hat{e}_m \cdot \vec{\Psi} = \sum_n C_n \hat{e}_m \cdot \hat{e}_n = \sum_n C_n \delta_{mn} = C_m. \tag{3.8-3}$$

$$|\Psi|^2 \equiv \vec{\Psi} \cdot \vec{\Psi} = \sum_{m,n} C_m^* C_n \hat{e}_m \cdot \hat{e}_n = \sum_n |C_n|^2, \tag{3.8-4}$$

so that the sum of the $|C_n|^2$ is unity if $\vec{\Psi}$ has unit length. It is useful to go to a *matrix representation*[12] by writing the unit vectors as columns,

$$\hat{e}_1 = \begin{pmatrix} 1 \\ 0 \\ 0 \\ 0 \\ \vdots \end{pmatrix}, \quad \hat{e}_2 = \begin{pmatrix} 0 \\ 1 \\ 0 \\ 0 \\ \vdots \end{pmatrix}, \quad \hat{e}_3 = \begin{pmatrix} 0 \\ 0 \\ 1 \\ 0 \\ \vdots \end{pmatrix}, \ldots \tag{3.8-5}$$

Then (2) is, with the vector sign omitted to conform to customary matrix notation,

$$\Psi = C_1 \begin{pmatrix} 1 \\ 0 \\ 0 \\ 0 \\ \vdots \end{pmatrix} + C_2 \begin{pmatrix} 0 \\ 1 \\ 0 \\ 0 \\ \vdots \end{pmatrix} + C_3 \begin{pmatrix} 0 \\ 0 \\ 1 \\ 0 \\ \vdots \end{pmatrix} + \cdots = \begin{pmatrix} C_1 \\ C_2 \\ C_3 \\ C_4 \\ \vdots \end{pmatrix}. \tag{3.8-6}$$

A matrix Ω is a rectangular array of numbers,

$$\Omega = \begin{pmatrix} \Omega_{11} & \Omega_{12} & \Omega_{13} & \cdot & \cdot \\ \Omega_{21} & \Omega_{22} & \Omega_{23} & \cdot & \cdot \\ \Omega_{31} & \Omega_{32} & \Omega_{33} & \cdot & \cdot \\ \cdot & \cdot & \cdot & \cdot & \cdot \end{pmatrix}, \tag{3.8-7}$$

[12] The essential properties of matrices are summarized here. For more details see, for example, B. Kolman, *Elementary Linear Algebra* (New York: The Macmillan Company, 1970). A compact summary of matrix algebra is in J. Mathews and R. L. Walker, *Mathematical Methods of Physics*, 2d ed. (New York: W. A. Benjamin, Inc., 1970), Chap. 6.

and is defined if every element Ω_{mn} is given. The number of rows and the number of columns may each have any value. Matrices are equal if their corresponding elements are equal, that is, $\Omega = \Lambda$ if $\Omega_{mn} = \Lambda_{mn}$ for all m and n.

Multiplication of a matrix by a scalar means multiplying each element by that scalar,

$$C\Omega = \begin{pmatrix} C\Omega_{11} & C\Omega_{12} & \cdot & \cdot \\ C\Omega_{21} & C\Omega_{22} & \cdot & \cdot \\ \cdot & \cdot & \cdot & \cdot \end{pmatrix}. \tag{3.8-8}$$

Addition of two matrices means adding their corresponding elements, that is,

$$\Gamma = \Omega + \Lambda \qquad \text{if} \qquad \Gamma_{mn} = \Omega_{mn} + \Lambda_{mn}. \tag{3.8-9}$$

The two matrices Ω and Λ must have the same number of rows and the same number of columns. In (6), (8) and (9) have already been used for the special case of matrices that have only one column, often called *column matrices.*

Matrix multiplication is defined by

$$\Gamma = \Omega\Lambda \qquad \text{if} \qquad \Gamma_{mn} = \sum_j \Omega_{mj}\Lambda_{jn}. \tag{3.8-10}$$

This operation has meaning only if the number of columns of Ω equals the number of rows of Λ. For example, if

$$\Omega = \begin{pmatrix} 0 & 1 \\ i & 2 \end{pmatrix}, \qquad \Lambda = \begin{pmatrix} i & 1 \\ 3 & 4 \end{pmatrix}, \tag{3.8-11}$$

then

$$\Omega + \Lambda = \begin{pmatrix} i & 2 \\ 3 + i & 6 \end{pmatrix},$$

and

$$\Omega\Lambda = \begin{pmatrix} 0 \cdot i + 1 \cdot 3 & 0 \cdot 1 + 1 \cdot 4 \\ i \cdot i + 2 \cdot 3 & i \cdot 1 + 2 \cdot 4 \end{pmatrix} = \begin{pmatrix} 3 & 4 \\ 5 & 8 + i \end{pmatrix}.$$

In general, matrices do not commute. The same matrices (11) multiplied in opposite order give

$$\Lambda\Omega = \begin{pmatrix} i & i + 2 \\ 4i & 11 \end{pmatrix}.$$

The identity matrix I has elements $I_{mn} = \delta_{mn}$,

$$I = \begin{pmatrix} 1 & 0 & 0 & \cdot \\ 0 & 1 & 0 & \cdot \\ 0 & 0 & 1 & \cdot \\ \cdot & \cdot & \cdot & \cdot \end{pmatrix}. \tag{3.8-12}$$

Note that for any Ω,

$$I\Omega = \Omega I = \Omega. \tag{3.8-13}$$

Taking the *complex conjugate* Ω^* of a matrix Ω means taking the complex conjugate of each element, so that if Ω is given by (7), then

$$\Omega^* = \begin{pmatrix} \Omega_{11}^* & \Omega_{12}^* & \cdot & \cdot \\ \Omega_{21}^* & \Omega_{21}^* & \cdot & \cdot \\ \cdot & \cdot & \cdot & \cdot \end{pmatrix}.$$

The *transpose* $\tilde{\Omega}$ of a matrix Ω is obtained by exchanging the rows and columns of Ω, so that if Ω is given by (7), then

$$\tilde{\Omega} = \begin{pmatrix} \Omega_{11} & \Omega_{21} & \cdot & \cdot \\ \Omega_{12} & \Omega_{22} & \cdot & \cdot \\ \cdot & \cdot & \cdot & \cdot \end{pmatrix}.$$

The *Hermitian conjugate* $\Omega^\dagger$ of a matrix Ω is obtained by taking its complex conjugate and transposing it,

$$\Omega^\dagger = \tilde{\Omega}^*, \tag{3.8-14}$$

so that

$$\Omega^\dagger = \begin{pmatrix} \Omega_{11}^* & \Omega_{21}^* & \cdot \\ \Omega_{12}^* & \Omega_{22}^* & \cdot \\ \cdot & \cdot & \cdot \end{pmatrix}.$$

A matrix is *Hermitian* if

$$\Omega^\dagger = \Omega, \qquad \text{or} \qquad \Omega_{nm}^* = \Omega_{mn}. \tag{3.8-15}$$

For a column matrix or vector Ψ such as (6),

$$\Psi = \begin{pmatrix} C_1 \\ C_2 \\ C_3 \\ \vdots \end{pmatrix}, \qquad \Psi^\dagger = (C_1^*, C_2^*, C_3^*, \ldots), \tag{3.8-16}$$

and the number called $\vec{\Psi} \cdot \vec{\Psi}$ in vector notation is

$$\Psi^\dagger \Psi = C_1^* C_1 + C_2^* C_2 + \cdots = \sum_n |C_n|^2. \tag{3.8-17}$$

A matrix operating on a vector produces another vector,

$$\Omega\Psi = \Phi, \tag{3.8-18}$$

where the elements of Φ, Ω, and Ψ are related by

$$\Phi_i = \sum_j \Omega_{ij} \Psi_j.$$

If $\Phi = \omega \Psi$, where ω is a scalar,

$$\Omega \Psi = \omega \Psi, \tag{3.8-19}$$

and Ψ is an *eigenvector* of the matrix Ω with eigenvalue ω. The structure of (19) is the same as that of (2.1-5). The difference is that there the emphasis was on ordinary functions $\Psi(\vec{r}, t)$ and operators, such as $-(\hbar^2/2M)\nabla^2 + V$, that produce other functions by operating on functions.

PROBLEM 3.24

Consider the three matrices

$$\sigma_x = \begin{pmatrix} 0 & 1 \\ 1 & 0 \end{pmatrix}, \qquad \sigma_y = \begin{pmatrix} 0 & -i \\ i & 0 \end{pmatrix}, \qquad \sigma_z = \begin{pmatrix} 1 & 0 \\ 0 & -1 \end{pmatrix}. \tag{3.8-20}$$

a. Compute $\sigma_x^2 + \sigma_y^2 + \sigma_z^2$.

b. Compute $\sigma_x \sigma_y$, $\sigma_z \sigma_x$, $\sigma_y \sigma_z$, and $\sigma_z \sigma_y$.

c. Obtain the eigenvectors and eigenvalues of each of the three matrices by writing (19) as a pair of linear homogeneous algebraic equations. Arrange for each of the eigenvectors to have unit length; see (4).

The three matrices σ_x, σ_y, and σ_z are the *Pauli spin matrices.* In Chapter 8, $S_x = (\hbar/2)\sigma_x$, $S_y = (\hbar/2)\sigma_y$, and $S_z = (\hbar/2)\sigma_z$ are used to describe the x, y, and z components of the spin of electrons and other particles.

PROBLEM 3.25

Show that for two matrices A and B,

$$(AB)^\dagger = B^\dagger A^\dagger. \tag{3.8-21}$$

Many of our concepts can now be translated into matrix language. First, choose the eigenfunctions Ψ_n of some Hermitian operator Ω as *basis functions,* and use them to expand Ψ, the state function, as in (3.6-1),

$$\Psi = \sum_n C_n \Psi_n,$$

where the C_n are given by (3.6-2). Because a definite set of Ψ_n has been selected, the set of C_n is a complete description of the state,

$$\Psi = \begin{pmatrix} C_1 \\ C_2 \\ C_3 \\ \vdots \end{pmatrix}, \tag{3.8-22}$$

as in (6). The range of n, that is, the number of dimensions of the complex vector space, can well be infinite. For example, if the basis functions are the harmonic oscillator eigenstates, it may be that only the lowest few states contribute appreciably in some particular expansion, but *all* states are needed to form a complete set. The sums involved in matrix multiplication are then infinite series, and questions of convergence can arise.

The expectation value of Ω is given by (3.7-1),

$$\begin{aligned}\langle\Omega\rangle &= \sum_{m,n} C_m^* \omega_n \delta_{mn} C_n \\ &= (C_1^*, C_2^*, C_3^* \ldots) \begin{pmatrix} \omega_1 & 0 & 0 & \cdot \\ 0 & \omega_2 & 0 & \cdot \\ 0 & 0 & \omega_3 & \cdot \\ \cdot & \cdot & \cdot & \cdot \end{pmatrix} \begin{pmatrix} C_1 \\ C_2 \\ C_3 \\ \cdot \end{pmatrix} \\ &= \Psi^\dagger \Omega \Psi.\end{aligned} \tag{3.8-23}$$

Now Ω is a *diagonal* matrix, that is, a matrix with nonzero elements only on the leading diagonal. These elements on the leading diagonal are the eigenvalues. The reason that Ω has the simple form summarized by

$$\Omega_{mn} = \omega_n \delta_{mn} \tag{3.8-24}$$

is that the basis functions are the eigenfunctions of Ω. For any other operator Λ,

$$\begin{aligned}\langle\Lambda\rangle &= \int \Psi^* \Lambda \Psi \\ &= \int \left(\sum_m C_m^* \Psi_m^*\right) \Lambda \left(\sum_n C_n \Psi_n\right) \\ &= \sum_{m,n} C_m{}^* \left(\int \Psi_m^* \Lambda \Psi_n\right) C_n \\ &= \sum_{m,n} C_m^* \Lambda_{mn} C_n,\end{aligned} \tag{3.8-25}$$

where the matrix elements of the operator Λ are given by

$$\Lambda_{mn} \equiv \int \Psi_m^* \Lambda \Psi_n. \tag{3.8-26}$$

With the matrix Λ defined as the array of the Λ_{mn}, (25) becomes the matrix equation

$$\langle\Lambda\rangle = \Psi^\dagger \Lambda \Psi. \tag{3.8-27}$$

Since the basis functions in the expansions are the eigenfunctions of Ω, and not necessarily those of Λ, the matrix Λ generally will not be diagonal.

PROBLEM 3.26

Show that if the basis functions are eigenfunctions of both Ω and Λ, then the matrices Ω and Λ commute.

If Λ is to represent a measurable quantity, then its expectation value must be real. Take the complex conjugate of (25).

$$\begin{aligned}\langle\Lambda\rangle^* &= \sum_{m,n} C_m \Lambda_{mn}^* C_n^* \\ &= \sum_{m,n} C_n^* \Lambda_{mn}^* C_m \\ &= \sum_{m,n} C_m^* \Lambda_{nm}^* C_n \\ &= \sum_{m,n} C_m^* (\Lambda^\dagger)_{mn} C_n \\ &= \Psi^\dagger \Lambda^\dagger \Psi.\end{aligned} \tag{3.8-28}$$

If Λ is Hermitian as defined in (15),

$$\Lambda = \Lambda^\dagger, \qquad \Lambda_{mn}^* = \Lambda_{nm},$$

then $\langle\Lambda\rangle^* = \langle\Lambda\rangle$ for any set of C_n, that is, for any state vector Ψ. It follows that a matrix must be Hermitian if it is to represent a measurable quantity. Notice how these ideas and language are nearly a repetition of Section 3.3.

PROBLEM 3.27

Show that $\langle\Lambda\rangle = \langle\Lambda\rangle^*$ if $\Lambda = \Lambda^\dagger$ by applying (21), the conclusion of Problem 3.25, to (27).

In 1925 W. Heisenberg developed a formulation of quantum mechanics that was based on matrix methods.[13] A central topic of the physics of that time was the discreteness of atomic states; spectroscopic studies had produced lists of discrete energy levels for a great variety of systems. Heisenberg begins his paper with the observation that these lists of energy levels contain the directly

[13] The original paper, W. Heisenberg, *Z. f. Phys.* **33**, 879 (1925), develops the physical ideas and its own notation for dealing with these ideas. M. Born and P. Jordan, *Z. f. Phys.* **34**, 858 (1925), pointed out that Heisenberg's calculations could be expressed in terms of matrix algebra. M. Born, W. Heisenberg, and P. Jordan, *Z. f. Phys.* **35**, 557 (1926) then extended the theory. J. Mehra, "The Birth of Quantum Mechanics," *CERN Report* **76** (10), (1976), gives an excellent history of these developments. One reason for the rapid advance of quantum theory was that, during the nineteen twenties, the absence of modern publishing procedures permitted journals to process manuscripts quickly.

available quantities, and that quantum theory should deal with them, rather than with unknown and possibly unknowable particle trajectories. One is led quite naturally to a description of states which, in our notation, uses

$$\begin{pmatrix} 1 \\ 0 \\ 0 \\ \vdots \end{pmatrix}, \quad \begin{pmatrix} 0 \\ 1 \\ 0 \\ \vdots \end{pmatrix}, \quad \begin{pmatrix} 1/\sqrt{2} \\ 1/\sqrt{2} \\ 0 \\ \vdots \end{pmatrix}$$

to say that the system is definitely in the first state, that it is definitely in the second state, and that it is described by a superposition of equal amounts of first and second states. It then follows that the operators of the theory can be written as matrices. The central task becomes that of finding the eigenvalues of the Hermitian matrices that represent measurable quantities, for these are the possible results of measurements.

In 1926 E. Schrödinger[14] developed the alternate approach in terms of a wave equation which we have emphasized, and which today dominates introductory treatments of the subject. Its techniques are more familiar to those with the usual mathematical education, and it can be argued that it is easier in an absolute sense. It is a formidable task to find the eigenvalues of even simple energy operators by matrix methods alone. Also, a large part of our effort concerns itself with scattering problems, and these are most naturally pictured in terms of waves.

The matrix approach and the wave equation approach are equivalent, as Schrödinger and others showed almost immediately.[15] We shall use matrices primarily in the next section to take another look at the harmonic oscillator, in the treatment of spin angular momentum, and in perturbation theory. Minor uses of matrix methods and language will occur in other sections; for example, in agreement with general custom, integrals of the kind given in (26) are called *matrix elements* even when matrix methods are not being used explicitly.

The equivalence of matrix and differential equation representations suggests correctly that the essentials of quantum theory can be expressed in ways that do not depend on particular choices of mathematical language. The full exploitation of this comment requires treatments that are more formal and advanced than this book. There is, however, a simple and important notation that does not depend on whether matrices or differential operators are used, the *Dirac bracket notation.*[16]

[14] E. Schrödinger, *Ann. d. Physik* **79**, 361, 489; **81**, 109 (1926).

[15] E. Schrödinger, *Ann. d. Physik* **79**, 734 (1926); C. Eckart, *Phys. Rev.* **28**, 711 (1926).

[16] The Dirac bracket notation is developed in P. A. M. Dirac, *The Principles of Quantum Mechanics*, 4th ed. (London: Oxford University Press, 1958), especially Chaps. I, II, and III. A discussion of this notation, and extensions of other topics of this section, are given in E. Merzbacher, *Quantum Mechanics*, 2d ed. (New York: John Wiley & Sons, Inc., 1970), especially Chap. 14.

A state Ψ_a is represented by a *ket vector* $|a\rangle$, where a stands for whatever quantum numbers describe the system. For example, since three quantum numbers are needed to describe an energy eigenstate of a three-dimensional harmonic oscillator, or of a particle in a three-dimensional rigid-walled box, such states $\Psi_{n_x n_y n_z}$ are represented by $|n_x n_y n_z\rangle$. The conjugate of a state, say $\Psi_b^*(\vec{r}, t)$ if it is viewed as a function, or $\Psi_b^\dagger$ if it is seen as the Hermitian conjugate of a vector Ψ_b, is represented by a *bra vector* $\langle b|$. The overlap integral, or the scalar product, is given by

$$\int \Psi_b^* \Psi_a \, dx = \Psi_b^\dagger \Psi_a \equiv \langle b||a\rangle \equiv \langle b|a\rangle. \tag{3.8-29}$$

The *bra* vector is placed in front of the *ket* vector to close the *bracket*; hence the name. The matrix element of an operator is written as

$$\int \Psi_b^* \Omega \Psi_a \, dx = \Psi_b^\dagger \Omega \Psi_a = \langle b|\Omega|a\rangle. \tag{3.8-30}$$

If Ω is Hermitian,

$$\langle a|\Omega|b\rangle = \langle b|\Omega|a\rangle^*. \tag{3.8-31}$$

The equation

$$\Omega|\omega\rangle = \omega|\omega\rangle \tag{3.8-32}$$

says that $|\omega\rangle$ is an eigenfunction, or eigenvector, or eigenstate of Ω with eigenvalue ω. A general state function $\Psi(x)$ can be written $|x\rangle$. Its expansion in eigenfunctions and the evaluation of the coefficients, as in (3.6-1) and (3.6-2), looks like this:

$$|x\rangle = \sum_n C_n|n\rangle, \tag{3.8-33}$$

$$\langle m|x\rangle = \langle m|\sum_n C_n|n\rangle = \sum_n C_n\langle m|n\rangle = C_m. \tag{3.8-34}$$

3.9 *The Harmonic Oscillator Again*

To provide an example of methods that do not depend on solving Schrödinger's differential equation, we shall look at the one-dimensional harmonic oscillator again. Begin by accepting that the energy operator is

$$H = \frac{p^2}{2M} + \frac{Kx^2}{2}, \tag{3.9-1}$$

and that the momentum and position satisfy the commutation relation

$$i[p, x] = \hbar. \tag{3.9-2}$$

In (3.3-6) and (3.4-9) this relation was seen as a result of the fact that $p = -i\hbar\, \partial/\partial x$ in position space. Problem (3.11) shows that it is valid in momentum space. Here it is promoted to a more exalted status; (2) has become a basic relation between the operators p and x which does not depend on the particular representation in which the calculations are done.[17]

We now use a clever trick that was developed by P. A. M. Dirac. Define the two operators

$$A \equiv \sqrt{\frac{M\omega_x}{2\hbar}}\, x + \frac{i}{\sqrt{2M\omega_x \hbar}}\, p, \tag{3.9-3}$$

$$A^\dagger \equiv \sqrt{\frac{M\omega_x}{2\hbar}}\, x - \frac{i}{\sqrt{2M\omega_x \hbar}}\, p, \tag{3.9-4}$$

where $\omega_x \equiv \sqrt{K/M}$. The notation is sensible because $A^\dagger$ is the Hermitian conjugate of A in a matrix representation.

PROBLEM 3.28

Prove that (4) is the Hermitian conjugate of (3).

Calculate $A^\dagger A$ and $AA^\dagger$.

$$A^\dagger A = \frac{1}{\hbar\omega_x}\left(\frac{p^2}{2M} + \frac{M\omega_x^2 x^2}{2}\right) - \frac{i}{2\hbar}[p, x] \tag{3.9-5}$$

$$= \frac{H}{\eta} - \frac{1}{2},$$

$$AA^\dagger = \frac{H}{\eta} + \frac{1}{2}, \tag{3.9-6}$$

where $\eta \equiv \hbar\omega_x$. Subtracting (5) from (6) gives the commutation relation

$$[A, A^\dagger] = 1. \tag{3.9-7}$$

The energy eigenvalue problem $H\Psi_E = E\Psi_E$ can be written

$$\eta(A^\dagger A + \tfrac{1}{2})\Psi_E = E\Psi_E, \tag{3.9-8}$$

or

$$\eta(AA^\dagger - \tfrac{1}{2})\Psi_E = E\Psi_E, \tag{3.9-9}$$

[17] Footnote 4 at the beginning of Section 3.4 indicates a general approach for obtaining such commutation relations.

or

$$\tfrac{1}{2}\eta(A^\dagger A + AA^\dagger)\Psi_E = E\Psi_E. \tag{3.9-10}$$

Suppose that there is a solution Ψ_E, with energy E, of (8). Then operating with A produces a solution with energy $E - \eta$:

$$\eta\left(AA^\dagger A + \frac{A}{2}\right)\Psi_E = EA\Psi_E.$$

Subtract $\eta A\Psi_E$ from both sides.

$$\eta(AA^\dagger - \tfrac{1}{2})A\Psi_E = (E - \eta)A\Psi_E. \tag{3.9-11}$$

Comparison with (9) then shows that $A\Psi_E$ is indeed an eigenfunction of H with eigenvalue $E - \eta$, and A is therefore called a *lowering operator* or *destruction operator*. This procedure can be repeated by applying A to produce the state $A^2\Psi_E$ with energy $E - 2\eta$, and, in general,

$$HA^n\Psi_E = (E - n\eta)A^n\Psi_E. \tag{3.9-12}$$

If the energy could be lowered in steps of η without limit, one would eventually reach a negative energy. However, H is the sum of squares of Hermitian operators, so that its expectation value, and therefore its eigenvalue, can never be negative; see Problem 3.12. The lowering process must eventually reach a lowest energy E_0 and corresponding Ψ_0, such that the process terminates there, that is, such that

$$A\Psi_0 = 0. \tag{3.9-13}$$

With this condition, (8) determines that $E_0 = \eta/2$. Since this value was reached by successive lowerings of the energy by η, we recover the familiar result of (2.7-12),

$$E_n = (n + \tfrac{1}{2})\eta.$$

Just as A is a *lowering* or *destruction* operator, $A^\dagger$ is a *raising* or *creation* operator. If Ψ_E has energy E, then $A^\dagger\Psi_E$ has energy $E + \eta$, and

$$HA^{\dagger n}\Psi_E = (E + n\eta)A^{\dagger n}\Psi_E. \tag{3.9-14}$$

PROBLEM 3.29

Obtain (14). The procedure is almost the same as that used to obtain (12).

All of the Section 2.7 results can be reproduced by going to the explicit representation used there, that is, by taking the operator x to be the multiplying variable x, p to be the operator $-i\hbar\,\partial/\partial x$, and the state function to be a function of x. Then (13) becomes a differential equation for the lowest state,

$$\left(\beta^2 x + \frac{\partial}{\partial x}\right)\Psi_0 = 0,$$

where $\beta \equiv \sqrt{M\omega_x/\hbar}$ as in (2.7-3). This equation is solved by

$$\Psi_0 \sim e^{-\beta^2 x^2/2}.$$

The normalization constant must be found in the usual way by integrating $|\Psi_0|^2$ from $-\infty$ to $+\infty$. The result is

$$\Psi_0 = \sqrt{\frac{\beta}{\sqrt{\pi}}}\,e^{-\beta^2 x^2/2}, \tag{3.9-15}$$

just as in Section 2.7. Now all of the higher states can be generated by repeated application of $A^\dagger$. With the notation Ψ_n, or $|n\rangle$, for the state with energy $(n + \frac{1}{2})\eta$,

$$\Psi_n(x) \sim (A^\dagger)^n \Psi_0(x),$$

or

$$|n\rangle \sim (A^\dagger)^n |0\rangle. \tag{3.9-16}$$

A proportionality sign rather than an equal sign appears because of an interesting complication. Start with (15) for $|0\rangle$ so that the ground state is normalized, $\langle 0|0\rangle = 1$, and generate $|1\rangle$ by operating with $A^\dagger$. Will $|1\rangle$ be normalized? There is no reason why it should be, and it is not. Equation (14) shows that $A^\dagger|n\rangle$ is a state with energy $(n + 1 + \frac{1}{2})\eta$, but normalization has been ignored.

Suppose that Ψ_n is normalized, and that the next higher state is

$$\Psi_{n+1} = N_{n+} A^\dagger \Psi_n, \tag{3.9-17}$$

with N_{n+} chosen so that Ψ_{n+1} is also normalized. What is the number N_{n+}? The answer can be found by requiring, in matrix notation, that

$$\Psi_{n+1}^\dagger \Psi_{n+1} = (N_{n+} A^\dagger \Psi_n)^\dagger (N_{n+} A^\dagger \Psi_n) = 1,$$

or

$$\Psi_n^\dagger A N_{n+}^* N_{n+} A^\dagger \Psi_n = |N_{n+}|^2 \Psi_n^\dagger A A^\dagger \Psi_n = 1.$$

From (9), because $E_n = (n + \frac{1}{2})\eta$,

$$AA^\dagger \Psi_n = \left(\frac{E_n}{\eta} + \frac{1}{2}\right)\Psi_n = (n + 1)\Psi_n.$$

$$\therefore \quad |N_{n+}|^2 \Psi_n^\dagger (n + 1)\Psi_n = 1,$$

or, since Ψ_n is normalized,

$$N_{n+} = \frac{1}{\sqrt{n+1}}, \tag{3.9-18}$$

within an arbitrary phase factor.

PROBLEM 3.30

Suppose that Ψ_n is a normalized state, and that the next lower state is

$$\Psi_{n-1} = N_{n-} A \Psi_n, \tag{3.9-19}$$

with N_{n-} chosen so that Ψ_{n-1} is normalized. Show that

$$N_{n-} = \frac{1}{\sqrt{n}} \tag{3.9-20}$$

within an arbitrary phase factor.

We now have an elegant method for generating the normalized harmonic oscillator wave functions. Begin with a normalized Ψ_0, which is given as a function of x in (15). Express $A^\dagger$ in the position representation by writing the coordinate x for the operator x, and $-i\hbar\, \partial/\partial x$ for the operator p. Then

$$\begin{aligned}
\Psi_1 &= \frac{1}{\sqrt{0+1}} A^\dagger \Psi_0, \\
\Psi_2 &= \frac{1}{\sqrt{1+1}} A^\dagger \Psi_1 = \frac{1}{\sqrt{(1+1)(0+1)}} (A^\dagger)^2 \Psi_0, \\
&\vdots \\
\Psi_n &= \frac{1}{\sqrt{n}} A^\dagger \Psi_{n-1} = \frac{1}{\sqrt{n!}} (A^\dagger)^n \Psi_0.
\end{aligned} \tag{3.9-21}$$

This discussion has emphasized the position representation, but these procedures are equally applicable in, for example, the momentum representation if Ψ_0 is the Fourier transform of (15), and, in $A^\dagger$, the operator p is replaced by the multiplier p, and x is replaced by $i\hbar\, \partial/\partial p$; see Problem 3.9.

PROBLEM 3.31

Use the prescription (21) to obtain Ψ_1 and Ψ_2, and compare with the results of Section 2.7.

Finally, we shall display the various operators explicitly in a matrix representation. From (18) and (20),

$$A\Psi_n = \sqrt{n}\,\Psi_{n-1}, \tag{3.9-22}$$

$$A^\dagger \Psi_n = \sqrt{n+1}\,\Psi_{n+1}. \tag{3.9-23}$$

As in Section 3.8, represent Ψ_n by a column vector which has 1 in the nth location and zeros elsewhere:

$$\Psi_0 = \begin{pmatrix} 1 \\ 0 \\ 0 \\ 0 \\ \vdots \end{pmatrix}, \qquad \Psi_1 = \begin{pmatrix} 0 \\ 1 \\ 0 \\ 0 \\ \vdots \end{pmatrix}, \qquad \Psi_2 = \begin{pmatrix} 0 \\ 0 \\ 1 \\ 0 \\ \vdots \end{pmatrix}, \ldots \tag{3.9-24}$$

Then A is given by

$$A = \begin{pmatrix} 0 & \sqrt{1} & 0 & 0 & \cdot \\ 0 & 0 & \sqrt{2} & 0 & \cdot \\ 0 & 0 & 0 & \sqrt{3} & \cdot \\ \cdot & \cdot & \cdot & \cdot & \cdot \end{pmatrix}. \tag{3.9-25}$$

For example,

$$A\Psi_2 = \begin{pmatrix} 0 & \sqrt{1} & 0 & 0 & \cdot \\ 0 & 0 & \sqrt{2} & 0 & \cdot \\ 0 & 0 & 0 & \sqrt{3} & \cdot \\ 0 & 0 & 0 & 0 & \cdot \\ \cdot & \cdot & \cdot & \cdot & \cdot \end{pmatrix} \begin{pmatrix} 0 \\ 0 \\ 1 \\ 0 \\ \cdot \end{pmatrix} = \begin{pmatrix} 0 \\ \sqrt{2} \\ 0 \\ 0 \\ \cdot \end{pmatrix} = \sqrt{2}\,\Psi_1.$$

The matrix $A^\dagger$ is obtained just by interchanging rows and columns because A is real.

$$A^\dagger = \begin{pmatrix} 0 & 0 & 0 & 0 & \cdot \\ \sqrt{1} & 0 & 0 & 0 & \cdot \\ 0 & \sqrt{2} & 0 & 0 & \cdot \\ 0 & 0 & \sqrt{3} & 0 & \cdot \\ \cdot & \cdot & \cdot & \cdot & \cdot \end{pmatrix}. \tag{3.9-26}$$

It also does its job correctly; check, for example, that $A^\dagger\Psi_2 = \sqrt{3}\Psi_3$. Once A and $A^\dagger$ are determined, the Hamiltonian can be obtained in several ways. From (8),

$$H = \eta(A^\dagger A + \tfrac{1}{2})$$

$$= \eta \begin{pmatrix} 0 & 0 & 0 & 0 & \cdot \\ \sqrt{1} & 0 & 0 & 0 & \cdot \\ 0 & \sqrt{2} & 0 & 0 & \cdot \\ 0 & 0 & \sqrt{3} & 0 & \cdot \\ \cdot & \cdot & \cdot & \cdot & \cdot \end{pmatrix} \begin{pmatrix} 0 & \sqrt{1} & 0 & 0 & \cdot \\ 0 & 0 & \sqrt{2} & 0 & \cdot \\ 0 & 0 & 0 & \sqrt{3} & \cdot \\ 0 & 0 & 0 & 0 & \cdot \\ \cdot & \cdot & \cdot & \cdot & \cdot \end{pmatrix}$$

$$+ \frac{\eta}{2} \begin{pmatrix} 1 & 0 & 0 & 0 & \cdot \\ 0 & 1 & 0 & 0 & \cdot \\ 0 & 0 & 1 & 0 & \cdot \\ 0 & 0 & 0 & 1 & \cdot \\ \cdot & \cdot & \cdot & \cdot & \cdot \end{pmatrix}$$

$$= \eta \begin{pmatrix} 0 & 0 & 0 & 0 & \cdot \\ 0 & 1 & 0 & 0 & \cdot \\ 0 & 0 & 2 & 0 & \cdot \\ 0 & 0 & 0 & 3 & \cdot \\ \cdot & \cdot & \cdot & \cdot & \cdot \end{pmatrix} + \frac{\eta}{2} \begin{pmatrix} 1 & 0 & 0 & 0 & \cdot \\ 0 & 1 & 0 & 0 & \cdot \\ 0 & 0 & 1 & 0 & \cdot \\ 0 & 0 & 0 & 1 & \cdot \\ \cdot & \cdot & \cdot & \cdot & \cdot \end{pmatrix} \tag{3.9-27}$$

$$= \begin{pmatrix} \eta/2 & 0 & 0 & 0 & \cdot \\ 0 & 3\eta/2 & 0 & 0 & \cdot \\ 0 & 0 & 5\eta/2 & 0 & \cdot \\ 0 & 0 & 0 & 7\eta/2 & \cdot \\ \cdot & \cdot & \cdot & \cdot & \cdot \end{pmatrix}.$$

The matrix representing the energy is diagonal, and the numbers on the leading diagonal are the eigenvalues. This result has to emerge because, in (24), the basis vectors were chosen to be the energy eigenfunctions; see (3.8-23) and (3.8-24), and the discussion there.

PROBLEM 3.32

a. Solve (3) and (4) for x and p. Then use (25) and (26) to construct matrices for x and p. Verify that these matrices satisfy (2).

b. Substitute the x and p obtained in (**a**) into expression (1) for the energy operator H and obtain the result of (27).

3.10 *Time Derivatives of Expectation Values*

Suppose that a charged particle is in an electric field. It is reasonable that its expectation value of momentum $\langle \vec{p} \rangle$ will change. If classical mechanics is a guide,

$$\frac{d}{dt}\langle \vec{p} \rangle = \langle \vec{F} \rangle = \langle -\vec{\nabla} V \rangle \tag{3.10-1}$$

seems like a good guess. It is plausible that the rate of change of the mean value of momentum equals the mean value of the applied force. Equation (1) is right, and this section shows how such results can be obtained.

Consider the general problem of calculating the time derivative of an expectation value, $(d/dt)\langle \Omega \rangle$. An ordinary derivative is used because $\langle \Omega \rangle$ depends at most on the time; the integrations that give $\langle \Omega \rangle$ remove the other variables. Since the limits of the integral do not depend on the time,

$$\begin{aligned}\frac{d}{dt}\langle \Omega \rangle &= \frac{d}{dt}\int \Psi^* \Omega \Psi = \int \frac{\partial}{\partial t}(\Psi^* \Omega \Psi) \\ &= \int \left\{ \frac{\partial \Psi^*}{\partial t} \Omega \Psi + \Psi^* \Omega \frac{\partial \Psi}{\partial t} + \Psi^* \frac{\partial \Omega}{\partial t} \Psi \right\}.\end{aligned} \tag{3.10-2}$$

Use the Schrödinger equation and its conjugate,

$$\frac{\partial \Psi}{\partial t} = -\frac{i}{\hbar} H \Psi, \qquad \frac{\partial \Psi^*}{\partial t} = \frac{i}{\hbar} H^* \Psi^*,$$

in the first two terms and recognize the third term as the expectation value of $\partial \Omega / \partial t$.

$$\frac{d}{dt}\langle \Omega \rangle = \frac{i}{\hbar}\int [(H^* \Psi^*)(\Omega \Psi) - \Psi^* \Omega H \Psi] + \left\langle \frac{\partial \Omega}{\partial t} \right\rangle.$$

Since H is Hermitian,

$$\int (H^* \Psi^*)(\Omega \Psi) = \int \Psi^* H \Omega \Psi.$$

$$\therefore \quad \frac{d}{dt}\langle \Omega \rangle = \left\langle \frac{i}{\hbar}[H, \Omega] \right\rangle + \left\langle \frac{d\Omega}{dt} \right\rangle. \tag{3.10-3}$$

The expectation value of the commutator $(i/\hbar)[H, \Omega]$ gives the time rate of change of $\langle \Omega \rangle$ caused by the change in Ψ. The term $\langle \partial \Omega / \partial t \rangle$ takes into account any explicit time dependence in Ω. For many important operators, such as those that are functions only of position and momentum, $\partial \Omega / \partial t$ is zero. Some applications of (3) follow.

EXAMPLE

Suppose that $V = gx$. In classical mechanics the force is then $-\vec{\nabla}V = -g\hat{e}_x$, and $d\vec{p}/dt = -g\hat{e}_x$. What is $d\langle\vec{p}\rangle/dt$ in quantum mechanics?

SOLUTION

Since $\partial p/\partial t = 0$,

$$\frac{d}{dt}\langle\vec{p}\rangle = \left\langle\frac{i}{\hbar}[H, \vec{p}]\right\rangle.$$

$$[H, \vec{p}] = \frac{1}{2M}[p^2, \vec{p}] + g[x, \vec{p}].$$

$$[p^2, \vec{p}] = 0, \qquad [x, p_x] = i\hbar, \qquad [x, p_y] = 0, \qquad [x, p_z] = 0.$$

$$\therefore \quad [H, \vec{p}] = i\hbar g$$

$$\therefore \quad \frac{d}{dt}\langle\vec{p}\rangle = \left\langle\frac{i}{\hbar}[H, \vec{p}]\right\rangle = \langle -g\hat{e}_x\rangle$$

$$= \int \Psi^*(-g\hat{e}_x)\Psi\, dx = -g\hat{e}_x,$$

for any Ψ. This result is a special case of *Ehrenfest's theorem*, which is proved below.

Suppose Ω is H. Then

$$\frac{d}{dt}\langle H\rangle = \left\langle\frac{i}{\hbar}[H, H]\right\rangle + \left\langle\frac{\partial H}{\partial t}\right\rangle. \tag{3.10-4}$$

The first term is zero because any operator commutes with itself. For a Hamiltonian that is the sum of the kinetic energy and a potential energy that may depend on the time,

$$\frac{\partial H}{\partial t} = \frac{\partial}{\partial t}\left[\frac{p^2}{2M} + V(\vec{r}, t)\right] = \frac{\partial V(\vec{r}, t)}{\partial t}. \tag{3.10-5}$$

This term is different from zero only if there is *explicit* time dependence in V. If a particle is moving through a region in which V depends just on position, and the expectation value $\langle V\rangle$ changes in time because $\langle\vec{r}\rangle$ is changing, then there is only an *implicit* time dependence, and expression (5) is zero. A nonzero

$\partial V/\partial t$ means that V at fixed $\vec{r}$ is changing because, for example, the particle is between capacitor plates that are being charged.

Equation (4) then becomes

$$\frac{d}{dt}\langle H\rangle = \left\langle \frac{\partial V}{\partial t}\right\rangle, \tag{3.10-6}$$

a statement of energy conservation in the expectation value sense.

PROBLEM 3.33

The conclusion

$$\frac{d}{dt}\langle H\rangle = 0 \qquad \text{if} \qquad \frac{\partial H}{\partial t} = \frac{\partial V}{\partial t} = 0 \tag{3.10-7}$$

was obtained in Problem 3.21 by an entirely different technique. It was also shown there that under these conditions ΔH was a constant. Use (3) to get this result.

Equation (3) also provides a quick and clever way to obtain the result (1.5-8),

$$\frac{d}{dt}\left(\int \Psi^*\Psi\right) = 0, \tag{3.10-8}$$

that is, to show that the normalization integral does not depend on the time. If Ω is chosen to be unity, then (3) becomes

$$\frac{d}{dt}\langle 1\rangle = 0$$

because the identity has no time dependence and commutes with everything. Since

$$\langle 1\rangle = \int \Psi^* 1 \Psi = \int \Psi^*\Psi,$$

(8) follows.

It is always pleasant and reassuring to see simple connections between classical and quantum physics. In 1927 P. Ehrenfest[18] showed that the expectation values of the position and momentum of a particle, and of the force on the

[18] P. Ehrenfest, *Z. f. Physik* **45**, 455 (1927).

particle, have the same relationships as in Newtonian mechanics. The time rate of change of the expectation value of position is

$$\frac{d}{dt}\langle\vec{r}\rangle = \left\langle \frac{i}{\hbar}\left[\frac{p_x^2 + p_y^2 + p_z^2}{2M} + V(\vec{r}, t), \vec{r}\right]\right\rangle. \tag{3.10-9}$$

Since

$$[V, \vec{r}] = 0, \qquad [p_x^2, y] = 0, \ldots, [p_z^2, x] = 0,$$

the only surviving terms are, with $\vec{r} = \hat{e}_x x + \hat{e}_y y + \hat{e}_z z$,

$$\frac{d}{dt}\langle\vec{r}\rangle = \left\langle \frac{i}{2M\hbar}(\hat{e}_x[p_x^2, x] + \hat{e}_y[p_y^2 y] + \hat{e}_z[p_z^2, z])\right\rangle.$$

Because $[p_x, x] = \hbar/i$,

$$[p_x^2, x] = p_x(p_x x - x p_x) + (p_x x - x p_x)p_x = \frac{2\hbar}{i} p_x. \tag{3.10-10}$$

$$\therefore \quad \frac{d}{dt}\langle\vec{r}\rangle = \left\langle \frac{i}{2M\hbar}\frac{2\hbar}{i}(\hat{e}_x p_x + \hat{e}_y p_y + \hat{e}_z p_z)\right\rangle = \frac{\langle\vec{p}\rangle}{M}. \tag{3.10-11}$$

The time rate of change of the expectation value of momentum is

$$\frac{d}{dt}\langle\vec{p}\rangle = \left\langle \frac{i}{\hbar}\left[\frac{p^2}{2M} + V(\vec{r}, t), \vec{p}\right]\right\rangle. \tag{3.10-12}$$

Since p^2 commutes with all components of p,

$$\frac{d}{dt}\langle\vec{p}\rangle = \left\langle \frac{i}{\hbar}[V(\vec{r}, t), \vec{p}]\right\rangle.$$

For any function $V(\vec{r}, t)$ and any Ψ,

$$[V, p_x]\Psi = V\frac{\hbar}{i}\frac{\partial}{\partial x}\Psi - \frac{\hbar}{i}\frac{\partial}{\partial x}(V\Psi) = -\frac{\hbar}{i}\left(\frac{\partial V}{\partial x}\right)\Psi,$$

so that one has the operator equations

$$[V, p_x] = -\frac{\hbar}{i}\frac{\partial V}{\partial x}, \qquad [V, \vec{p}] = -\frac{\hbar}{i}\vec{\nabla} V. \tag{3.10-13}$$

$$\therefore \quad \frac{d}{dt}\langle\vec{p}\rangle = -\langle\vec{\nabla} V\rangle. \tag{3.10-14}$$

Equation (11) should be compared with the classical relation $\dot{\vec{r}} = \vec{p}/M$, while (14) should be compared with the classical relation $\dot{\vec{p}} = -\vec{\nabla} V = \vec{F}$. Of course, these results come out of the theory because they were put into the theory. Sections 1.2 and 1.4 use the correspondence principle repeatedly to guide the construction of the Schrödinger equation.

PROBLEM 3.34

At $t = 0$ a particle with potential energy zero everywhere has the wave function

$$\Psi(x, 0) = \sqrt{\frac{\beta}{\sqrt{\pi}}}\, e^{ip_0x/\hbar - \beta^2x^2/2}.$$

a. Calculate $\langle x \rangle$ and $\langle p \rangle$.

b. Use (3) directly to evaluate $(d/dt)\langle x \rangle$ and $(d/dt)\langle p \rangle$.

c. In this problem the wave function is given only at $t = 0$. How does it happen that the time derivatives of various quantities are determined?

PROBLEM 3.35

The discussion here has been in terms of functions $\Psi(\vec{r}, t)$ and operators on such functions. Equation (3), like most of the basic equations, is also valid if everything is expressed in terms of matrices. Review the matrix representation of the harmonic oscillator given at the end of Section 3.9. Use (3.9-27) for H and the matrices for x and p obtained in Problem 3.32. Calculate the matrix $(i/\hbar)[H, x]$, and compare it with the matrix for p.

Section 3.4 ends with the remark that the $\Delta E\, \Delta t$ uncertainty principle does not follow from the theorem (3.4-1) because time is a parameter, not an operator. This uncertainty principle can, however, be obtained from (3). Any system is described by measurables such as momentum, position, angular momentum, and so forth that are represented by operators, $P, Q, R, \ldots$.[19] Assume that these operators do not have explicit time dependence, so that (3) becomes

$$\hbar \frac{d}{dt}\langle P \rangle = \langle i[H, P] \rangle.$$

The uncertainty principle theorem (3.4-1) does apply to H and P,

$$\Delta H\, \Delta P \geq \tfrac{1}{2}|\langle i[H, P] \rangle|.$$

Therefore

$$\Delta H \frac{\Delta P}{\left|\dfrac{d\langle P \rangle}{dt}\right|} \geq \frac{\hbar}{2}. \tag{3.10-15}$$

[19] This argument is an adaptation of that given by A. Messiah, *Quantum Mechanics*, Vol. I, trans. G. M. Temmer (Amsterdam: North-Holland Publishing Company, 1962), pp. 319–320.

The ratio

$$\frac{\Delta P}{\left|\dfrac{d\langle P\rangle}{dt}\right|} \equiv \Delta t_P \tag{3.10-16}$$

is the time required for $\langle P\rangle$ to change by ΔP, the width of the distribution. It is therefore the time that must elapse before one can deduce easily from measurements of P that the system has really changed.

Now calculate $\Delta t_Q, \Delta t_R, \ldots$ in the same way, and call the shortest of these time intervals Δt. This Δt is the least time that must elapse before one can deduce easily from any measurement that the system has really changed, and with it, the largest lower bound on ΔH is obtained from (15),

$$\Delta H\,\Delta t \geq \frac{\hbar}{2}.$$

Since H is the name of the operator, ΔH is really the right name for the width of the energy distribution, but the special role of energy has led to the custom of writing ΔE, as well as $\langle E\rangle$ rather than $\langle H\rangle$. The energy-time uncertainty principle is therefore

$$\Delta E\,\Delta t \geq \frac{\hbar}{2}. \tag{3.10-17}$$

Here ΔE is the usual root mean squared width of a distribution as defined by (3.1-3) and used in Section 3.4, but Δt is something quite different. It is the time required for the expectation value of a dynamical variable to change by the width of its distribution.

PROBLEM 3.36

For the $\Psi(x, 0)$ of Problem 3.34, obtain Δt_x, the time for the center of the packet to move by a distance equal to its root mean squared width Δx. Compute ΔE, and show that

$$\Delta E\,\Delta t_x = \frac{\hbar}{2}\sqrt{1 + \frac{\hbar^2\beta^2}{4p_0^2}}.$$

3.11 *Unstable States*

Excited molecules, atoms, and nuclei, and many elementary particles, are unstable systems that decay with the emission of some form of radiation. Section 2.6 describes such systems rather artificially by introducing complex potential energies. This section uses a more fundamental approach that illustrates many of the topics of this chapter.

The most familiar unstable systems follow an approximately exponential decay law and live for a long time in terms of their own "natural time" T_n. The size of the hydrogen atom, 10^{-8} cm, divided by the expectation value of the speed of its electron, 10^8 cm/sec, gives a T_n of 10^{-16} sec. The mean life of an excited state is about 10^{-8} sec, or $10^8\ T_n$. As another example, consider unstable nuclei. Inside them, nucleons, and clusters of nucleons such as alpha particles, have kinetic energy $\simeq 10$ MeV/nucleon and speeds $\simeq 10^9$ cm/sec. Since nuclear sizes $\simeq 10^{-12}$ cm, $T_n \simeq 10^{-21}$ sec. Most heavy alpha-emitting nuclei have mean lives in the range from 10^{-6} sec through 10^{10} years, or $10^{15}\ T_n$ through $10^{38}\ T_n$. Human reaction time is around 0.1 sec, and human life expectancy is around $10^{10} \cdot 0.1$ sec, so one could argue that even a 10^{-6} sec alpha emitter lives 10^5 times longer than people do. Most systems of this kind are, like most people, only slightly influenced in their ordinary activities by the fact of their mortality. Their wave functions can be obtained to high accuracy as if they were completely stable, and they are often called *quasi-stationary states.*

As discussed in Sections 1.3 and 3.10, $\Delta E\, \Delta t \geq \hbar/2$ suggests that a system with mean life τ has an energy uncertainty, or natural width, $\Delta E \simeq \hbar/\tau$. Quasi-stationary states generally have a ΔE that is much less than their decay energy. For example, if a particle is in a well of size a, the energies of the two lowest states are $E_1 \simeq \hbar^2/2Ma^2$ and $E_2 \simeq$ a few $\hbar^2/2Ma^2$; the precise values would depend on the shape of the well. The natural time is the size divided by an estimate of the mean speed,

$$T_n \simeq \frac{a}{\sqrt{2E/M}} \simeq \frac{Ma^2}{\hbar}.$$

The ratio of the mean life to T_n, the number that is about 10^8 for the excited hydrogen atom and 10^{15} to 10^{38} for typical alpha emitting nuclei, is

$$\frac{\tau}{T_n} \simeq \frac{\hbar/\Delta E}{Ma^2/\hbar} = \frac{\hbar^2/Ma^2}{\Delta E} \simeq \frac{E_2 - E_1}{\Delta E}. \tag{3.11-1}$$

This argument deals with a particular model, but the conclusion is generally valid: The ratio τ/T_n is likely to be of the same magnitude as the ratio of the decay energy to the natural width. For the excited hydrogen atom $E/\Delta E \simeq 10$ eV/$(\hbar/10^{-8}$ sec), or 10^8. For a 10^{-6}-sec alpha emitter, $E/\Delta E \simeq 10$ MeV/$(\hbar/10^{-6}$ sec), or 10^{16}. Much of the following deals with states that are not quite eigenstates of energy because they do decay, but that have a very small ΔE.

The general problem can be stated as follows. At $t = 0$ there is a nonstationary state $\Psi(x, 0)$. It might, for example, be a hydrogen atom with energy around 12.09 eV above its ground state. The Schrödinger equation determines the time development of Ψ. The Hamiltonian H is complicated because it must describe the particle, the radiation, and the emission of the radiation by the particle. At any time $t > 0$, $\Psi(x, t)$ will be a superposition of the original undecayed state and the various possible outcomes of decay. For the hydrogen atom example, $\Psi(x, t)$ will be a superposition of states conveniently characterized by the number of photons that are flying away from the atom; the undecayed state is characterized by zero photons.

The undecayed state will be orthogonal to all the other components in $\Psi(x, t)$. For the hydrogen atom example, think in terms of a number-of-photons operator, with eigenvalue 0 for the undecayed state and eigenvalue 1 or 2 or larger integers for the various decayed states. The operator is Hermitian because it represents something that can be measured with a necessarily real outcome. Since the undecayed state is an eigenstate with eigenvalue zero, and each decayed state has eigenvalue not zero, (3.5-2) guarantees the claimed orthogonality.

For any system, no matter whether it is a radioactive nucleus, an unstable meson, or an excited molecule, there is some quantity that can distinguish between the undecayed system and all possible outcomes of decay. This quantity can be represented by a Hermitian operator, the $\Psi(x, 0)$ will have one eigenvalue, the decayed states will have different eigenvalues, and orthogonality is assured. What has really been used here is the universally applicable statement,

> *If there is a measurement that must give one result for one state and a different result for another state, then the two states are orthogonal.*

PROBLEM 3.37

Examine the previous three paragraphs carefully and then prove the concluding statement in your own words. Does anything need to be said about degeneracy?

The expansion postulate, Section 3.6, guarantees that $\Psi(x, t)$ can be written as a superposition of $\Psi(x, 0)$, the undecayed state, and all the states Φ_n that describe the possible outcomes of decay.

$$\Psi(x, t) = u(t)\Psi(x, 0) + \sum_n b_n(t)\Phi_n(x). \qquad (3.11\text{-}2)$$

The coefficient of $\Psi(x, 0)$ has the special name $u(t)$, for *undecayed*, but it is an expansion coefficient of the same kind as the $b_n(t)$.

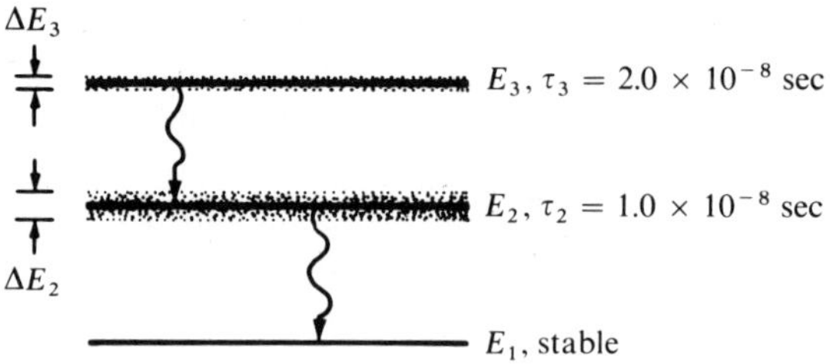

Fig. 3.11-1. A typical atomic decay scheme.

PROBLEM 3.38

The states of an atom have a small $\Delta E/E$ and decay to the next lower state through photon emission (Figure 3.11-1). The decay is very nearly exponential. The initial state $\Psi(x, 0)$ is Ψ_3, with energy near E_3 and no photons. It decays with mean life $\tau_3 = 2.0 \times 10^{-8}$ sec to the state Ψ_2, in which the atom's energy is near E_2 and a photon γ_{32} with energy near $E_3 - E_2$ has been emitted. The state Ψ_2 then decays with mean life $\tau_2 = 1.0 \times 10^{-8}$ sec to the state Ψ_1, which consists of two photons, γ_{32} and γ_{21}, and the atomic ground state. To a very good approximation, (2) for this system can be written in the form

$$\Psi(x, t) = u(t)\Psi_3 + b_2(t)\Psi_2 + b_1(t)\Psi_1.$$

Calculate and plot as a function of time the probabilities of the system being in Ψ_3, in Ψ_2, and in Ψ_1, that is $|u(t)|^2$, $|b_2(t)|^2$, and $|b_1(t)|^2$. (Actually Ψ_2 and Ψ_1 each represent many states because the photons can be emitted in many different ways.)

The coefficient $u(t)$ in (2) can be evaluated by the standard use of (3.6-2), which works for all expansions in terms of orthogonal functions. Multiply by $\Psi^*(x, 0)$, integrate, and use the fact that $\Psi(x, 0)$ is orthogonal to all Φ_n.

$$\int \Psi^*(x, 0)\Psi(x, t)dx = u(t) \int |\Psi(x, 0)|^2\, dx + \sum_n b_n(t) \int \Psi^*(x, 0)\Phi_n(x)dx$$
$$= u(t). \tag{3.11-3}$$

The probability of finding the undecayed system is

$$|u(t)|^2 = \left| \int \Psi^*(x, 0)\Psi(x, t)dx \right|^2. \tag{3.11-4}$$

This reasonable result says that the overlap integral $u(t)$ measures the amplitude of $\Psi(x, 0)$ that is still left in $\Psi(x, t)$, and that the square of this amplitude gives the probability of still finding the undecayed system at time t.

One can get a useful expression for $u(t)$ by expanding the wave function in terms of the exact energy eigenfunctions $\psi_E(x)$, as in (3.6-8). With H as the exact Hamiltonian of the system,

$$H\psi_E(x) = E\psi_E(x). \tag{3.11-5}$$

The expansion in terms of the $\psi_E(x)$ can be a sum as in (3.6-4) and (3.6-8) if the possible energies are discrete, or an integral if the possible energies are continuously distributed. The usual situation involves continuously distributed energy eigenvalues because final states typically include a free particle. In alpha decay there is no boundary condition that restricts the energy of the alpha particle. The expansion of the initial state analogous to (3.6-6) is

$$\Psi(x, 0) = \int C(E')\psi_{E'}(x)dE'. \tag{3.11-6}$$

The energy eigenfunctions are orthonormal in the delta function sense, as in (3.5-10):

$$\int \psi_E^*(x)\psi_{E'}(x)dx = \delta(E - E'). \tag{3.11-7}$$

To obtain the coefficients $C(E)$, multiply (6) by ψ_E^* and integrate.

$$\begin{aligned}\int \psi_E^*(x)\Psi(x, 0)dx &= \int C(E')\left[\int \psi_E^*(x)\psi_{E'}(x)dx\right]dE' \\ &= \int C(E')\delta(E - E')dE' = C(E).\end{aligned} \tag{3.11-8}$$

The wave function at any t can then be written as

$$\Psi(x, t) = \int C(E)\psi_E(x)e^{-iEt/\hbar}\, dE, \tag{3.11-9}$$

which is (3.6-8) translated into language appropriate for continuously distributed energies.

With (6) and (9), expression (3) for the amplitude of the undecayed state becomes

$$\begin{aligned}u(t) &= \int\left[\int C^*(E')\psi_{E'}^*(x)dE'\right]\left[\int C(E)\psi_E(x)e^{-iEt/\hbar}\, dE\right]dx \\ &= \iint C^*(E')C(E)e^{-iEt/\hbar}\left[\int \psi_{E'}^*(x)\psi_E(x)dx\right]dE'\, dE \\ &= \iint C^*(E')C(E)e^{-iEt/\hbar}\delta(E' - E)dE'\, dE \\ &= \int |C(E)|^2 e^{-iEt/\hbar}\, dE.\end{aligned} \tag{3.11-10}$$

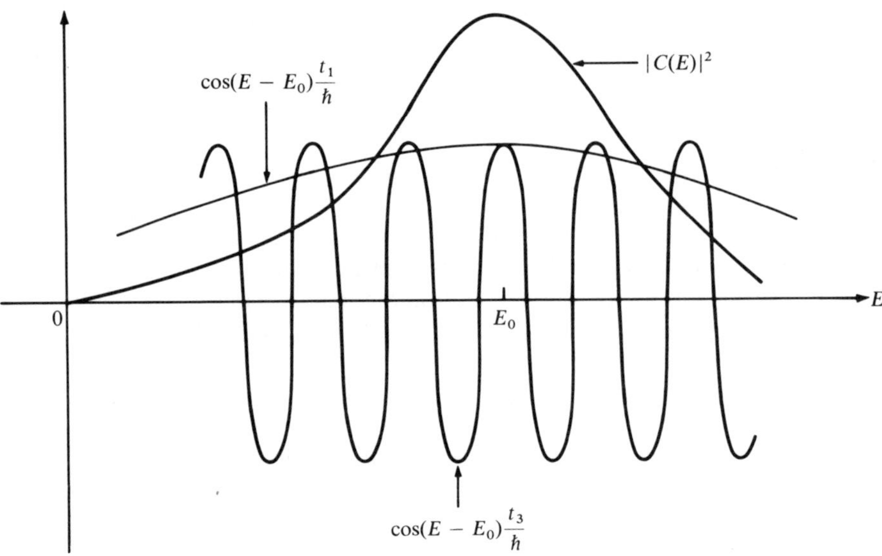

Fig. 3.11-2. The time development of the integral in (11). For large t the oscillations in the integrand give near cancellation of the positive and negative contributions.

Except for a numerical factor, $u(t)$ is the Fourier transform of $|C(E)|^2$. The quantity $|C(E)|^2$ is the energy spectrum of the unstable state, in the sense that $|C(E)|^2\,dE$ is the probability of finding the energy between E and $E + dE$. Equation (10) says that if the energy spectrum is known, then the probability of finding the undecayed system is determined.

PROBLEM 3.39

Write the development (5) through (10) for a discrete rather than a continuous energy spectrum.

Suppose that the energy spectrum $|C(E)|^2$ is any smooth peak with maximum at E_0 and width ΔE (Figure 3.11-2). Equation (10) can be written as

$$u(t) = e^{-iE_0t/\hbar} \int |C(E)|^2 e^{-i(E-E_0)t/\hbar}\, dE. \qquad (3.11\text{-}11)$$

At $t = 0$ the integrand is simply $|C(E)|^2$, and $|u(0)|^2 = 1$. Now consider the integrand at three times,

$$t_1 \ll \frac{\hbar}{\Delta E}, \qquad t_2 \simeq \frac{\hbar}{\Delta E}, \qquad t_3 \gg \frac{\hbar}{\Delta E}.$$

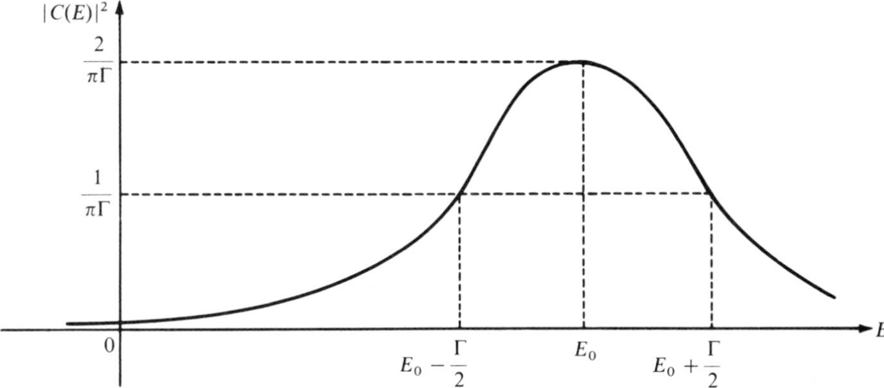

Fig. 3.11-3. The Lorentz or Breit–Wigner form (12). The full width at half maximum is Γ.

At t_1 the oscillations in $\exp[-i(E - E_0)t/\hbar]$ as a function of E have caused only negligible reduction in the integral because the exponential is close to unity in the neighborhood of the peak. At t_2 the oscillations begin to cause a substantial reduction. At t_3 there is very little left because the integrand has become a large number of alternately positive and negative contributions that very nearly cancel each other. It follows that the mean life must be of the order of $\hbar/\Delta E$.

The typical energy spectrum of a quasi-stationary state is a narrow peak, of width $\Delta E \simeq \hbar/\tau$, centered at some energy E_0 that one casually calls "the energy of the decaying state" because ΔE is so small. A fairly realistic form for such a spectrum is

$$|C(E)|^2 = \frac{\Gamma/2\pi}{(E - E_0)^2 + \Gamma^2/4}. \tag{3.11-12}$$

This expression is known as the *Lorentz line shape* when it occurs in atomic spectroscopy, and as the *Breit–Wigner resonance* form when it occurs in nuclear physics. It also arises in the study of the resonance of damped oscillators, mechanical or electrical, with frequency rather than energy as the variable.[20] The constant in the numerator, $\Gamma/2\pi$, has been chosen so that

$$\int_{-\infty}^{+\infty} |C(E)|^2 \, dE = 1 \tag{3.11-13}$$

because the probability of finding the system at some energy must be unity; see Problem 3.20. The maximum of $|C(E)|^2$ is at E_0, and the full width at half maximum equals Γ (Figure 3.11-3).

[20] F. S. Crawford, Jr., *Waves, Berkeley Physics Course*, Vol. 3 (New York: McGraw-Hill Book Company, 1968), Secs. 3.2 and 6.4. Pages 303–307 contain for a classical system nearly the same calculation as is presented here.

An excited atom might be described by $E_0 \simeq 10$ eV, $\Gamma \simeq \hbar/10^{-8}$ sec $\simeq 10^{-7}$ eV. For all quasi-stationary states, $|C(E)|^2$ is a very tall and narrow spike. The limit $\Gamma \to 0$ turns expression (12) into a way of writing the Dirac delta function $\delta(E - E_0)$, and the state $\Psi(x, t)$ into a truly stationary state of definite energy.

If (12) is used as the energy spectrum, (10) becomes

$$u(t) = \int_{-\infty}^{+\infty} \frac{\Gamma/2\pi}{(E - E_0)^2 + \Gamma^2/4} e^{-iEt/\hbar}\, dE. \tag{3.11-14}$$

With

$$w \equiv E - E_0, \qquad e^{-iwt/\hbar} = \cos\frac{wt}{\hbar} - i\sin\frac{wt}{\hbar},$$

$$u(t) = \frac{\Gamma}{2\pi} e^{-iE_0t/\hbar} \int_{-\infty}^{+\infty} \frac{\cos wt/\hbar - i\sin wt/\hbar}{w^2 + \Gamma^2/4}\, dw.$$

The part of the integrand that contains $\sin wt/\hbar$ contributes nothing because it is odd in w. The rest is given in standard tables of definite integrals. The result is[21]

$$u(t) = e^{-iE_0t/\hbar} e^{-\Gamma t/2\hbar}. \tag{3.11-15}$$

[21] This footnote is for those who know about complex contour integration. Equation (14) can be written

$$u(t) = \frac{\Gamma}{2\pi} \int_{-\infty}^{+\infty} \frac{e^{-iEt/\hbar}}{(E - E_0 + i\Gamma/2)(E - E_0 - i\Gamma/2)}.$$

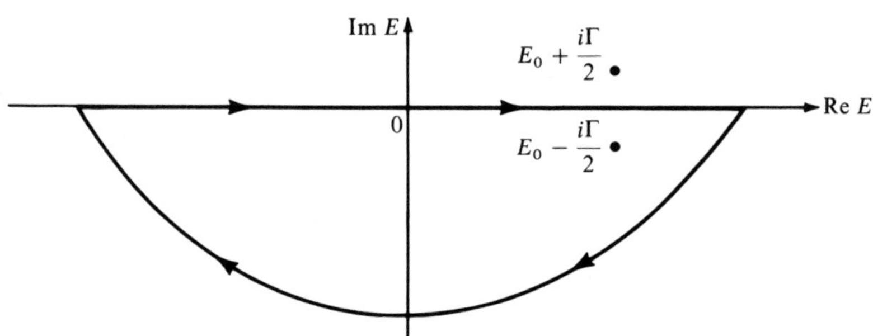

In the complex E plane the integrand has two poles, one at $E_0 - i\Gamma/2$ and the other at $E_0 + i\Gamma/2$. Because $e^{-iEt/\hbar} \to 0$ on a large semicircle in the lower half plane, take the contour shown. Since the pole at $E_0 - i\Gamma/2$ is encircled in the negative direction,

$$u(t) = -2\pi i \cdot \text{Residue at } \left(E_0 - \frac{i\Gamma}{2}\right)$$

$$= -2\pi i \frac{\Gamma}{2\pi} \frac{e^{-i(E_0 - i\Gamma/2)t/\hbar}}{E_0 - i\Gamma/2 - E_0 - i\Gamma/2} = e^{-iE_0t/\hbar} e^{-\Gamma t/2\hbar}.$$

Therefore, the probability of finding the undecayed system at time t is

$$|u(t)|^2 = e^{-\Gamma t/\hbar} = e^{-t/\tau}, \tag{3.11-16}$$

where

$$\tau \equiv \frac{\hbar}{\Gamma}. \tag{3.11-17}$$

The conclusion is that if the energy spectrum has the shape (12) with full width at half maximum Γ, then the decay must be exponential with mean life $\tau = \hbar/\Gamma$. The uncertainty principle and the discussion that accompanies Figure 3.11-2 imply that $\Gamma\tau$ should be of the order of $\hbar$, but now we know the exact relation between full width at half maximum and the mean life for the line shape (12).

The results should be compared with those of Section 2.6, where decay with mean life τ was described by a complex energy, $E = E_0 - i\Gamma/2$, with, according to (2.6-7), $\tau = \hbar/\Gamma$ as here. The time dependence obtained there for $\Psi(x, t)$ is precisely the same as that contained in $u(t)$ according to (15).

The argument can be turned around. If the decay is exponential with mean life τ, then, if one accepts a few reasonable assumptions, it follows that the energy spectrum must be Lorentzian with full width at half maximum $\Gamma = \hbar/\tau$. Exponential decay has been verified to high precision for many systems, and the spectrum (12) must therefore be an excellent description of these systems.

Is this treatment a realistic model of unstable states? The answer is a qualified yes. No deviation from exponential decay has ever been observed in radioactive nuclei, excited atoms, and similar systems. However, the energy spectrum $|C(E)|^2$ cannot be precisely Lorentzian for all energies. If nothing else, there must be a deviation from the line shape (12) for E sufficiently far below E_0 because $|C(E)|^2$ must go to zero at some finite $E_{\min}$. There is no such thing as a negative energy alpha particle coming out of a nucleus. The actual situation is that the far wings of most energy distributions, that is, the parts of $|C(E)|^2$ that lie above or below E_0 by many times Γ, do differ substantially from the form (12), but that so little of the area under the curve is out there that its effect is negligible. It can be shown that there should be deviations from exponential decay at very early times, just after the unstable state is formed, and then again at very large times, after many mean lives have elapsed.[22] Both of these domains are beyond experimental reach for ordinary systems.

PROBLEM 3.40

Suppose that the energy spectrum of some state has a Gaussian shape,

$$|C(E)|^2 = Ae^{-\gamma^2(E-E_0)^2}.$$

[22] A general treatment of these effects requires several theorems regarding functions of a complex variable. See L. Fonda, G. C. Ghirardi, A. Rimidi, and T. Weber, *Nuovo Cimento* **15A**, 689 (1973), and references there for the mathematical development and for the various physical interpretations that have been discussed. R. G. Winter, *Phys. Rev.* **123**, 1503 (1961) gives numerical results for an idealized model.

Determine A. Then find the decay law for this state; that is, find, as a function of the time, the probability that the system has not yet decayed.

PROBLEM 3.41

In 1904 F. Soddy[23] wrote, "It may be pointed out that the actual life of the different atoms of the same unstable element has all values between zero and infinity. Some break up during the first second of existence, and since only a fraction of the total changes per second, the quantity is, theoretically, never reduced to zero, and some persist indefinitely. This constitutes the first difference in properties between the individual atoms of the same element that has ever been discovered. It may be likened to the individual differences of velocity that exist between the molecules of a gas at constant temperature according to the kinetic theory. It is open to question whether all atomic properties are not really average properties, the individual atoms continually passing with great rapidity through phases varying widely among themselves in chemical and physical nature." Invent an explanation of the exponential decay of radioactive nuclei that would have made sense to prequantum scientists. During the early days of nuclear physics, such explanations were, of course, proposed.[24]

3.12 *Additional Chapter 3 Problems*

PROBLEM 3.42

a. For a one-dimensional harmonic oscillator, sketch a graph of the probability density $\rho(x)$, and calculate Δx if the oscillator

i. is in the ground state ψ_0 (see Problem 3.3).

ii. is in the second excited state ψ_2.

iii. is described by classical mechanics.

Mark the positions $\pm\Delta x$ on the graphs.

b. *Guess* $|\psi_{10}|^2$ and Δx_{10}, the probability density and root mean squared deviation in x for the tenth state.

PROBLEM 3.43

a. For a one-dimensional infinite square well, $V(x) = 0$, $0 \leq x \leq a$, $V(x) = \infty$ otherwise, obtain Δp_x for the general energy eigenstate $\psi_n(x)$.

b. Use the results of Problem 3.2 to obtain $\Delta x \, \Delta p_x$ for the ψ_n.

[23] F. Soddy, *Manchester Memoirs* **48** (8), 21 (1904).

[24] F. A. Lindemann, *Phil. Mag.* **30**, 560 (1915); N. R. Campbell, *The Structure of the Atom* (London: Cambridge University Press, 1923), Sec. 14.

c. For a particle in a box, $V(x, y, z) = 0$ if $0 \leq x \leq a$ *and* $0 \leq y \leq b$ *and* $0 \leq z \leq c$, $V(x, y, z) = \infty$ otherwise, calculate $\langle \vec{p} \cdot \vec{p} \rangle$.

PROBLEM 3.44

Section 3.6 mentions the parity operator $\mathscr{P}$, defined by $\mathscr{P}\Psi(\vec{r}, t) = \Psi(-\vec{r}, t)$, and the complex conjugation operator K, defined by $K\Psi(\vec{r}, t) = \Psi^*(\vec{r}, t)$.

a. Is $\mathscr{P}$ linear? Prove that it is Hermitian.

b. Is K linear? Prove that it is not Hermitian.

PROBLEM 3.45

$V = 0, 0 \leq x \leq a$; $V = \infty$ otherwise. The wave function is a superposition of the two lowest energy eigenstates, and $\langle H \rangle = 3\pi^2\hbar^2/2Ma^2$.

a. Find the most general wave function that satisfies these conditions.

b. The probability of finding the particle between $a/2$ and a is as large at $t = 0$ as the other conditions permit. Find the most general wave function that satisfies this additional condition.

PROBLEM 3.46

$V = 0$, $0 \leq x \leq a$; $V = \infty$ otherwise, and the particle is in the lowest energy eigenstate. Suddenly the potential well expands and becomes $V = 0$, $0 \leq x \leq 3a$; $V = \infty$ otherwise. Immediately after the well has expanded, the energy is measured. Obtain the possible results of the energy measurement and their probabilities. Evaluate numerically the probabilities for the lowest eight possible results, and calculate their sum.

PROBLEM 3.47

Obtain the uncertainty principle for $H = p^2/2M + V(\vec{r})$ and p_x. What does it say for a particle in a force-free region? For a particle in a harmonic oscillator bound state? Qualitatively, for a wave packet moving through the linear potential discussed in Section 2.9?

PROBLEM 3.48

$\Psi = C_0\Psi_0 + C_1\Psi_1$, where Ψ_0 and Ψ_1 are the two lowest energy eigenfunctions for the one-dimensional harmonic oscillator. Obtain $\langle v \rangle \equiv d\langle x \rangle/dt$. What choice of C_0 and C_1 gives the largest amplitude for the oscillations in $\langle v \rangle$?

PROBLEM 3.49

The energy of a system has only two eigenvalues, E_1 and E_2. Develop a matrix representation by defining the two energy eigenstates as column vectors and using them as the basis states. Write the Hamiltonian matrix. Write the most general state that has $\langle E \rangle = (3E_1 + 2E_2)/5$, and check by formally calculating $\langle E \rangle$. Obtain a raising operator that changes a state with energy E_1 into a state with energy E_2, and that gives zero when it operates on a state with energy E_2. Obtain the corresponding lowering operator. This formalism is applied to spin angular momentum in Chapter 8.

PROBLEM 3.50

An alpha particle emitter has a mean life of 1.0 seconds. The most probable energy of the alpha is precisely 6 MeV. What is the probability per disintegration for an alpha to be emitted with energy greater than 6.0000001 MeV?

AIP Niels Bohr Library,
Stein Collection

AIP Niels Bohr Library

Maria Goeppert Mayer and J. Hans D. Jensen made key contributions to our understanding of the shell *structure of nuclei. Atoms are exceptionally stable if there are 2, 10, 18, 36, 54, or 86 electrons, the right number to complete a shell in the atomic potential. Similarly, if the neutron number N or the proton number Z equals one of the so-called* magic numbers *2, 8, 20, 28, 50, 82, or 126, a shell is filled in the nuclear potential. The plot gives the measured binding energy* B_{exp}*, subtracted from* B_{th}*, the value given by a theoretical formula that does not include shell effects, and shows the particularly strong binding produced by* $N = 50, 82,$ *and 126, The fact that nucleons fill successive shells shows that they obey the Pauli exclusion principle, and therefore that they are* fermions.

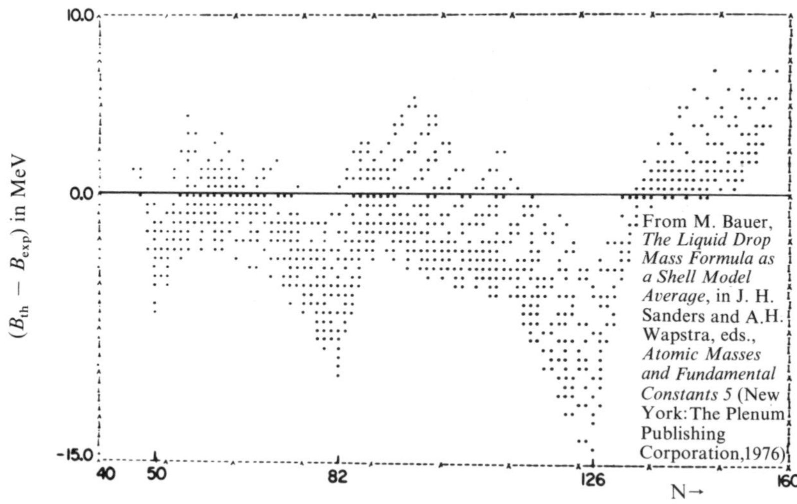

From M. Bauer, *The Liquid Drop Mass Formula as a Shell Model Average*, in J. H. Sanders and A.H. Wapstra, eds., *Atomic Masses and Fundamental Constants 5* (New York: The Plenum Publishing Corporation, 1976)

4

Systems of Several Particles

Most of Chapter 2 deals with the behavior of a single particle in a given environment. The wave function is treated unambiguously as a function of x, y, z and of the time. Much of Chapter 3, however, is worded so that the results do not depend on the number and kinds of coordinates that describe the system, for two reasons. One is the need to leave room in the theory for variables other than x, y, and z that are needed to describe what a particle is doing; spin is the most important and familiar example. The other reason is that we want to discuss systems of two or more particles, and therefore want to use wave functions that, for N particles, depend on $3N$ spatial coordinates and the time.

4.1 *Extension of the Theory*

Consider a system of N particles. There might be just two, such as an electron and a proton that have formed a hydrogen atom, or a neutron and a proton that have formed a deuterium nucleus, or a neutron and a proton that are undergoing a collision. There might be 10^{23}, such as the conduction electrons in a block of metal. There might be three particles, such as the nucleus and the two electrons of a helium atom.

PROBLEM 4.1

The nucleus of the common isotope of helium contains two neutrons and two protons. Can one really consider the helium atom as a three-particle system? In all experiments? In some experiments? Why?

If a state function Ψ describes an N-particle system, it must depend on the time and on $3N$ spatial coordinates, x_1, y_1, z_1, the position of the first particle, ..., x_N, y_N, z_N, the position of the Nth particle.

$$\Psi = \Psi(x_1, y_1, z_1, x_2, y_2, z_2, \ldots, x_N, y_N, z_N, t) = \Psi(\vec{r}_1, \vec{r}_2, \ldots, \vec{r}_N, t). \quad (4.1\text{-}1)$$

The energy operator must include the kinetic energies of all particles, and a potential energy that may depend on all coordinates.

$$H = \frac{p_{x1}^2 + p_{y1}^2 + p_{z1}^2}{2M_1} + \cdots + \frac{p_{xN}^2 + p_{yN}^2 + p_{zN}^2}{2M_N} + V(\vec{r}_1, \ldots, \vec{r}_N, t). \quad (4.1\text{-}2)$$

In the usual position representation, with

$$p_{x1} = \frac{\hbar}{i}\frac{\partial}{\partial x_1}, \ldots, p_{zN} = \frac{\hbar}{i}\frac{\partial}{\partial z_N},$$

$$p_1^2 = p_{x1}^2 + p_{y1}^2 + p_{z1}^2 = -\hbar^2\left(\frac{\partial^2}{\partial x_1^2} + \frac{\partial^2}{\partial y_1^2} + \frac{\partial^2}{\partial z_1^2}\right) = -\hbar^2\nabla_1^2, \quad (4.1\text{-}3)$$

$$p_2^2 = -\hbar^2\nabla_2^2, \ldots, p_N^2 = -\hbar^2\nabla_N^2,$$

the energy operator becomes

$$H = -\frac{\hbar^2}{2M_1}\nabla_1^2 - \cdots - \frac{\hbar^2}{2M_N}\nabla_N^2 + V(\vec{r}_1, \ldots, \vec{r}_N, t). \quad (4.1\text{-}4)$$

Since a partial derivative of x_1 with respect to x_2 is zero,

$$[p_{xj}, x_k] = [p_{yj}, y_k] = [p_{zj}, z_k] = \frac{\hbar}{i}\delta_{jk}. \quad (4.1\text{-}5)$$

With Ψ and H given by expressions (1) and (2), the Schrödinger equation is

$$H(\vec{r}_1, \ldots, \vec{r}_N, t)\Psi(\vec{r}_1, \ldots, \vec{r}_N, t) = i\hbar\frac{\partial}{\partial t}\Psi(\vec{r}_1, \ldots, \vec{r}_N, t). \quad (4.1\text{-}6)$$

This statement cannot be derived; there is no way to deduce uniquely a two-particle theory from a one-particle theory. We are again at one of those points where reasonableness, simplicity, and agreement with experiment suggest the form of the theory, but where an additional assumption is really being made.

PROBLEM 4.2

If the N particles have nothing to do with each other, then (6) should imply N single-particle Schrödinger equations. Verify that it does by requiring that

$$V(\vec{r}_1, \vec{r}_2, \ldots, \vec{r}_N, t) = V_1(\vec{r}_1, t) + V_2(\vec{r}_2, t) + \cdots + V_N(\vec{r}_N, t),$$

substituting a trial solution

$$\Psi(\vec{r}_1, \vec{r}_2, \ldots, \vec{r}_N, t) = \Phi_1(\vec{r}_1, t)\Phi_2(\vec{r}_2, t)\cdots\Phi_N(\vec{r}_N, t),$$

and separating variables.

With $dx_j\,dy_j\,dz_j \equiv d\tau_j$, the quantity

$$|\Psi(\vec{r}_1, \vec{r}_2, \ldots, \vec{r}_N, t)|^2\, d\tau_1\, d\tau_2 \cdots d\tau_N \tag{4.1-7}$$

is the probability that, all at the same time t, particle 1 is in the volume element $d\tau_1$ at $\vec{r}_1$, particle 2 is in the volume element $d\tau_2$ at $\vec{r}_2, \ldots,$ particle N is in the volume element $d\tau_N$ at $\vec{r}_N$. Since every particle has to be somewhere, normalization means that

$$\iiiint \cdots \int |\Psi(\vec{r}_1, \vec{r}_2, \ldots, \vec{r}_N, t)|^2\, d\tau_1\, d\tau_2 \cdots d\tau_N = 1, \tag{4.1-8}$$

where $3N$ integrations over the full range of the variables are required. The expression

$$\rho_1(\vec{r}_1) = \int \cdots \int |\Psi(\vec{r}_1, \vec{r}_2, \ldots, \vec{r}_N, t)|^2\, d\tau_2 \cdots d\tau_N, \tag{4.1-9}$$

with $3N - 3$ integrations *not* including those over x_1, y_1, and z_1, gives the probability distribution for particle 1 alone.

EXAMPLE

A neutron and a proton are in a box, that is, each has a potential energy as in Section 2.4,

$$\begin{aligned} V(x, y, z) &= 0 \qquad \text{if } 0 \le x \le a \text{ and } 0 \le y \le b \text{ and } 0 \le z \le c, \\ &= \infty \qquad \text{otherwise.} \end{aligned}$$

With $\vec{r}_n$ and $\vec{r}_p$ as the neutron and proton positions, the wave function is $\Psi(\vec{r}_n, \vec{r}_p, t)$. In terms of integrals over $|\Psi|^2$, obtain the probability that

a. the neutron is in the upper third of the box, $\frac{2}{3}c \le z_n \le c$, if the proton can be anywhere.

b. both the neutron and the proton are in the upper third of the box.

c. the neutron is in the upper third and the proton is in the lower two-thirds of the box.

SOLUTION

Let $dx_n\, dy_n\, dz_n \equiv d\tau_n$, $dx_p\, dy_p\, dz_p \equiv d\tau_p$.

$$\textbf{a.} \quad \int_{x_n=0}^{a} \int_{y_n=0}^{b} \int_{z_n=2c/3}^{c} \int_{x_p=0}^{a} \int_{y_p=0}^{b} \int_{z_p=0}^{c} |\Psi(\vec{r}_n, \vec{r}_p, t)|^2\, d\tau_n\, d\tau_p$$

$$\textbf{b.} \quad \int_0^a \int_0^b \int_{2c/3}^{c} \int_0^a \int_0^b \int_{2c/3}^{c} |\Psi(\vec{r}_n, \vec{r}_p, t)|^2\, d\tau_n\, d\tau_p$$

$$\textbf{c.} \quad \int_0^a \int_0^b \int_{2c/3}^{c} \int_0^a \int_0^b \int_0^{2c/3} |\Psi(\vec{r}_n, \vec{r}_p, t)|^2\, d\tau_n\, d\tau_p$$

Note that **(b)** + **(c)** = **(a)**.

The statement of this example does not imply that the wave function can be factored in the sense that

$$\Psi(\vec{r}_n, \vec{r}_p, t) = \Phi_n(\vec{r}_n, t)\Phi_p(\vec{r}_p, t).$$

If, however, Ψ can be written in this form, then the answer to (**a**) becomes

$$\left[\int_0^a \int_0^b \int_{2c/3}^{c} |\Phi_n(\vec{r}_n, t)|^2\, d\tau_n\right]\left[\int_0^a \int_0^b \int_0^{c} |\Phi_p(\vec{r}_p, t)|^2\, d\tau_p\right].$$

The second term gives unity because it is the normalization integral for Φ_p.

PROBLEM 4.3

For the example given above, obtain actual numbers for (**a**), (**b**), and (**c**) if the neutron and proton are each in the lowest possible energy eigenstate.

As announced at the beginning of this chapter, the major results of Chapter 3 can be applied to systems of several particles with, at most, a modest adjustment of the language. Review Chapter 3 to verify this statement.

PROBLEM 4.4

A neutron and proton do not interact with each other, but move in the same harmonic oscillator potential, so that

$$V = V(\vec{r}_n, \vec{r}_p) = \frac{K}{2}(x_n^2 + y_n^2 + z_n^2) + \frac{K}{2}(x_p^2 + y_p^2 + z_p^2).$$

Use the single-particle harmonic oscillator wave functions that are called Ψ_{000} and Ψ_{001} in Section 2.7 to construct a normalized two-particle wave function with the following properties. Both $\langle z_n \rangle$ and $\langle z_p \rangle$ oscillate in time. When $\langle z_n \rangle$ is at its extreme in the $+z$ direction, $\langle z_p \rangle$ is at its extreme in the $-z$ direction.

Verify explicitly that the wave function satisfies the Schrödinger equation.

Various conservation laws for systems of particles can be obtained from (3.10-3). If the Hamiltonian does not contain explicit time dependence, then (3.10-4) shows without modification that the expectation value of the total energy is constant. Of course, it is not true that the expectation value of the energy of any one particle must be constant. Suppose, for example, that

$$H = -\frac{\hbar^2}{2M_1}\nabla_1^2 - \frac{\hbar^2}{2M_2}\nabla_2^2 + V(\vec{r}_1 - \vec{r}_2). \tag{4.1-10}$$

The potential energy depends on the separation between the particles, as in the hydrogen atom, where $V \sim 1/|\vec{r}_1 - \vec{r}_2|$. Now the energy of particle 1 alone,

$$H_1 = -\frac{\hbar^2}{2M_1}\nabla_1^2 + V(\vec{r}_1 - \vec{r}_2),$$

does not commute with H;

$$[H, H_1] = \left[-\frac{\hbar^2}{2M_2}\nabla_2^2, V(\vec{r}_1 - \vec{r}_2)\right] \neq 0.$$

For the system described by the Hamiltonian (10), the expectation value of the total momentum is constant. The x component of the total momentum is given by

$$p_x = p_{x1} + p_{x2} = \frac{\hbar}{i}\frac{\partial}{\partial x_1} + \frac{\hbar}{i}\frac{\partial}{\partial x_2}.$$

The kinetic energy terms in H commute with p_x. Since V depends on the difference between $\vec{r}_1$ and $\vec{r}_2$,

$$\frac{\partial}{\partial x_1} V(\vec{r}_1 - \vec{r}_2) = -\frac{\partial}{\partial x_2} V(\vec{r}_1 - \vec{r}_2),$$

so that, for any Ψ,

$$(p_{x1} + p_{x2})V\Psi = V(p_{x1} + p_{x2})\Psi, \tag{4.1-11}$$

and $[p_x, H] = 0$. Since $\vec{p}$ does not depend explicitly on the time, (3.10-3) assures that $\langle \vec{p} \rangle$ is conserved.

PROBLEM 4.5

a. Review the argument leading to (11) and show that $\langle \vec{p}_1 \rangle$ alone will not generally be a constant.

b. Generalize the argument to N interacting particles. Begin by justifying that

$$V = \sum_{j>k} V_{jk}(\vec{r}_j - \vec{r}_k)$$

is a sensible expression for the total potential energy. Then show that

$$p_x = \sum_{j=1}^{N} p_{xj}$$

commutes with the Hamiltonian.

PROBLEM 4.6

Suppose that two particles interact because of a $V(\vec{r}_1 - \vec{r}_2)$, and in addition are in an externally caused force field. The Hamiltonian is given by

$$H = -\frac{\hbar^2}{2M_1}\nabla_1^2 - \frac{\hbar^2}{2M_2}\nabla_2^2 + V(\vec{r}_1 - \vec{r}_2) + V_1(\vec{r}_1) + V_2(\vec{r}_2). \tag{4.1-12}$$

Devise a concrete example that shows that $\langle \vec{p} \rangle$ should not be constant under these circumstances. Then show that $[\vec{p}, H] \neq 0$.

The quantum results regarding the total energy and momentum of systems are similar to the corresponding classical results.[1] Compare the classical and quantal statements and methods of proof.

[1] K. R. Symon, *Mechanics*, 3d ed. (Reading, Mass.: Addison-Wesley Publishing Co., Inc., 1971), Chap. 4.

If a two-particle Hamiltonian is given by (10), then the Schrödinger equation can be separated into two equations, one that describes the motion of the center of mass, and another that describes the relative motion. Define the relative coordinates $\vec{r} = (x, y, z)$ by

$$\vec{r} \equiv \vec{r}_1 - \vec{r}_2, \qquad \text{or} \qquad x \equiv x_1 - x_2, \qquad y \equiv y_1 - y_2, \qquad z \equiv z_1 - z_2. \tag{4.1-12}$$

Define the center-of-mass coordinates $\vec{r}_c = (x_c, y_c, z_c)$ by

$$\vec{r}_c \equiv \frac{M_1\vec{r}_1 + M_2\vec{r}_2}{M}, \qquad M \equiv M_1 + M_2. \tag{4.1-13}$$

To express the kinetic energy in terms of the relative and center-of-mass coordinates, note that

$$\frac{1}{M_1}\frac{\partial^2}{\partial x_1^2} + \frac{1}{M_2}\frac{\partial^2}{\partial x_2^2} = \frac{1}{M}\frac{\partial^2}{\partial x_c^2} + \frac{1}{\mu}\frac{\partial^2}{\partial x^2}, \tag{4.1-14}$$

where the *reduced mass*

$$\mu \equiv \frac{M_1 M_2}{M_1 + M_2}, \tag{4.1-15}$$

just as in classical mechanics.

PROBLEM 4.7

Obtain (14).

The Schrödinger equation becomes

$$\left[-\frac{\hbar^2}{2M}\nabla_c^2 - \frac{\hbar^2}{2\mu}\nabla^2 + V(\vec{r})\right]\Psi = i\hbar\frac{\partial \Psi}{\partial t}, \tag{4.1-16}$$

where ∇_c^2 and ∇^2 are the Laplacians in terms of the center-of-mass and relative coordinates. Let

$$\Psi(\vec{r}_1, \vec{r}_2, t) = \Psi_c(\vec{r}_c, t)\Psi_r(\vec{r}, t), \tag{4.1-17}$$

substitute, and divide by Ψ.

$$\frac{1}{\Psi_c}\left[-\frac{\hbar^2}{2M}\nabla_c^2\Psi_c - i\hbar\frac{\partial \Psi_c}{\partial t}\right] = -\frac{1}{\Psi_r}\left[-\frac{\hbar^2}{2\mu}\nabla^2\Psi_r + V(\vec{r})\Psi_r - i\hbar\frac{\partial \Psi_r}{\partial t}\right]. \tag{4.1-18}$$

Since the left side depends on $\vec{r}_c$ and t, and the right side depends on $\vec{r}$ and t, the two sides must equal something that is at most a function of t. Choosing anything different from zero would raise the origin of energy in one of the two equations and lower it in the other by the same amount. With both sides set equal to zero,

$$-\frac{\hbar^2}{2M}\nabla_c^2\Psi_c(\vec{r}_c, t) = i\hbar\frac{\partial\Psi_c(\vec{r}_c, t)}{\partial t} \tag{4.1-19}$$

and

$$-\frac{\hbar^2}{2\mu}\nabla^2\Psi_r(\vec{r}, t) + V(\vec{r})\Psi_r(\vec{r}, t) = i\hbar\frac{\partial\Psi_r(\vec{r}, t)}{\partial t}. \tag{4.1-20}$$

Equation (19) describes a free particle with mass M, the sum of the two masses. Equation (20) describes a single particle at $\vec{r}$ with mass μ, the reduced mass, moving under the influence of $V(\vec{r})$ as if the source of the interaction were fixed at the origin. Calculations dealing with two isolated interacting particles therefore fall apart into two problems. One is trivial; it merely describes the free motion of the center of mass. The other may involve hard work, but it is in every way equivalent to a single-particle problem. These results are exactly the same as those obtained in classical mechanics.

4.2 *Identical Particles*

Think about interacting macroscopic objects that are similar. They might be three planets of the same mass that are circling a star, or two colliding billiard balls. According to classical mechanics, nothing much depends on whether the objects are "similar" or "identical," and it is always possible to name and distinguish them. The centers of the colliding billiard balls follow sharp trajectories. If the two balls are called #1 and #2 before collision, it is clear which is #1 and which is #2 after the collision to anyone who has been watching (Figure 4.2-1). Even if the collision itself is shrouded in fog, it is possible

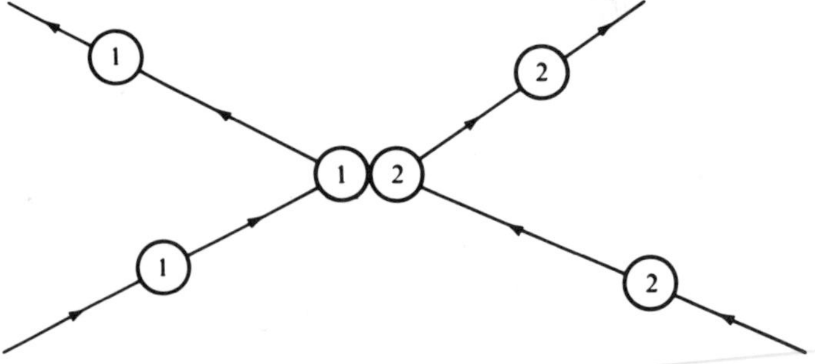

Fig. 4.2-1. Two colliding billiard balls.

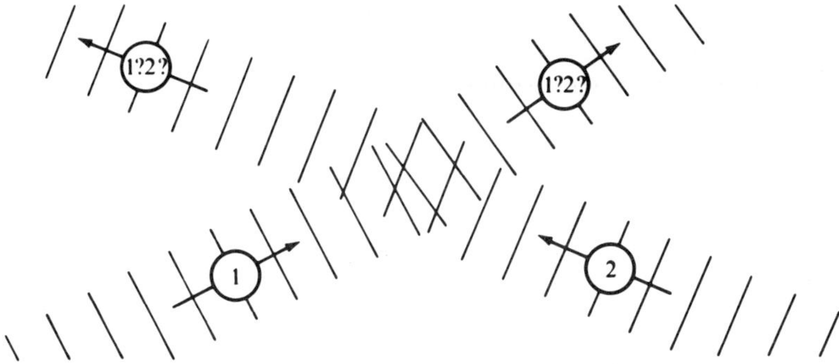

Fig. 4.2-2. Two colliding electrons.

to know which outgoing ball is #1 and which is #2 if the balls are marked with a negligible bit of paint, and all of classical experience shows that the presence of the paint does not affect the outcome appreciably.

Now think about interacting electrons. They might be the three electrons of a lithium atom, or two free colliding electrons. They are necessarily identical, and there is no way to put distinguishing labels on them without changing them into something entirely different. This fact is a consequence of the quantization of mass and charge. If matter were infinitely divisible, there might be something that could be used as a negligible bit of paint for marking electrons. The granularity of matter implies that changes in objects must have a minimum size, and, by definition, this minimum size is not negligible at the atomic level.

PROBLEM 4.8

G. W. von Leibniz (1646–1716) proved that atoms did not exist.[2] If grass were made of atoms, then, within some reasonable range of sizes and shapes, the number of possible different blades of grass is finite. At a party an acquaintance of Leibniz hunted all over the lawn for two identical blades of grass, but found none. Use what is known now about atomic sizes to estimate very roughly the number of blades that one should examine if one wants to have a good chance of seeing two that are identical.

If there were a good chance of finding two identical blades of grass during a party, how large would atoms be?

It is then impossible to distinguish electrons from each other by labeling them. If they interact appreciably, it is also impossible to keep track of them by following their motions (Figure 4.2-2). In a scattering experiment with

[2] P. Morrison, *Am. J. Phys.* **26**, 358 (1958).

reasonably well-defined momenta, the uncertainty principle assures that the electrons are represented by broad wave packets, and that the individual trajectories cannot be traced through a close collision. The wave functions of the three electrons bound in the same lithium atom overlap, and one cannot localize one of the electrons in a small part of the atom's volume without disrupting the system. Identity in quantum mechanics means much more than the classical "so similar that for present purposes the differences are negligible." It means that no measurement of any kind can be affected if the roles of two electrons are exchanged, and this fact has fascinating consequences.

In this section, 1, 2, ..., N are shorthand for all the coordinates of particles #1, #2, ..., #N. If particle #2 is an electron, then 2 stands for three spatial coordinates and a description of the spin orientation. The Schrödinger equation,

$$H(1, 2, \ldots, N, t)\Psi(1, 2, \ldots, N, t) = i\hbar \frac{\partial}{\partial t} \Psi(1, 2, \ldots, N, t), \qquad (4.2\text{-}1)$$

is the same as (4.1-6), except for a compact notation that leaves room for coordinates other than x, y, and z. Now suppose that the N particles are identical; they are all electrons, or they are all neutrons, or they are all pions. The Hamiltonian must then be symmetric under the exchange of particle labels. In a lithium atom there are three electrons, with negative contributions to the potential energy because of the attraction of a nucleus at the origin, and with positive contributions to the potential energy because of their mutual Coulomb repulsion. The general form (4.1-4) becomes[3]

$$H(1, 2, 3) = -\frac{\hbar^2}{2M}(\nabla_1^2 + \nabla_2^2 + \nabla_3^2) - \frac{3e^2}{4\pi\varepsilon_0}\left(\frac{1}{r_1} + \frac{1}{r_2} + \frac{1}{r_3}\right) + \frac{e^2}{4\pi\varepsilon_0}\left(\frac{1}{|\vec{r}_1 - \vec{r}_2|} + \frac{1}{|\vec{r}_1 - \vec{r}_3|} + \frac{1}{|\vec{r}_2 - \vec{r}_3|}\right). \qquad (4.2\text{-}2)$$

Exchanging the roles of 1 and 2, or of 1 and 3, or of 2 and 3, does not change H. For two colliding electrons with Coulomb interaction,

$$H(1, 2) = -\frac{\hbar^2}{2M}(\nabla_1^2 + \nabla_2^2) + \frac{e^2}{4\pi\varepsilon_0|\vec{r}_1 - \vec{r}_2|} \qquad (4.2\text{-}3)$$

is invariant under the exchange of 1 and 2. What do such exchanges do to the wave function?

Suppose that the situation described by (3) is a typical scattering experiment, with one electron incident with 5 keV kinetic energy, and one electron initially bound in an atom. If the electrons were distinguishable, the wave function $\Xi(1, 2, t)$ before collision would then depend on 1 and on 2 in very

[3] If spin were included, there would be additional small terms that are also unchanged by the exchange of particle labels.

different ways. The coordinates of one electron would appear in the description of a broad wave packet, nearly a plane wave, while the coordinates of the other electron would appear in some function that is piled up around its atom. One might try writing terms like

$$\Xi(1, 2, t) = e^{ikz_1}e^{-(1/a)\sqrt{x_2^2+y_2^2+z_2^2}}e^{-iEt/\hbar}, \tag{4.2-4}$$

where the target atom is at the origin, $(\hbar k)^2/2M = 5$ keV, $a \simeq 10^{-8}$ cm, and E is the total energy. This expression has no simple behavior under the exchange of labels 1 and 2, and also there is no compelling reason for writing it rather than

$$\Xi(2, 1, t) = e^{ikz_2}e^{-(1/a)\sqrt{x_1^2+y_1^2+z_1^2}}e^{-iEt/\hbar}. \tag{4.2-5}$$

If either of these two expressions is used to calculate something measurable, the result will generally depend on the placement of the labels. According to (4), the expectation value of the momentum of electron 1 is $\hbar k\hat{e}_z$, while according to (5), it is zero.

Since however the electrons are identical, the wave function must be arranged so that the demands of the identity of the particles are satisfied; all expectation values must be invariant under the exchange of particle labels. Clearly, a way to achieve this goal is to take neither (4) nor (5), but their sum or difference,

$$\Xi_\pm = \Xi(1, 2, t) \pm \Xi(2, 1, t). \tag{4.2-6}$$

Under the exchange of 1 and 2, Ξ_+ is unchanged while Ξ_- changes sign, and expectation values are unchanged in either case.

To treat the general problem of identical particles, it is useful to define the *exchange operator* P_{12} by

$$P_{12}\Psi(1, 2, t) = \Psi(2, 1, t). \tag{4.2-7}$$

Since all measurements must have the same expectation value for $\Psi(1, 2, t)$ and $\Psi(2, 1, t)$ if the particles are identical,

$$\Psi(2, 1, t) = C\Psi(1, 2, t).$$

A second application of P_{12} to (7) gives

$$P_{12}^2\Psi(1, 2, t) = P_{12}\Psi(2, 1, t) = C^2\Psi(1, 2, t).$$

But two exchanges return Ψ to its original form, so that P_{12}^2 is the identity, and P_{12} has the eigenvalues

$$C = +1 \qquad \text{or} \qquad -1. \tag{4.2-8}$$

The corresponding eigenfunctions behave like the $\Xi_\pm$ of (6). Although the exchange operator P_{12} and the parity operator are entirely different in their meanings, the structure of this argument is much like that used for parity in Section 2.5; review and compare the discussion given there.

Regardless of whether the two particles are identical, any $\Psi(1, 2, t)$ can be written as a linear combination of functions that are even and that are odd under exchange,

$$\Psi(1, 2, t) = C_+ \Psi_+(1, 2, t) + C_- \Psi_-(1, 2, t). \tag{4.2-9}$$

If the two particles are different, then both terms in this expression can be present. There is no simple symmetry associated with exchanging the coordinates of the electron and the proton in a hydrogen atom. If, however, the particles are identical, then the argument that leads to (8) applies and Ψ must be either Ψ_+ or Ψ_-. A state that describes two identical particles must be an eigenfunction of the exchange operator P_{12} with eigenvalue $+1$ or -1.

For two electrons, is the eigenvalue of P_{12} equal to $+1$ or -1? One might guess that, just as with parity, the answer depends on the circumstances, but that guess is wrong. Suppose that the Hamiltonian of the entire universe is written as H(1, 2, everything else). Since 1 and 2 are identical, H depends on their coordinates in the same way, as illustrated by (2), and therefore P_{12} commutes with H.

$$[P_{12}, H(1, 2, \text{everything else})] = 0. \tag{4.2-10}$$

From (3.10-3),

$$\frac{d}{dt}\langle P_{12}\rangle = 0, \tag{4.2-11}$$

so that if the eigenvalue, and therefore $\langle P_{12}\rangle$, is -1 at one time, it is -1 for all time. Equation (3.10-3) is not really needed for this conclusion. If $\Psi(1, 2, t_0)$ is the state at t_0, then the Schrödinger equation gives Ψ at $t_0 + \Delta t$, where Δt is small, as

$$\begin{aligned}\Psi(1, 2, t_0 + \Delta t) &= \Psi(1, 2, t_0) + \frac{\partial \Psi}{\partial t}\Delta t \\ &= \Psi(1, 2, t_0) + \frac{1}{i\hbar} H\Psi(1, 2, t_0)\,\Delta t. \end{aligned} \tag{4.2-12}$$

Since H is necessarily even under exchange of 1 and 2, the increment that is added to Ψ in each Δt has the same symmetry as Ψ itself.

Any pair of electrons can be given the names 1 and 2, and therefore any test of the symmetry of Ψ, in any system, will answer the question for all systems and all electrons. There is abundant evidence that the answer is -1, so that for any system Ψ changes sign if two electrons are exchanged. Such evidence can be obtained from scattering experiments, where the results depend on whether Ψ is even or odd under the exchange of the incident and target particles; we return to this question in Chapter 11. The answer is also implied by a variety of observations on atomic and molecular systems, and in particular by the *Pauli exclusion principle.*

In the ground state of the lithium atom, two electrons with opposing spins are close to the nucleus, tightly bound in the K shell. The third electron is in the L shell; it is on the average much further from the nucleus and is rather loosely bound.[4] No two electrons are in the same state, and the L shell electron cannot lower its energy by radiating and jumping into the K shell. The Pauli exclusion principle summarizes this and similar situations in the assertion, *at most one electron can be in a given state.* This assertion is right and useful, and is essential for understanding the special case of quantum physics called chemistry. The principle follows directly from the statement that, for any pair of electrons, the eigenvalue of P_{12} must be -1, and that therefore the eigenfunction of P_{12} can be written in the form

$$\Psi_- = \Psi(1, 2, t) - \Psi(2, 1, t). \tag{4.2-13}$$

If 1 and 2 are in the same state, the dependence of Ψ on the coordinates 1 and 2 is the same. The two terms cancel, Ψ_- is identically zero and is therefore never found, and the Pauli principle has been derived.

It is often true that the interaction between electrons is less important than some force, such as the Coulomb force of a nucleus, in which both electrons move. Then, as shown in Problem 4.2,

$$\Psi(1, 2, t) \simeq \Phi_a(1, t)\Phi_b(2, t). \tag{4.2-14}$$

For example, Φ_a might describe a K shell electron with spin up, and Φ_b might describe an L shell electron with spin down. The antisymmetric state (13) becomes

$$\Psi_- = \frac{1}{\sqrt{2}}[\Phi_a(1, t)\Phi_b(2, t) - \Phi_a(2, t)\Phi_b(1, t)]. \tag{4.2-15}$$

PROBLEM 4.9

If Φ_a and Φ_b are orthonormal, is Ψ_- normalized?

In the form (15) it is particularly clear why Ψ_- is zero if 1 and 2 are in the same state, for then Φ_a and Φ_b are the same functions. Even if the interactions

[4] Review at this time an introduction to atomic structure such as that in K. F. Ford, *Classical and Modern Physics*, Vol. 3 (Lexington, Mass.: Xerox College Publishing, 1972), Chap. 24, or E. H. Wichmann, *Quantum Physics, Berkeley Physics Course*, Vol. 4 (New York: McGraw-Hill Book Company, 1971), Chap. 3.

between the electrons are not negligible, one can write Ψ_- as a sum of terms of the form (15),

$$\Psi_- = \frac{1}{\sqrt{2}} \sum_{m>n} C_{mn}[\Phi_m(1, t)\Phi_n(2, t) - \Phi_m(2, t)\Phi_n(1, t)] \qquad (4.2\text{-}16)$$

if the Φ_n are a complete set of functions for a single particle, such as the set of all energy eigenfunctions of a single electron in the Coulomb field of a nucleus. The sum is restricted to $m > n$ to avoid counting the same term twice. This expression is a perfectly general way to write an eigenfunction of P_{12} with eigenvalue -1.

PROBLEM 4.10

If in (16) Ψ_- is normalized and all of the Φ_n are orthonormal, what is $\sum_{m>n} |C_{mn}|^2$? What is the physical meaning of $|C_{mn}|^2$?

PROBLEM 4.11

Is any Ψ_+ orthogonal to any Ψ_-?

PROBLEM 4.12

Is P_{12} Hermitian?

If there are three or more electrons, then the entire system must be described by a Ψ_- that changes sign if any pair of electrons is exchanged. Expression (15) suggests how such a Ψ_- can be constructed, because it can be written as a determinant,

$$\Psi_- = \frac{1}{\sqrt{2}} \begin{vmatrix} \Phi_a(1, t) & \Phi_a(2, t) \\ \Phi_b(1, t) & \Phi_b(2, t) \end{vmatrix}. \qquad (4.2\text{-}17)$$

For three electrons[5] a successful guess is

$$\Psi_- = \frac{1}{\sqrt{3!}} \begin{vmatrix} \Phi_a(1, t) & \Phi_a(2, t) & \Phi_a(3, t) \\ \Phi_b(1, t) & \Phi_b(2, t) & \Phi_b(3, t) \\ \Phi_c(1, t) & \Phi_c(2, t) & \Phi_c(3, t) \end{vmatrix}, \qquad (4.2\text{-}18)$$

and for N electrons, Ψ_- can be written as an N by N determinant, or a sum of such determinants. Determinants do the job because the exchange of any two electrons means exchanging two columns, which changes the sign of the result. The Pauli exclusion principle is built in because, if two states are equal, two rows

[5] A good treatment of these topics that includes a discussion of three-particle systems is given by D. S. Saxon, *Elementary Quantum Mechanics* (San Francisco: Holden-Day, Inc., 1968), Chap. VIII.

are equal and the determinant is zero. Determinants of this kind were applied to the study of atomic structure by J. C. Slater, and are called Slater determinants.

What can be said about the exchange symmetries applicable to protons, neutrons, pions, and all the other particles? Within the confines of nonrelativistic quantum theory, each kind of particle raises a separate question that must be answered by experiment. From the study of scattering processes and nuclear structure, one learns that protons, neutrons, and other particles with half-integral spin have, in the language of (8), $C = -1$. For example, neutrons and protons inside nuclei do not all fall into the lowest available state, but fill successively higher levels, or "shells," qualitatively similar to those occupied by electrons in atoms. This observation alone shows that protons and neutrons obey the Pauli exclusion principle. The requirement that the wave function changes sign under particle exchange affects the statistical treatment of these particles, and the appropriate formalism is called *Fermi-Dirac statistics.* All particles that obey such statistics are called *fermions.*

Photons, pions, and other particles with integral spin have $C = +1$, and are described by state functions that do not change sign under particle exchange. Their statistics is called *Bose-Einstein statistics*, and the particles are called *bosons.*

As stated above, the connection between spin and statistics, the observation that $C = -1$ for half integral spin particles and $+1$ for integral spin particles, has to be seen as an unexplained experimental result in nonrelativistic theory. However, through the use of relativistic quantum theory, this connection can be proved from general and plausible assumptions.

To summarize and illustrate the ideas of this section, we shall look at two-particle wave functions in a simple one-dimensional model. Assume that both particles are subject to the potential energy

$$
\begin{aligned}
V &= 0, \qquad -\frac{a}{2} \le x \le +\frac{a}{2}, \\
&= \infty, \qquad x < -\frac{a}{2} \quad \text{or} \quad +\frac{a}{2} < x,
\end{aligned}
$$

that they have the same mass, and that they do not interact with each other. The two lowest single-particle states are

$$
\begin{aligned}
\Phi_A(x_j, t) &= \sqrt{\frac{2}{a}} \cos \frac{\pi x_j}{a} e^{-iE_A t/\hbar}, \qquad E_A = \frac{\pi^2 \hbar^2}{2Ma^2}, \\
\Phi_B(x_j, t) &= \sqrt{\frac{2}{a}} \sin \frac{2\pi x_j}{a} e^{-iE_B t/\hbar}, \qquad E_B = \frac{4\pi^2 \hbar^2}{2Ma^2},
\end{aligned}
$$

where x_j is x_1 for particle 1 and x_2 for particle 2.

Suppose first that the two particles are distinguishable; perhaps one is a neutron and the other is an antineutron. Then there is no exchange symmetry limitation, and the four lowest-energy eigenstates of the system are

$$\Phi_A(x_1, t)\Phi_A(x_2, t), \qquad E = E_A + E_A, \tag{4.2-19}$$

$$\Phi_A(x_1, t)\Phi_B(x_2, t), \qquad E = E_A + E_B, \tag{4.2-20}$$

$$\Phi_B(x_1, t)\Phi_A(x_2, t), \qquad E = E_B + E_A, \tag{4.2-21}$$

$$\Phi_B(x_1, t)\Phi_B(x_2, t), \qquad E = E_B + E_B. \tag{4.2-22}$$

The second and third states are degenerate since both have $E = 5\pi^2\hbar^2/2Ma^2$. They are, however, different states because the particles are distinguishable. Now calculate the mean squared separation between the two particles in the second state. With all integrals taken from $-a/2$ to $+a/2$,

$$\begin{aligned}
\langle(x_1 - x_2)^2\rangle &= \iint \Phi_A^*(x_1, t)\Phi_B^*(x_2, t)(x_1 - x_2)^2\Phi_A(x_1, t)\Phi_B(x_2, t)dx_1\, dx_2 \\
&= \left(\frac{2}{a}\int x_1^2 \cos^2\frac{\pi x_1}{a}\, dx_1\right)\left(\int |\Phi_B(x_2, t)|^2\, dx_2\right) \\
&\quad + \left(\int |\Phi_A(x_1, t)|^2\, dx_1\right)\left(\frac{2}{a}\int x_2^2 \sin^2\frac{2\pi x_2}{a}\, dx_2\right) \\
&\quad - 2\left(\frac{2}{a}\int x_1 \cos^2\frac{\pi x_1}{a}\, dx_1\right)\left(\frac{2}{a}\int x_2 \sin^2\frac{2\pi x_2}{a}\, dx_2\right) \\
&= \left(\frac{a^2}{12} - \frac{a^2}{2\pi^2}\right)(1) + (1)\left(\frac{a^2}{12} - \frac{a^2}{8\pi^2}\right) - 2(0)(0) \\
&= a^2\left(\frac{1}{6} - \frac{5}{8\pi^2}\right).
\end{aligned} \tag{4.2-23}$$

The term in x_1x_2 contributes nothing because it leads to the product of two integrals with odd integrands. This result means that the particles are uncorrelated. Neither particle cares where the other one is, x_1x_2 is as often positive as it is negative, and

$$\langle(x_1 - x_2)^2\rangle = \langle x_1^2 + x_2^2 - 2x_1x_2\rangle = \langle x_1^2\rangle + \langle x_2^2\rangle + 0. \tag{4.2-24}$$

The root mean squared separation is then

$$\sqrt{\langle(x_1 - x_2)^2\rangle} = a\sqrt{\frac{1}{6} - \frac{5}{8\pi^2}} = 0.321a. \tag{4.2-25}$$

Next, suppose that the two particles are indistinguishable *fermions*; perhaps they are both electrons, or they are both neutrons. Suppose also that the spin orientation of the two particles is the same; any other assumption would require the spin formalism of Chapter 8. Then the Pauli exclusion principle forbids states in which both particles are in the same single-particle states. With Φ_A and Φ_B as the available single-particle states, only one energy, $E_A + E_B = 5\pi^2\hbar^2/2Ma^2$, can be reached. Neither expression (20) nor (21) alone can be the wave function because neither is antisymmetric under exchange of the two fermions, the wave functions must have the structure of (15) or (17), and instead of the four wave functions (19), (20), (21), and (22), there is only one,

$$\Psi_- = \frac{1}{\sqrt{2}}[\Phi_A(x_1, t)\Phi_B(x_2, t) - \Phi_B(x_1, t)\Phi_A(x_2, t)]. \tag{4.2-26}$$

Notice that if $x_1 = x_2$, then this expression is zero, so that there is zero probability of finding the two particles at the same place. Even though there is no interaction between the particles, they have to stay away from each other because of the required antisymmetry. For the state (26) the mean squared separation between the particles is

$$\langle(x_1 - x_2)^2\rangle = \iint \Psi_-^*(x_1 - x_2)^2\Psi_-\, dx_1\, dx_2. \tag{4.2-27}$$

The calculation is less tedious than it might seem. Many of the integrals obtained by multiplying out (27) are zero because their integrands are odd, and others are unity because they are normalization integrals for Φ_A or Φ_B. Further simplification occurs because

$$\int f(x_1)dx_1 = \int f(x_2)dx_2.$$

Therefore

$$\begin{aligned}\langle(x_1 - x_2)^2\rangle &= \frac{2}{a}\int x_1^2 \cos^2\frac{\pi x_1}{a}\, dx_1 + \frac{2}{a}\int x_2^2 \sin^2\frac{2\pi x_2}{a}\, dx_2 \\ &\quad + 2\left(\frac{2}{a}\int x_1 \cos\frac{\pi x_1}{a}\sin\frac{2\pi x_1}{a}\, dx_1\right)^2 \\ &= \left(\frac{a^2}{12} - \frac{a^2}{2\pi^2}\right) + \left(\frac{a^2}{12} - \frac{a^2}{8\pi^2}\right) + 2\left(\frac{16a}{9\pi^2}\right)^2.\end{aligned} \tag{4.2-28}$$

The first two terms are the same as those found in (23) for distinguishable particles, while the last term is $\langle 2x_1x_2\rangle$. Even though there is no interaction between the particles, the demands of antisymmetry tend to keep them apart and so correlate their positions. The root mean squared separation becomes

$$\sqrt{\langle(x_1 - x_2)^2\rangle} = 0.410a \tag{4.2-29}$$

rather than the $0.321a$ found in (25).

Last, suppose that the two particles are indistinguishable *bosons*; perhaps they are both positive pions, π^+. Pions have zero spin, so that only spatial coordinates need be considered. For identical bosons the state must be symmetric under exchange of the particles. With Φ_A and Φ_B as the available single-particle states, there are now three possibilities. Expressions (19) and (22) can be taken over directly because they are symmetric. Expressions (20) and (21) will not do because they are not symmetric, but their sum,

$$\Psi_+ = \frac{1}{\sqrt{2}}[\Phi_A(x_1, t)\Phi_B(x_2, t) + \Phi_B(x_1, t)\Phi_A(x_2, t)], \tag{4.2-30}$$

is acceptable. This state has the same energy, $E_A + E_B = 5\pi^2\hbar^2/2Ma^2$, as expressions (20), (21), and (26). For this state the mean squared separation between the particles is given by

$$\langle(x_1 - x_2)^2\rangle = \iint \Psi_+^*(x_1 - x_2)^2\Psi_+\, dx_1\, dx_2. \tag{4.2-31}$$

The result is

$$\langle(x_1 - x_2)^2\rangle = \left(\frac{a^2}{12} - \frac{a^2}{2\pi^2}\right) + \left(\frac{a^2}{12} - \frac{a^2}{8\pi^2}\right) - 2\left(\frac{16a}{9\pi^2}\right)^2. \tag{4.2-32}$$

The particles are correlated so as to bring them closer together, and

$$\sqrt{\langle(x_1 - x_2)^2\rangle} = 0.196a. \tag{4.2-33}$$

PROBLEM 4.13

Go through the details of obtaining (28) from (27), and of obtaining (32) from (31). Arrange the work so that the kinds of integrals that occur, and the way many of these integrals combine or cancel, can be seen clearly.

In classical mechanics, particles that do not interact are uncorrelated, and nothing like the difference between the results (25), (29), and (33) makes any sense. That there is a difference should be seen as an illustration of the somewhat unprecise but useful quantal statement:

Indistinguishable possibilities can interfere.

Section 1.2 treats an idealized two-slit electron diffraction experiment. Interference occurs because there is no way to distinguish between two possibilities, namely between the electron going through the upper slit, and the electron going through the lower slit. If something is done to find out through which slit the electron goes, the two possibilities become distinguishable and there is no interference. Here the two indistinguishable possibilities are $\Psi(1, 2, t)$ and $\Psi(2, 1, t)$ if 1 and 2 are identical particles. The unclassical correlations are the result of the interference between these possibilities.

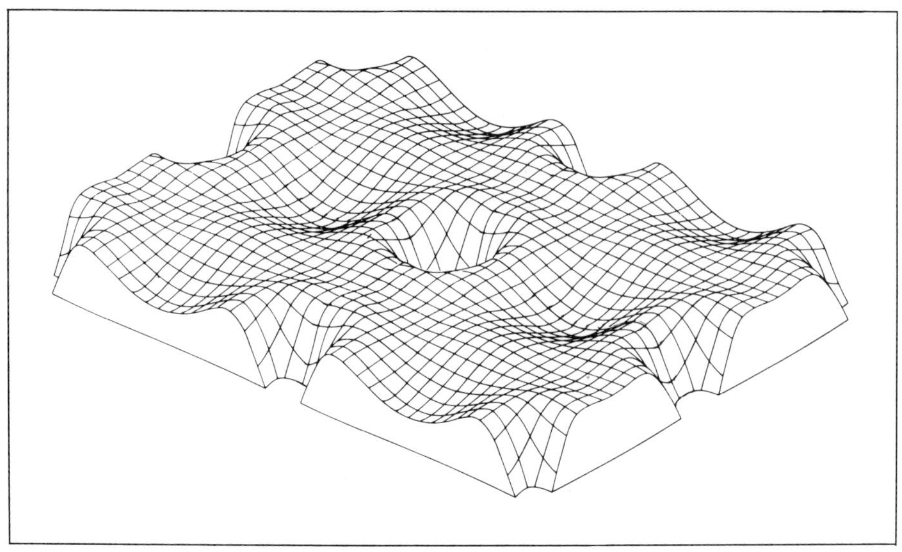

The conduction electron density in aluminum, in one of the principal planes of the face-centered cubic structure, where the nuclei are in a 2.86 Å × 4.04 Å rectangular array. There is only a rather modest variation except in the immediate neighborhood of the ion cores. The plot shows the results of an approximate calculation, and the contribution is small, but not actually zero, near the ions. (*Reproduced from* Pseudopotentials in the Theory of Metals, *1966, by Walter A. Harrison, with permission of the publishers, Benjamin/Cummings, Inc., Reading, Massachusetts, U.S.A.*)

5

Electrons in Solids

A short chapter cannot pretend to give a crystal-clear summary of a large and amorphous subject like solid state physics.[1] Here, and in the other chapters that present applications, there is no attempt to provide surveys of various fields. Rather, the purpose is to discuss a few topics that use the tools that we have developed, that provide a picture of important phenomena, and that carry forward the development of general quantum ideas. A discussion of electrons in solids provides several such topics. Accurate calculations in this area require a variety of approximation methods and computational techniques, but very simple models can provide useful insights.

A lithium atom is really a rather stable ion consisting of the nucleus and the two inner, or K shell electrons, and a loosely bound valence electron in the L shell. The two K shell electrons hover close to the nucleus and are tightly bound, while the valence electron tends to be near the edge of the atom and has a binding energy of only 5.39 eV. For comparison, note that it costs 75.6 eV to remove the second electron, and even more energy, that you should be able

[1] Useful introductory treatments include N. W. Ashcroft and N. D. Mermin, *Solid State Physics* (New York: Holt, Rinehart and Winston, 1976) and C. Kittel, *Introduction to Solid State Physics*, 5th ed. (New York: John Wiley & Sons, Inc., 1976).

to calculate to three significant figures on 1 cm^2 of paper, to remove the third electron after the other two are gone.

PROBLEM 5.1

a. Calculate, on 1 cm^2 of paper, to three significant figures, the energy necessary to remove from an isolated Li atom the third electron after the other two are gone.

b. What effects would have to be considered if much better than three-figure accuracy were required?

c. Invent a crude picture that makes the 5.39-eV binding of the valence electron seem reasonable.

Lithium metal has a body-centered cubic crystal structure with 3.04 Å between nearest neighbors. The potential energy seen by an electron halfway between two isolated Li^+ ions separated by this distance (Figure 5-1) is given by

$$2\left(\frac{-e^2}{4\pi\varepsilon_0 \cdot 1.52\,\text{Å}}\right) = -19.\text{ eV},$$

an energy well below the 5.39-eV valence electron binding. In the actual Li lattice the potential that an electron sees halfway between nearest neighbors is, on the one hand, lowered substantially more by the presence of other ions that are a little further away; on the other, the potential is raised substantially by the

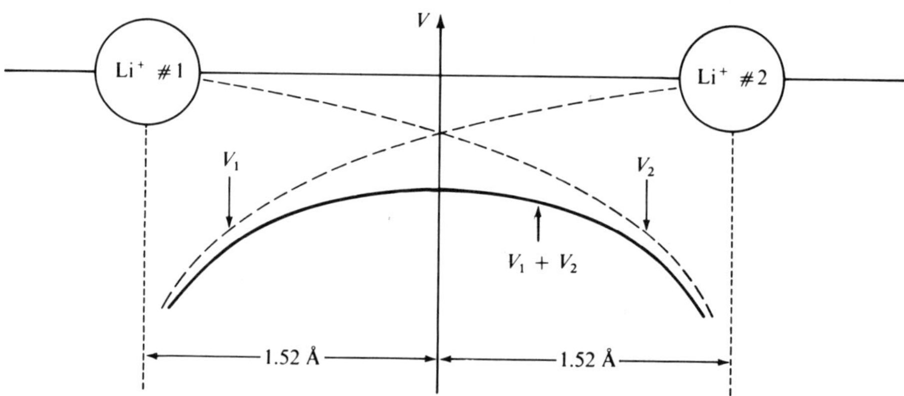

Fig. 5-1. The potential energy of an electron between two Li^+ ions.

presence of the many valence electrons that are drifting around. Nevertheless, it is true that the potential barrier between the Li^+ ions is pulled down enough to permit valence electrons to roam freely through the entire piece of metal. Extremely strong electric forces arise if there is even a small deviation from electrical neutrality, and there will in each neighborhood be an average density of one electron per positive ion, but the 10^{24} valence electrons in 20 cm^3 of Li have to be viewed as belonging to all the 10^{24} ions.

We are going to discuss the behavior of the electrons at two levels of sophistication. First, in the *free electron model*, the potential bumps caused by the ions are ignored, and one pretends that the electrons feel zero force as long as they stay away from the edge of the metal. Second, we will recognize that there actually is a *periodic potential* that modifies the results of the free electron model, and examine these modifications by studying a simple one-dimensional picture.

5.1 *The Free Electron Model*

Here a macroscopic piece of metal is treated as a box that contains many electrons with negligible interactions. The essential results do not depend on the shape and size of the box if its dimensions are very large compared to atomic diameters. We will use an $a \times a \times a$ region, with a of the order of a cm, in which $V = 0$, with $V = \infty$ outside. The number of free electrons is the number of atoms times the valence, so that the box contains about 10^{24} electrons.

The 10^{24} electrons are identical fermions, and the considerations of Section 4.2 apply. The wave function of the system is a $10^{24} \times 10^{24}$ Slater determinant. Fortunately, it will not be necessary to write out this determinant; it will suffice to recognize that the Pauli exclusion principle asks that, if 10^{24} electrons are to be accommodated, then 10^{24} different single-particle states must be available. Except for emphasis on a large number of highly excited states rather than the few lowest states, the problem is the one discussed in Section 2.4. With $a = b = c$, (2.4-3) and (2.4-4) give the single-particle energy eigenfunctions and eigenvalues,

$$\psi_{n_x n_y n_z}(x, y, z) = |n_x n_y n_z\rangle = \sqrt{\frac{8}{a^3}}\sin\left(\frac{n_x \pi x}{a}\right)\sin\left(\frac{n_y \pi y}{a}\right)\sin\left(\frac{n_z \pi z}{a}\right), \quad (5.1\text{-}1)$$

$$E_{n_x n_y n_z} = \frac{\pi^2 \hbar^2}{2Ma^2}(n_x^2 + n_y^2 + n_z^2). \quad (5.1\text{-}2)$$

Each of the states $|n_x n_y n_z\rangle$ can accommodate two electrons because of spin. As described in any introductory modern physics text, electrons have intrinsic spin angular momentum that can be given two and only two independent orientations. This idea is explored in Chapter 8. The result needed now is that

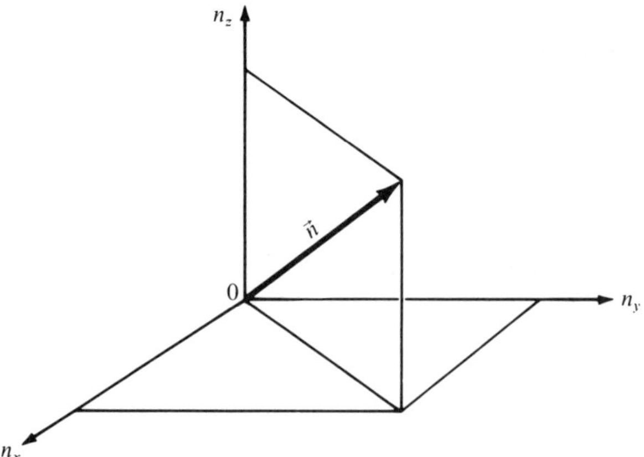

Fig. 5.1-1. The (n_x, n_y, n_z) space used for counting states.

each $|n_x n_y n_z\rangle$ can be occupied by one electron that has spin "up" and another electron "down," where "up" and "down" refer to orientations with respect to some convenient axis.

How many states are there with an energy less than some energy E? If E is small, so that $n_x^2 + n_y^2 + n_z^2$ is not a very large number, then the answer depends on E in a detailed and messy way, as Problem 2.7 shows. However, if $n_x^2 + n_y^2 + n_z^2$ ranges up to large numbers, as it must to accommodate 10^{24} electrons, the details and discontinuities seen in Problem 2.7 become insignificant, and one can count as follows.

In a space with the three Cartesian axes (n_x, n_y, n_z), each $|n_x n_y n_z\rangle$ corresponds to a lattice point with integral coordinates, and there is one such point per unit volume (Figure 5.1-1). Define q by

$$E = \frac{\pi^2 \hbar^2}{2Ma^2} q^2. \tag{5.1-3}$$

Comparison of (2) and (3) then turns the question, "How many spatial states are there with energy less than E?" into the question, "How many ways can $n_x^2 + n_y^2 + n_z^2$ be less than q^2?" Changing the sign of n_x or n_y or n_z does not produce a new state because it only changes the sign of the wave function (1). The answer to the question is therefore given by the number of points with integral coordinates, inside a sphere of radius

$$q = \frac{a}{\pi\hbar} \sqrt{2ME}, \tag{5.1-4}$$

with $n_x > 0$, $n_y > 0$, and $n_z > 0$, that is, in $\frac{1}{8}$ of the volume of the sphere. With a factor of two to account for spin, the number of states with energy less than E is

$$2 \cdot \frac{1}{8} \cdot \frac{4\pi}{3} q^3 = \frac{a^3}{3\pi^2 \hbar^3} (2ME)^{3/2}. \tag{5.1-5}$$

The answer is proportional to the volume a^3, and it follows that the $a \times a \times a$ cube was only a convenient device: Any large volume $\mathscr{V}$ can be approximated by an assembly of cubes of various sizes, each cube has a number of states with energy less than E proportional to its own volume, and the total number of states is the sum of the separate contributions. The result (5) can therefore be generalized to

$$\frac{\mathscr{V}}{3\pi^2 \hbar^3} (2ME)^{3/2}, \tag{5.1-6}$$

where $\mathscr{V}$ is any volume that is large in all directions.

PROBLEM 5.2

The argument that leads to (6) presumes that there are so many states that it is not important whether the sphere of radius q includes the few highest states correctly or not. Explore the validity of this approach numerically. Take $\mathscr{V} = 1\ \text{cm}^3$ and $E = 5$ eV.

a. How many states are there with energy less than E? What is the number q defined by (4)?

b. Suppose that q, the radius of the sphere in (n_x, n_y, n_z) space, were increased by one unit. By what fraction would the total number of states given by (6) change? Does the answer justify the method of counting that was used to obtain (6)?

EXAMPLE

Suppose that $\mathscr{V}$ is not "large in all directions" because the metal is in the form of a very thin rod, 1 cm × 4.2 Å × 4.2 Å. Obtain the number of states with energy less than E, for $0 < E < 10$ eV.

SOLUTION

The energy levels in a box 1 cm × 4.2 Å × 4.2 Å are

$$E_{n_x n_y n_z} = \frac{\pi^2 \hbar^2}{2M} \left[\frac{n_x^2}{(1\ \text{cm})^2} + \frac{n_y^2 + n_z^2}{(4.2\ \text{Å})^2} \right].$$

$$\frac{\pi^2 \hbar^2}{2M(4.2\ \text{Å})^2} = 2.13\ \text{eV}.$$

Since $n_y \geq 1$ and $n_z \geq 1$, there are *no* states if $E < 4.26$ eV. For $E > 4.26$ eV but less than $(1 + 2^2)$ 2.13 eV = 10.65 eV, n_y and n_z both have to be 1, and $E_{n_x} = E_{n_x n_y n_z} - 4.26$ eV is the energy contributed by the x dependence. How many states are there with E_{n_x} less than $E - 4.26$ eV? Let

$$E' = \frac{\pi^2 \hbar^2}{2M(1\text{ cm})^2} q^2.$$

Then the question becomes, "How many ways can n_x^2 be less than q^2?" Only $n_x > 0$ must be counted, and the answer is

$$q = \frac{1\text{ cm}}{\pi\hbar}\sqrt{2ME'}.$$

With a factor of two to account for spin, the number of states with energy $< E$ is zero for $E < 4.26$ eV, and

$$\frac{2\text{ cm}}{\pi\hbar}\sqrt{2M(E - 4.26\text{ eV})}$$

if $4.26\text{ eV} < E < 10.65$ eV. At 10 eV this number is 7.8×10^7, roughly the cube root of the number that one gets for a 1 cm × 1 cm × 1 cm piece, as in Problem 5.2.

PROBLEM 5.3

Suppose that $\mathscr{V}$ is not "large in all directions," because the metal is in the form of a very thin film, 1 cm × 1 cm × 4.2 Å. Obtain the number of states with energy less than E, for $0 < E < 8$ eV, and compare the answer with (5) and with the preceding example.

The number of states with energy less than E *per unit volume*,

$$\eta(E) = \frac{(2ME)^{3/2}}{3\pi^2\hbar^3}, \tag{5.1-7}$$

is an *intrinsic* property that does not depend on the nature of the boundaries if they are sufficiently far away, and we shall now obtain (7) again by a different approach.

Suppose that the physical boundaries are "at infinity," that is, so far away that they impose no constraints on wave functions. Although there are no real boundaries, it is useful to think again in terms of a large but finite $a \times a \times a$ region and to decide that everything important happens only inside it. With this view, the results of measurements should depend only on the behavior of Ψ

inside the box. For example, one would like to calculate the mean of measurements of p_x from

$$\langle p_x \rangle_a = \int_0^a \int_0^a \int_0^a \Psi^* \frac{\hbar}{i} \frac{\partial}{\partial x} \Psi \, dx \, dy \, dz, \tag{5.1-8}$$

rather than from the integrations over infinite space that are used in Section 3.3 to define expectation values.

In Section 3.3 a partial integration shows that $\langle p_x \rangle$ is real if $\Psi \to 0$ as $x \to \pm\infty$. If (8) is used, is $\langle p_x \rangle_a$ real? To answer this question, write $\langle p_x \rangle_a^*$ and integrate over x by parts.

$$\begin{aligned}
\langle p_x \rangle_a^* &= \int_0^a \int_0^a \left\{ \int_0^a \Psi \left(\frac{\hbar}{-i} \frac{\partial}{\partial x} \Psi^* \right) dx \right\} dy \, dz \\
&= \int_0^a \int_0^a \left\{ \int_0^a \left(\frac{\hbar}{i} \frac{\partial}{\partial x} \Psi \right) \Psi^* \, dx - \frac{\hbar}{i} [\Psi^* \Psi]_{x=0}^{x=a} \right\} dy \, dz \\
&= \langle p_x \rangle_a - \frac{\hbar}{i} \int_0^a \int_0^a [|\Psi a, y, z, t)|^2 - |\Psi(0, y, z, t)|^2] dy \, dz. \qquad (5.1\text{-}9)
\end{aligned}$$

All is well if the second term is zero, but one cannot conclude immediately that it is; the points $x = 0$ and $x = a$ are not at infinity. To study the question, write the wave function as a superposition of eigenfunctions of p_x,

$$\Psi(x, y, z, t) = \sum_n C_n(y, z, t) e^{ik_{xn}x}. \tag{5.1-10}$$

Then the offending term in (9) becomes

$$|\Psi(a, y, z, t)|^2 - |\Psi(0, y, z, t)|^2 = \sum_{n, m} C_m^* C_n [e^{i(k_{xn} - k_{xm})a} - 1] \tag{5.1-11}$$

and is zero if, for all n and m,

$$(k_{xn} - k_{xm})a = 2\pi j, \tag{5.1-12}$$

where j is any integer. A convenient choice is to take the lowest k_{xn} to be zero, for then (12) is satisfied if each of the k_{xn} is given by

$$k_{xn} = \frac{2\pi n_x}{a}, \qquad n_x = 0, 1, 2, \ldots. \tag{5.1-13}$$

With this choice,

$$\Psi = \sum_n C_n(y, z, t) e^{2\pi i n_x x/a}, \tag{5.1-14}$$

and satisfies *periodic boundary conditions*, that is,

$$\Psi(x + a, y, z, t) = \Psi(x, y, z, t). \tag{5.1-15}$$

Exactly the same argument can be applied to the y and z behavior. The result is that the wave function must have the form

$$\Psi = \sum_{n_x, n_y, n_z} b_{n_x n_y n_z}(t) e^{2\pi i(n_x x + n_y y + n_z z)/a}. \tag{5.1-16}$$

Each term is a momentum eigenfunction with

$$\frac{\vec{p}}{\hbar} \equiv \vec{k} = \frac{2\pi}{a}(n_x \hat{e}_x + n_y \hat{e}_y + n_z \hat{e}_z). \tag{5.1-17}$$

The energy for such a term is given by

$$E_{n_x n_y n_z} = \frac{p^2}{2M} = \frac{2\pi^2\hbar^2}{Ma^2}(n_x^2 + n_y^2 + n_z^2), \tag{5.1-18}$$

so that the time dependence is determined, and (16) becomes

$$\Psi = \sum_{n_x, n_y, n_z} g_{n_x n_y n_z} \Phi_{n_x n_y n_z}, \tag{5.1-19}$$

where the $g_{n_x n_y n_z}$ are expansion coefficients, and

$$\Phi_{n_x n_y n_z} \equiv \frac{1}{a^{3/2}} \exp\left[\frac{2\pi i}{a}(n_x x + n_y y + n_z z) - \frac{i}{\hbar} E_{n_x n_y n_z} t\right]. \tag{5.1-20}$$

These plane waves are simultaneous eigenfunctions of momentum and energy and satisfy periodic boundary conditions in all three directions. With the factor $1/a^{3/2}$, they are normalized in the sense that

$$\int_0^a \int_0^a \int_0^a |\Phi_{n_x n_y n_z}|^2 \, dx \, dy \, dz = 1.$$

Using the waves (20) as a basis, that is, using periodic boundary conditions for all wave functions, is equivalent to pretending that all of space is filled by stacked $a \times a \times a$ boxes in which the same things happen at the same time.

The question "How many states are there with energy less than E?" can be discussed in terms of the plane waves given by (20). The argument is almost the same as that used before, and also uses the (n_x, n_y, n_z) space. Because the energy is now given by (18), define q' by

$$E = \frac{2\pi^2\hbar^2}{Ma^2} q'^2 \tag{5.1-21}$$

instead of using (3). Comparison of (18) and (21) shows that the question is, "How many ways can $n_x^2 + n_y^2 + n_z^2$ be less than q'^2?" Now a change in the sign of n_x or n_y or n_z does produce a new state, because such changes produce states with different momentum directions. The answer is therefore given by the number of points with integral coordinates anywhere inside a sphere of radius

$$q' = \frac{a}{2\pi\hbar}\sqrt{2ME}. \tag{5.1-22}$$

With a factor of two to account for spin, the number of states with energy less than E is

$$2 \cdot \frac{4\pi}{3} q'^3 = \frac{a^3}{3\pi^2\hbar^3}(2ME)^{3/2}, \tag{5.1-23}$$

the same as expression (5). Because in this method of bookkeeping, integral, rather than half-integral, wavelengths have to fit into the interval a, the number of states has been reduced by 2^3. However, because positive and negative n_x, n_y, and n_z give physically different states, the number of states has been increased by 2^3.

The argument that leads to (23) may seem like a lot of work that does nothing except reproduce (5). It is, however, important because it shows that states can be counted either in terms of standing waves, sin $\vec{k} \cdot \vec{r}$, or in terms of traveling waves, $e^{i\vec{k}\cdot\vec{r}}$. The traveling wave picture will turn out to be simpler and more physical for some applications in this and in subsequent chapters.

Suppose that the number of valence electrons per unit volume is N, and that they have the lowest possible energy because the system has a temperature of 0°K. The electrons will fill states, starting from the lowest, until they have all found a home in such a way that the Pauli exclusion principle is obeyed. Then there will be a well-defined highest single-particle energy that is reached, and this energy is called the *Fermi energy* E_F. From (7),

$$N = \frac{(2ME_F)^{3/2}}{3\pi^2\hbar^3},$$

$$E_F = \frac{\hbar^2}{2M}(3\pi^2 N)^{2/3}. \tag{5.1-24}$$

PROBLEM 5.4

a. Use (24) to compute E_F, in electron volts, for Li, K, Be, and Al. You will need to use the valence and the density of these metals. How does the thermal energy of electrons at room temperature compare with these numbers?

b. Use (24) to estimate E_F, in MeV, for the 146 neutrons in a U^{238} nucleus if they are free to move in the nuclear volume. Why are you asked to *estimate*, rather than to *compute* as in (**a**)?

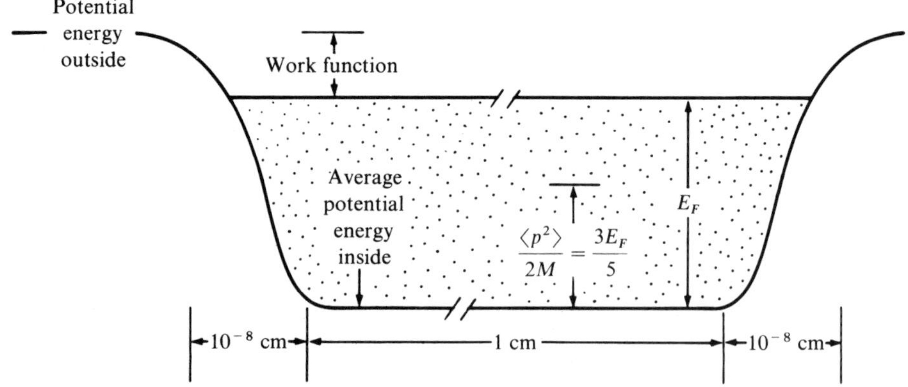

Fig. 5.1-2. The important energies in the free electron model.

Problem 5.4 shows that Fermi energies in metals are a few eV, about a hundred times thermal energy at room temperature. It follows that the actual distribution of electron energies is very close to the 0°K distribution, that is, to the distribution that has the lowest possible energy: All states up to E_F are filled, while all those above E_F are empty.

PROBLEM 5.5

In counting states to obtain (7) we assumed that the electrons are confined in a rigid-walled box, i.e., a box with an infinite potential rise at its surface. Actually, E_F and the work function are each a few eV, and there should be appreciable spilling of the electron wave functions into the classically inaccessible region. How does this fact affect the count of number of states?

Differentiating (7) gives the number of states per unit volume between E and $E + dE$,

$$\frac{d\eta(E)}{dE}\,dE = \frac{(2M)^{3/2}}{2\pi^2\hbar^3}\,E^{1/2}\,dE. \tag{5.1-25}$$

At absolute zero temperature the number of electrons per unit volume per unit energy is given by this expression up to E_F, and is zero above E_F. At higher temperatures, thermal excitation empties some states below E_F and fills some above E_F. Since $E_F \simeq$ a few eV, the dotted curve in Figure 5.1-3, if drawn to

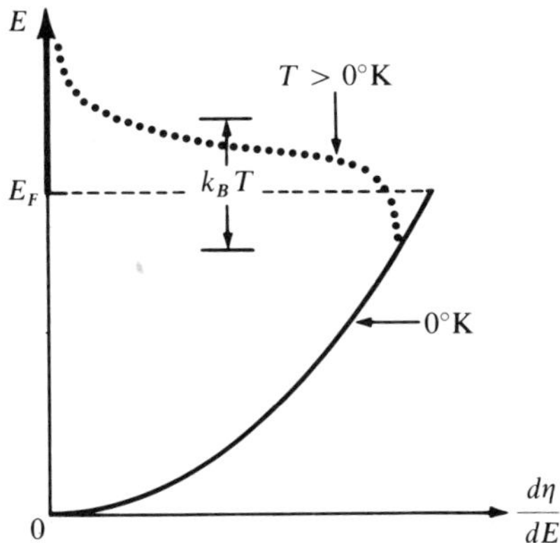

Fig. 5.1-3. The density of electrons per unit energy, $d\eta/dE$, plotted horizontally, versus energy. At 0°K the result is given by (25) up to E_F, and is zero for higher energies, as shown by the solid curve. At higher temperatures $d\eta/dE$ deviates from the 0°K curve, primarily in the interval $E_F \pm k_B T$, as shown by the dotted curve.

scale, corresponds to $k_B T \simeq 1$ eV, and $T \simeq 10^4$°K, which is above the boiling point of all elements.

PROBLEM 5.6

What is the total kinetic energy of the valence electrons in 1 cm^3 of Li at 0°K? Suggestion: First calculate the mean kinetic energy of electrons with the distribution given by (25); it is $\frac{3}{5}E_F$. Then multiply by the number of electrons. To get some feeling for magnitudes, express the result in calories.

As an application we shall calculate the *electronic paramagnetic susceptibility* of a typical metal. Each electron has a magnetic moment, directed opposite to its spin, with magnitude $M_s = 0.579 \times 10^{-8}$ eV/gauss, or 0.579×10^{-4} eV/(weber/m^2). The energy of orientation, $2BM_s$, in a magnetic field B of one weber/m^2 is then about 10^{-4} eV. Suppose that the field and the temperature satisfy $k_B T \ll M_s B \ll E_F$. This condition requires a very low temperature; how low? If there were no Pauli exclusion principle, most electron spins would line up with the field. There would be a magnetic moment per unit volume $\simeq M_s N$, and so classical physics predicts a susceptibility

$$\chi \simeq \frac{M_s N}{B},$$

a value vastly greater than one observes.

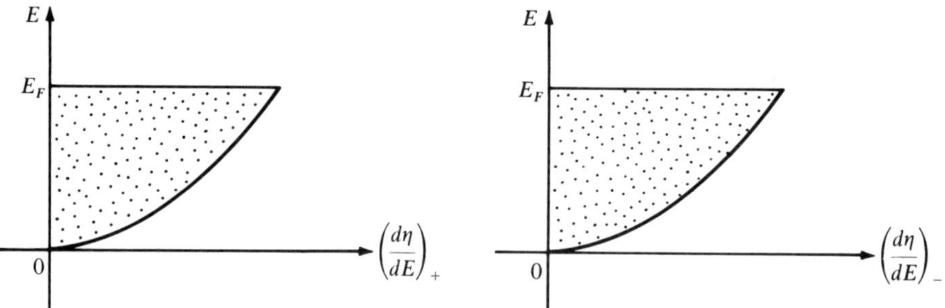

Fig. 5.1-4. The densities of occupied states for the two spin directions before B is turned on.

According to quantum physics, most electrons cannot flip their spins; if they did, they would go into states that are already occupied. Figure 5.1-4 shows the densities of occupied states with moments along (+) and opposite (−) the field direction plotted separately, before the field is turned on. The two graphs are identical because there is nothing yet that distinguishes between the directions.

Figure 5.1-5 shows the densities after the field is turned on, but before any electrons have had time to turn their spins around. One distribution is shifted up in energy by $M_s B$, while the other is shifted down by $M_s B$. Now the electrons at the top of the (−) distribution can flip their spins, occupy available states at the top of the (+) distribution, and so minimize the energy of the system. Figure 5.1-6 shows how, in the $k_B T \ll M_s B \ll E_F$ approximation, the tops of both

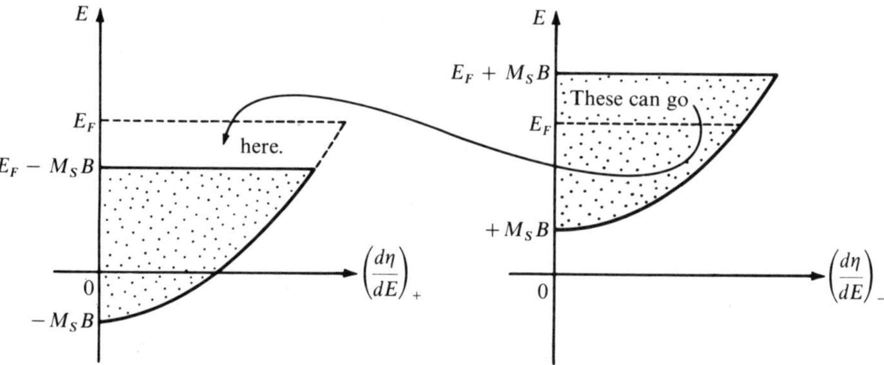

Fig. 5.1-5. The densities of occupied states for the two spin directions after B is turned on, but before any spins have flipped.

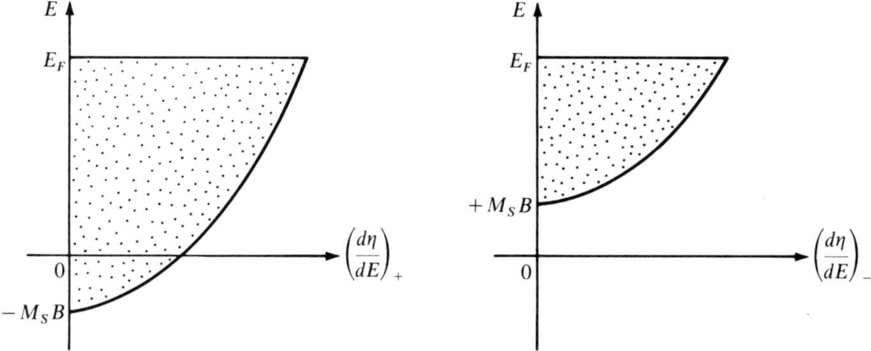

Fig. 5.1-6. The densities of occupied states for the two spin directions after B is turned on and after the spins have flipped to minimize the total energy.

distributions return to E_F because $(d\eta/dE)_- M_s B$ electrons flip their spins. Then

(Number of e^- aligned *with* $\vec{B}$) − (Number of e^- aligned *opposite* $\vec{B}$)

$$= 2M_s B\left(\frac{d\eta}{dE}\right)_{-,E_F} = M_s B\left(\frac{d\eta}{dE}\right)_{E_F}.$$

The net magnetic moment per unit volume becomes

$$M_s\left[M_s B\left(\frac{d\eta}{dE}\right)_{E_F}\right]$$

and the susceptibility

$$\chi = M_s^2\left(\frac{d\eta}{dE}\right)_{E_F}.$$

From (25) and (24),

$$\left(\frac{d\eta}{dE}\right)_{E_F} = \frac{(2M)^{3/2}}{2\pi^2\hbar^3} E_F^{1/2} = \frac{3N}{2E_F}. \tag{5.1-26}$$

$$\therefore \quad \chi = \frac{3M_s^2 N}{2E_F}. \tag{5.1-27}$$

The magnitude of the result is right, but other effects have to be considered if detailed agreement is wanted. The plane wave states of the electrons are affected by $\vec{B}$, exchange and correlation effects are important, and, for all except the lightest atoms, the ion cores give a substantial contribution. These matters, the experimentally more accessible case $M_s B \ll k_B T \ll E_F$, and numerical results are treated in various texts.[2]

One aspect of this calculation is very general and worth emphasizing. All electrons except those at the top of the Fermi distribution are locked in by the Pauli principle. There is nothing that most of the electrons can do except stay in their states, because all states near them in energy are filled. The interesting property, in this case the susceptibility, depends entirely on the relatively few electrons at the top of the distribution that had vacant states available. The same reasoning shows that electrons contribute little to the specific heat of metals. Near room temperature, only about 0.025 eV/E_F of the electrons are close enough to E_F for thermal excitation to unoccupied states to be likely.

The free electron model also provides a rough estimate of the *compressibility*

$$C = -\frac{1}{\mathscr{V}}\frac{d\mathscr{V}}{dP}, \tag{5.1-28}$$

where $\mathscr{V}$ is the volume and P is the pressure. The forces that resist compression are due to a variety of causes, but the pressure of the "gas" of valence electrons is a major contributor.

Let $\mathscr{N} = N\mathscr{V}$ be the total number of electrons in the piece of metal. The total kinetic energy of the electrons is $3\mathscr{N}E_F/5$; see Problem 5.6. In a change of volume $d\mathscr{V}$, the work done on the electron gas is $-P\,d\mathscr{V}$, and this work must equal the change in the total kinetic energy,

$$-P\,d\mathscr{V} = \tfrac{3}{5}\mathscr{N}\,dE_F.$$

With $N = \mathscr{N}/\mathscr{V}$ in (24),

$$P = -\tfrac{3}{5}\mathscr{N}\frac{dE_F}{d\mathscr{V}} = \frac{\hbar^2(3\pi^2)^{2/3}}{5M}\left(\frac{\mathscr{N}}{\mathscr{V}}\right)^{5/3},$$

$$\frac{dP}{d\mathscr{V}} = -\frac{2\mathscr{N}E_F}{3\mathscr{V}^2},$$

and (28) becomes

$$C = \frac{3}{2NE_F}. \tag{5.1-29}$$

For Li, $N = 4.63 \times 10^{22}/\text{cm}^3$ and $E_F = 4.73$ eV give $C = 4.3 \times 10^{-12}$ cm^2/dyne, while the measured value is 8.7×10^{-12} cm^2/dyne. The heavier alkali metals show better agreement with (29), while some other metals show greater

[2] C. Kittel, *Introduction to Solid State Physics*, 5th ed. (New York: John Wiley & Sons, Inc., 1976), Chap. 14; N. S. Ashcroft and N. D. Mermin, *Solid State Physics* (New York: Holt, Rinehart and Winston, 1976) Chap. 31.

discrepancies.[3] Detailed agreement would be a misleading accident. The zero pressure size is determined by a balance between forces trying to expand the volume and forces trying to reduce it. Any approach that deals with only one of these forces must be incomplete.

What does hold a piece of metal together? A serious answer would be an accurate calculation of the energy released when initially separated atoms are assembled to build the piece. Such a calculation would take us at least to the frontiers of solid state theory. Some of the physical ideas can, however, be seen in an absurdly simple model.

Pretend that each separated atom is a little $b_0 \times b_0 \times b_0$ box that contains an electron with energy

$$E_0 = \frac{3\pi^2\hbar^2}{2Mb_0^2}.$$

To build the piece of metal, assemble many little boxes. If the boxes were simply stacked together, then N, the number of electrons per unit volume, would be $1/b_0^3$. However, the boxes should be pushed into each other so that the electrons have overlapping wave functions and can easily get from one neighborhood to another. Perhaps the average distance between electrons should be halved, so that $N = 8/b_0^3$. The electrons then form an electron gas in which the average energy is

$$\begin{aligned}\tfrac{3}{5}E_F &= \frac{3}{5}\frac{\hbar^2}{2M}\left(3\pi^2 \cdot \frac{8}{b_0^3}\right)^{2/3}\\ &= \frac{12}{5(3\pi^2)^{1/3}}\frac{3\pi^2\hbar^2}{2Mb_0^2} = 0.8E_0.\end{aligned}$$

Together the electrons are 20% lower in energy than when they are apart. If in the separated atoms E_0 is a few eV, this result suggests a binding energy of an eV per atom, a reasonable amount.

In its details this model is almost silly. Also, it makes no mention of the very significant Coulomb interactions. What is correct is the idea that assembling metallic atoms permits the electrons to roam throughout the macroscopic piece, and therefore permits them to reduce their average energy below the single atom value. If an electron is given more room, it can lengthen its wavelength and lower its energy. If the electrons were bosons, then, in the assembled piece of metal, they could all collapse into the lowest state, with wavelength of about a centimeter and nearly zero energy. Since, however, they are fermions, they must fill a range of energy levels, and binding energies of about an eV per atom are typical of the alkali metals.

[3] N. W. Ashcroft and N. D. Mermin, *Solid State Physics* (New York: Holt, Rinehart and Winston, 1976) Chap. 2.

5.2 *Electrons in a Periodic Potential*

In Section 5.1 the potential energy of electrons in the interior of the metal was taken to be constant. In this section we continue to neglect the interactions between electrons, but recognize the presence of the positive ions. Figure 5.2-1 describes qualitatively the potential energy of an electron in a lattice with spacing b.

The consequences of such a periodic potential can be discussed from several points of view.[4] Detailed numerical agreement with experiment requires a three-dimensional description and a proper choice of the potential shape, but a qualitative understanding requires far less. Much of the interesting physics is determined just by the fact that the potential is periodic, and one can learn a lot by looking at a very simple one-dimensional model.

Suppose then that the wave functions and the potential depend only on x, and that the potential is periodic with a period $b \simeq 10^{-8}$ cm;

$$V(x + b) = V(x). \tag{5.2-1}$$

This statement is strictly right only for an infinite piece of material. Any real sample will have a length Qb, where $Q \simeq 10^8$, and (1) is wrong near the ends. However, (1) can be made to be right everywhere by the imposition of periodic boundary conditions, as described in Section 5.1, with the a of (5.1-15) equal to Qb. Alternatively, one can pretend that the Q atoms are arranged around a ring of circumference Qb. Either point of view makes (1) valid everywhere, and forces the wave function to satisfy

$$\psi(x + Qb) = \psi(x). \tag{5.2-2}$$

The imposition of this condition is a trick that will make anyone who has been raised carefully feel a little queasy. It simply is not true that the wave function is the same at the two ends of a 1-cm-long piece of metal; (2) is a boundary condition invented to make the problem tractable. The justification comes from the fact that if Q is large, then its actual size does not appear in physically interesting results, such as conductivities, which can be expressed as something per unit mass or per unit volume. The results (5.1-7) and (5.1-25) are relevant examples: To obtain them, it was useful to think in terms of some volume $\mathscr{V}$, but the important conclusions about the number and density of states *per unit volume* turned out to be independent of whether rigid walls or periodic boundary conditions were used, for large $\mathscr{V}$. Macroscopic boundary conditions can be arranged in many different ways without any change in the essential conclusions.

[4] C. Kittel, *Introduction to Solid State Physics*, 5th ed. (New York: John Wiley & Sons, Inc., 1976), Chap. 7; N. W. Ashcroft and N. D. Mermin, *Solid State Physics* (New York: Holt, Rinehart and Winston, 1976), Chap. 8; J. S. Blakemore, *Solid State Physics*, 2d ed. (Philadelphia: W. B. Saunders Co., 1974), Sec. 3.4.

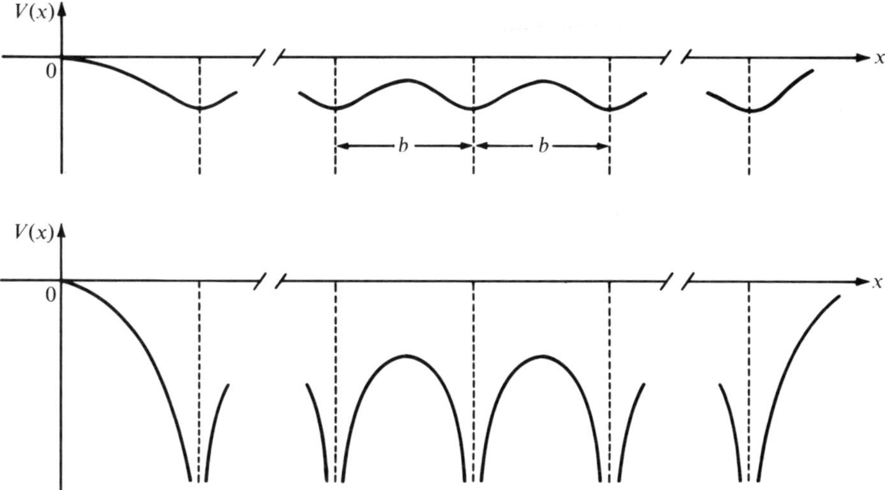

Fig. 5.2-1. The potential energy of an electron in a lattice. The lower curve shows $V(x)$ along a line that passes through adjacent nuclei. The upper curve shows $V(x)$ along a parallel line that misses the nuclei.

The boundary condition (2) specifies what happens to the wave function if the position is shifted through the large distance Qb. What happens if the position is shifted by only one lattice spacing b; that is, what is the relation between $\psi(x + b)$ and $\psi(x)$? The relation $\psi(x + b) = \psi(x)$ is one possibility, but there are many others. To examine this problem, it is useful to define the *displacement operator* D_b by

$$D_b\psi(x) = \psi(x + b). \tag{5.2-3}$$

The kinetic energy is independent of the choice of origin, and $V(x + b) = V(x)$. Therefore the Hamiltonian is unchanged by a shift of x through a distance b, and it commutes with D_b;

$$[D_b, H] = 0. \tag{5.2-4}$$

As shown in Section 3.7, it follows that there exist simultaneous eigenfunctions of D_b and H. To find these eigenfunctions, require that

$$D_b\psi(x) = e^{\alpha b}\psi(x),$$

or

$$\psi(x + b) = e^{\alpha b}\psi(x) \tag{5.2-5}$$

for any x, where the eigenvalue has for convenience been called $e^{\alpha b}$, with α some constant. Then, from the definition (3) of D_b and the relation (5),

$$D_b[e^{-\alpha x}\psi(x)] = e^{-\alpha(x+b)}\psi(x+b) = e^{-\alpha x}\psi(x).$$

It follows that

$$e^{-\alpha x}\psi(x) \equiv u(x)$$

is unchanged by the shift operator:

$$u(x+b) = u(x).$$

The eigenfunctions of D_b must then have the form

$$\psi(x) = e^{\alpha x}u(x), \tag{5.2-6}$$

where $u(x)$ is any function with periodicity b, and α is any complex constant.[5]

Now impose the single-valuedness requirement (2) on expression (6).

$$\begin{aligned} \psi(x+Qb) &= D_b^Q\psi(x) \\ &= e^{Q\alpha b}\psi(x) = \psi(x), \\ e^{Q\alpha b} &= 1, \\ Q\alpha b &= 2\pi i j, \qquad j = 0, \pm 1, \pm 2, \ldots. \end{aligned}$$

With the definition

$$\alpha = i\frac{2\pi j}{Qb} \equiv iK, \tag{5.2-7}$$

the eigenfunctions of the displacement operator are

$$\psi(x) = e^{iKx}u_K(x). \tag{5.2-8}$$

These are, in one dimension, the so-called Bloch functions.[6] Be careful to remember that $\hbar K$ is not the momentum of these states because $u_K(x)$ also depends on position.

PROBLEM 5.7

Is the displacement operator D_b Hermitian?

---

[5] That $\psi(x)$ must have the form (6) is known in solid state physics as Bloch's theorem. It is essentially the same as a result in the theory of differential equations called Floquet's theorem. See for example H. Margenau and G. M. Murphy, *The Mathematics of Physics and Chemistry*, 2d ed. (Princeton: D. Van Nostrand Co., Inc., 1956), Sect. 2.17.

[6] F. Bloch, *Z. Physik* **52**, 555 (1928). Note the date. Applications of quantum theory came very quickly after its construction.

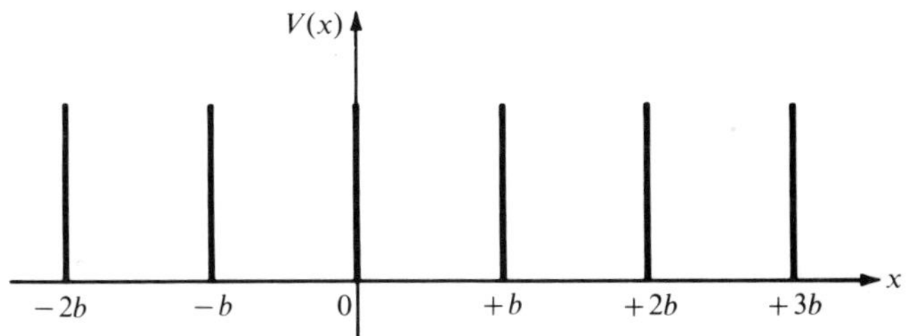

Fig. 5.2-2. A Dirac delta function model of a periodic potential.

To learn more, one can choose a particular $V(x)$ to carry out the calculation. The worst thing we did in this section was to go to one dimension, so, with the realization that it is hard to do much more harm, we choose a form for $V(x)$ that makes the algebra simple. As mentioned above, several important results follow from the periodicity, and the qualitative picture is not sensitive to the precise form of $V(x)$. Choose an array of Dirac delta functions (Figure 5.2-2),

$$V(x) = U[\delta(x) + \delta(x - b) + \delta(x + b) + \delta(x - 2b) + \delta(x + 2b) + \cdots].$$

It will turn out not to matter whether U is positive or negative. Now look at the way delta function potentials were handled in Section 2.3. The condition (2.3-6) on the derivatives at $x = b$ becomes

$$\left(\frac{d\psi}{dx}\right)_{b+\varepsilon} - \left(\frac{d\psi}{dx}\right)_{b-\varepsilon} = \frac{2MU}{\hbar^2}\psi(b). \tag{5.2-9}$$

For $0 < x < b$ the potential is zero and the general solution is

$$\psi(x) = Ae^{ikx} + Be^{-ikx},$$

where k must not be confused with K. The energy is $\hbar^2k^2/2M$ because the $\psi(x)$ are eigenstates of H, and the E that satisfies $H\psi = E\psi$ between the barriers satisfies it everywhere. Since

$$\psi(x) = D_b\psi(x - b) = e^{iKb}\psi(x - b),$$

the solution in the next cell is

$$\psi(x) = e^{iKb}[Ae^{ik(x-b)} + Be^{-ik(x-b)}], \qquad b < x < 2b.$$

Now require at b that $\psi(x)$ is continuous and that the condition (9) is satisfied. Consistency of the two resulting linear homogeneous equations for A and B then imposes the following requirement:

$$L \equiv \sqrt{1 + \frac{v^2}{k^2}}\cos(kb - \phi) = \cos Kb, \tag{5.2-10}$$

where

$$v \equiv \frac{MU}{\hbar^2}, \qquad \phi \equiv \tan^{-1}\left(\frac{v}{k}\right).$$

PROBLEM 5.8

a. Complete the derivation of (10).

b. Show that $L \to 1 + vb$ as $k \to 0$.

As k changes, the left side L of (10) oscillates with an amplitude

$$\sqrt{1 + \frac{v^2}{k^2}} > 1;$$

see Figure 5.2-3. The right side $\cos Kb$ is always between -1 and $+1$. Therefore (10) cannot be satisfied for those values of k that give $kb - \phi$ near $n\pi$. Wherever $|L| < 1$, a solution is arbitrarily close because j can be selected to bring $\cos Kb = \cos(2\pi j/Q)$ arbitrarily close to any number between -1 and $+1$. The allowed energies therefore occur in *bands*,

$$\frac{\hbar^2 k_1^2}{2M} < E < \frac{\hbar^2 k_2^2}{2M}, \qquad \frac{\hbar^2 k_3^2}{2M} < E < \frac{\hbar^2 k_4^2}{2M}, \ldots,$$

where $k_1, k_2, k_3, \ldots$ are the values where $L = \pm 1$, as shown in Figure 5.2-3, and electrons cannot have energies in the intervening gaps.

There is much that is incomplete in this model, but merely the existence of energy bands explains things that would otherwise be totally mysterious. The various kinds of electrical conductivity form one such topic.

Suppose that the valence electrons just fill a band. In other words, suppose that the number of electrons just equals the number of states needed to get to, say, k_4, and that the temperature is low and the energy gap is large, so that

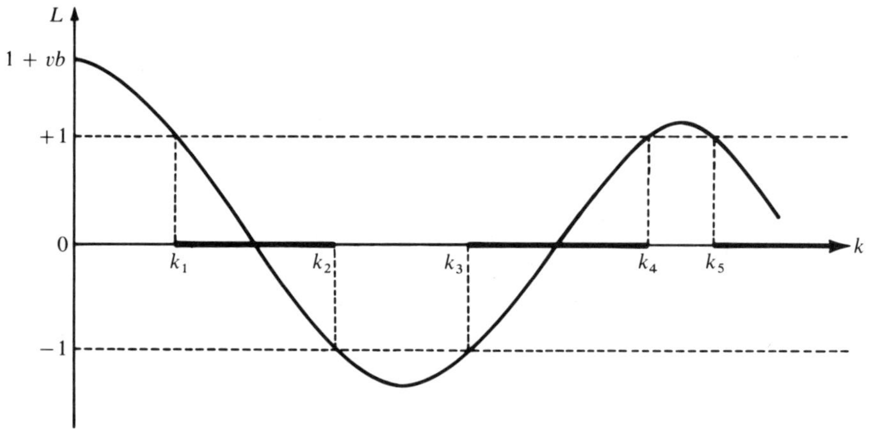

Fig. 5.2-3. A graph of the expression L defined in (10). The allowed values of k occur wherever $|L| < 1$.

virtually no electrons are excited into the next band. The application of an electric field then produces virtually no net current, because one cannot turn around the momentum of an unexcited electron without asking it to occupy a state that is already filled. The material is an *insulator*. At high temperatures there is some conductivity because a few electrons are thermally excited from the top of the filled band to the bottom of the empty band. These electrons can respond to a field because they have many empty states available. Further, the holes left in the originally filled band provide the freedom for a few electrons in this band to respond also.

Next, suppose that the valence electrons fill a substantial fraction, but not nearly all, of a band. Then the electrons near the Fermi energy are close to a large number of empty states, even a small electric field will turn around the momentum of many electrons to produce a substantial current, and the material is a good conductor, a *metal*. The conductivity is governed primarily by the ease with which these electrons move through the lattice. Thermal jiggling of the positive ions impedes the motion of the electrons. In the neighborhood of room temperature the conductivity varies roughly like the reciprocal of the temperature.

If a material has a completely filled band at absolute zero temperature, but has a small gap between the filled band and the next highest band, even a modest temperature rise promotes many electrons to the next band. Then there is a moderate conductivity that increases rapidly with temperature, and one calls the material a *semiconductor*. Examples include Si, Ge, and Te. The distinction between semiconductors and insulators is not sharp, and is really a matter of the applications for which the material is usually used.

If at absolute zero one has nearly filled or nearly empty bands, then a few electrons are available to conduct electricity, and one calls the material a *semimetal*. Examples include Bi, Sb, and As.

PROBLEM 5.9

Make an estimate of the gap E_g between the highest filled band and the lowest empty band in silicon as follows. The conductivity is governed primarily by the number of electrons that can have their state changed by a modest electric field, although the mobility of these electrons also affects the result. The number of electrons that are thermally excited across E_g is given reasonably well by $\exp(-E_g/2k_B T)$.[7] Use the following data.

T, °K:	500	1000
Conductivity, (ohm cm)$^{-1}$:	.03	20.

---

[7] C. Kittel, *Introduction to Solid State Physics*, 5th ed. (New York: John Wiley & Sons, Inc., 1976), Chap. 8; N. W. Ashcroft and N. D. Mermin, *Solid State Physics* (New York: Holt, Rinehart and Winston, 1976), Chap. 28.

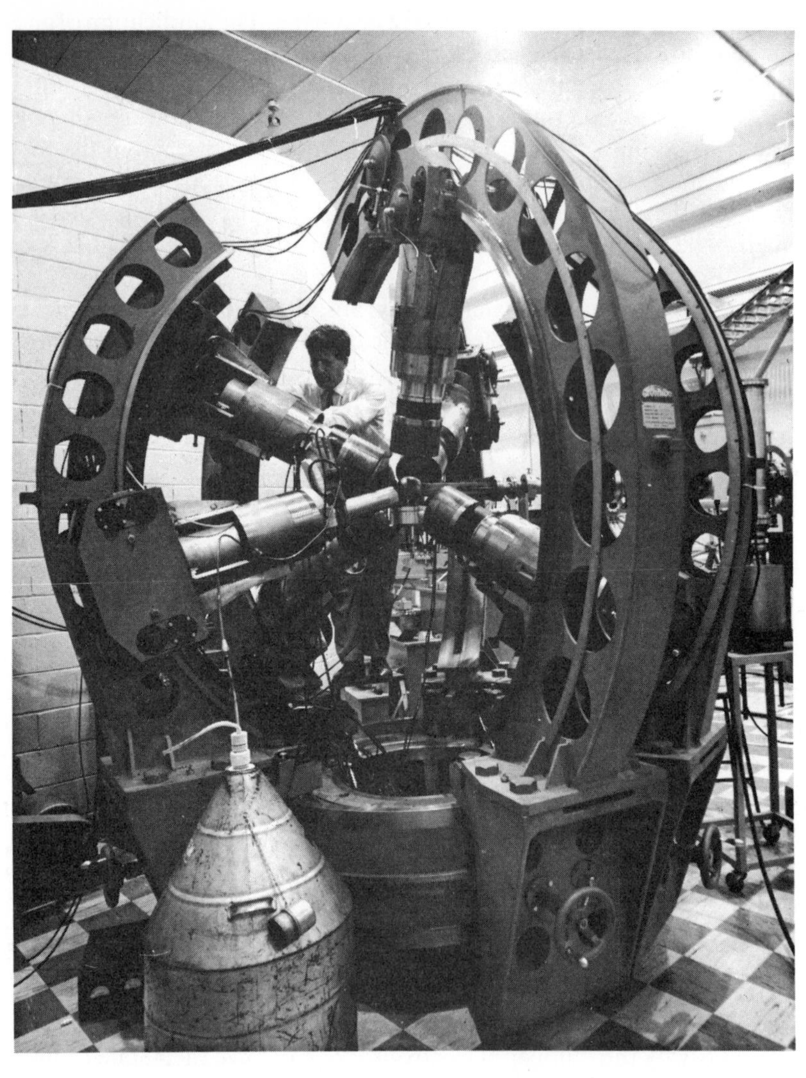

A device for determining the angular momenta and parities of nuclei in excited states. The beam from an accelerator comes in from the right and strikes the target at the center. The reactions produce excited nuclei, which decay by emitting gamma rays. The angular distributions of these gamma rays are measured with counters that can be moved about in the framework. The properties of the nuclear states can then be deduced from these distributions. For details and basic references, see A. J. Ferguson, Angular Correlation Methods, *in J. Cerny, ed.,* Nuclear Spectroscopy and Reactions, *part C (New York: Academic Press, 1974). (Courtesy of Chalk River Nuclear Laboratories, Atomic Energy of Canada Limited)*

6

Angular Momentum

In quantum as in classical mechanics the angular momentum concept facilitates the understanding and the solution of some of the most important problems. Section 6.1 deals with general properties of quantal angular momentum. Section 6.2 shows how orbital angular momentum operators and eigenfunctions appear when the Schrödinger equation is expressed in spherical polar coordinates.

6.1 *General Properties of Angular Momentum*

In classical mechanics the orbital angular momentum vector for a particle with position $\vec{r}$ and momentum $\vec{p}$ is

$$\vec{L} = \vec{r} \times \vec{p}. \tag{6.1-1}$$

Quantal orbital angular momentum is defined by the same expression, with $\vec{r}$ and $\vec{p}$ taken to be the position and momentum operators. This choice for $\vec{L}$ is the simplest thing that is Hermitian and that gives the classical result in the classical limit.

EXAMPLE

Should one use $\vec{L} = \vec{r} \times \vec{p}$ or $\vec{L} = -\vec{p} \times \vec{r}$?

SOLUTION

In classical mechanics the question would be silly because there it is obvious that $\vec{r} \times \vec{p} = -\vec{p} \times \vec{r}$. In quantum mechanics, however, $\vec{r}$ and $\vec{p}$ are operators that do not commute, and the order in which they are written must be examined. The z component is

$$L_z = (\vec{r} \times \vec{p})_z = xp_y - yp_x.$$
$$(-\vec{p} \times \vec{r})_z = -p_x y + p_y x.$$

Since x commutes with p_y and y commutes with p_x,

$$(-\vec{p} \times \vec{r})_z = -yp_x + xp_y = L_z;$$

the two expressions are the same. The vector product involves the terms xp_y, xp_z, yp_x, yp_z, zp_x, and zp_y, in which the factors commute, and not xp_x, yp_y, and zp_z, in which the factors do *not* commute. For this reason

$$\vec{r} \times \vec{p} = -\vec{p} \times \vec{r}$$

in quantum mechanics also.

Suppose that the Hamiltonian is invariant under arbitrary rotations. For example, the system might be an electron in a spherically symmetric potential. If classical experience is a useful guide, angular momentum should be conserved. It is, and here is a proof.

Consider a rotation through a very small angle φ about one of the axes, say the z axis. Since $\cos\varphi \simeq 1$ and $\sin\varphi \simeq \varphi$, the position (x, y, z) becomes

$$x' = x - \varphi y,$$
$$y' = y + \varphi x,$$
$$z' = z.$$

Any function of position changes, for small Δx, Δy, and Δz, according to

$$F(\vec{r}') = F(\vec{r}) + \frac{\partial F}{\partial x}\Delta x + \frac{\partial F}{\partial y}\Delta y + \frac{\partial F}{\partial z}\Delta z.$$

The increments in the coordinates are $\Delta x = -\varphi y$, $\Delta y = \varphi x$, and $\Delta z = 0$.

$$\therefore \quad F(\vec{r}') = F(\vec{r}) + \varphi\left(x\frac{\partial F}{\partial y} - y\frac{\partial F}{\partial x}\right) = R_z F(\vec{r}),$$

where R_z, the operator for infinitesimal rotations φ about the z axis, is given by

$$\begin{aligned} R_z &= 1 + \varphi\left(x\frac{\partial}{\partial y} - y\frac{\partial}{\partial x}\right) = 1 + \frac{i\varphi}{\hbar}(xp_y - yp_x) \\ &= 1 + \frac{i\varphi}{\hbar}(\vec{r} \times \vec{p})_z = 1 + \frac{i\varphi}{\hbar}L_z. \end{aligned} \tag{6.1-2}$$

Since H is invariant under rotations, it makes no difference whether one first rotates and then applies H to a wave function, or whether one first applies H and then rotates; the Hamiltonian and R_z commute,

$$[H, R_z] = 0. \tag{6.1-3}$$

Since H commutes with all constants, (2) and (3) show that H commutes with the z component of angular momentum,

$$[H, L_z] = 0. \tag{6.1-4}$$

Since L_z has no explicit time dependence, it follows from (3.10-3) that $d\langle L_z\rangle/dt = 0$. There is nothing sacred about the z axis, and the same argument shows that $\langle L_x\rangle$ and $\langle L_y\rangle$ are constant. The result is the quantum-mechanical theorem regarding the conservation of orbital angular momentum: If H is rotation invariant, $\langle \vec{L}\rangle$ is a constant, where by $\vec{L}$ is meant $\hat{e}_x L_x + \hat{e}_y L_y + \hat{e}_z L_z$.

To continue, we need some properties of the angular momentum operators. Since $[x, p_x] = i\hbar$, $[y, p_x] = 0, \ldots,$

$$\begin{aligned} [L_x, L_y] &= (yp_z - zp_y)(zp_x - xp_z) - (zp_x - xp_z)(yp_z - zp_y) \\ &= (zp_z - p_z z)(xp_y - yp_x) = i\hbar L_z, \\ [L_z, L_x] &= i\hbar L_y, \\ [L_y, L_z] &= i\hbar L_x. \end{aligned} \tag{6.1-5}$$

These three commutation relations are summarized by

$$\vec{L} \times \vec{L} = i\hbar\vec{L}. \tag{6.1-6}$$

This equation probably makes you uncomfortable because you "know" that the cross product of a vector with itself is zero. However, one necessarily gets zero only if the components of the vector commute with each other. Here they do not, and the three commutation relations (5) are contained in (6), that is, in

$$\begin{vmatrix} \hat{e}_x & \hat{e}_y & \hat{e}_z \\ L_x & L_y & L_z \\ L_x & L_y & L_z \end{vmatrix} = i\hbar\vec{L}.$$

These commutation relations among the components of angular momentum are extremely important; all basic properties of quantal angular momentum can be derived from them. We obtained (5), or (6), by translating the classically familiar $\vec{r} \times \vec{p}$ into quantum mechanics and then looking at the behavior of its components. For many purposes it is desirable to promote these commutation relations to the status of the starting point of the development by saying, "By definition, angular momentum is a vector that obeys (6)." This attitude will become essential a little later in the discussion of the intrinsic spin of particles, where there is no known $\vec{r} \times \vec{p}$ associated with internal structure.

Suppose that there are several kinds of angular momentum in a system. Think, for example, about an ^{14}N nucleus, which has an angular momentum that is the resultant of the spins of seven protons and of seven neutrons, plus the contributions of the orbital angular momenta that these particles have inside the nucleus. Suppose that each of the constituent angular momenta satisfies (6); what can be said about the resultant? The answer is, fortunately, that the components of the resultant also satisfy (6). Each of the contributions has its own coordinates and momenta which commute with the coordinates and momenta of all other contributions. If $\vec{L} = \vec{L}_1 + \vec{L}_2 + \cdots$,

$$\begin{aligned}[L_x, L_y] &= [L_{x1} + L_{x2} + \cdots, L_{y1} + L_{y2} + \cdots] \\ &= [L_{x1}, L_{y1}] + [L_{x2}, L_{y2}] + \cdots + [L_{x1}, L_{y2}] + \cdots.\end{aligned}$$

But components labeled with different indices commute;

$$[L_{x1}, L_{y2}] = [y_1 p_{z1} - z_1 p_{y1}, z_2 p_{x2} - x_2 p_{z2}] = 0$$

because $[p_{z1}, z_2] = 0$.

$$\therefore \quad [L_x, L_y] = i\hbar L_{z1} + i\hbar L_{z2} + \cdots = i\hbar L_z.$$

Without this result there would be a basic and horrendous difficulty. To do physics as we know it, one needs the option of treating a system as a particle or of analyzing its parts. For example, an atomic spectroscopist wants to treat a nucleus of ^{14}N as a single particle with a single angular momentum, a nuclear physicist wants to think about the contributions of the individual nucleons, and a particle physicist wants to speculate about the structure of each nucleon. All of them want to use the general theory of angular momentum without asking whether their system is composite or elementary.

The square of the magnitude of the angular momentum is

$$L^2 \equiv \vec{L} \cdot \vec{L} = L_x^2 + L_y^2 + L_z^2.$$

This quantity commutes with all three components of $\vec{L}$. For the x component,

$$[L_x, L^2] = [L_x, L_x^2] + [L_x, L_y^2] + [L_x, L_z^2].$$

The first commutator is zero. To work out the other two, note that for any operators Ω and Λ,

$$[\Omega, \Lambda^2] = \Omega\Lambda^2 + \Lambda\Omega\Lambda - \Lambda\Omega\Lambda - \Lambda^2\Omega$$
$$= [\Omega, \Lambda]\Lambda + \Lambda[\Omega, \Lambda].$$

$$\therefore \quad [L_x, L_y^2] = [L_x, L_y]L_y + L_y[L_x, L_y] = i\hbar(L_z L_y + L_y L_z),$$
$$[L_x, L_z^2] = [L_x, L_z]L_z + L_z[L_x, L_z] = -i\hbar(L_y L_z + L_z L_y).$$

These two commutators cancel, and the result is zero. The same argument works for the other two components, so that

$$[L_x, L^2] = 0, \qquad [L_y, L^2] = 0, \qquad [L_z, L^2] = 0,$$

or

$$[\vec{L}, L^2] = 0. \tag{6.1-7}$$

It follows from Section 3.4 that there is no uncertainty principle that prevents our knowing L^2 and any component of $\vec{L}$ simultaneously. However, (6) shows that, except for the special case of $\vec{L} = 0$, two components of $\vec{L}$ cannot be known simultaneously. We are going to find simultaneous eigenfunctions of L^2 and of L_z, and recognize that they will not be eigenfunctions of L_x or of L_y. Singling out the z axis is, of course, totally arbitrary; any direction will do. A scheme in which one constructs simultaneous eigenfunctions of L^2 and of L_x, and in which L_y and L_z are indefinite, would be equally valid. The z axis is, however, universally popular because of its special position in the definition of spherical polar coordinates.

PROBLEM 6.1

Are the commutation relations (6) and (7) valid in *momentum space*?

PROBLEM 6.2

Suppose that ψ is spherically symmetric; $\psi = \psi(r)$ only. Show that $\psi(r)$ is an eigenfunction of L_z and of L^2. What are the eigenvalues? (This problem can be worked by writing the angular momentum operators in Cartesian coordinates and using chain rule differentiation. It is a special case of results obtained in the next section.)

In the next section orbital angular momentum eigenfunctions will emerge from a separation of the Schrödinger equation in spherical polar coordinates. First, however, we shall look at an algebraic approach that depends only on the commutation relations (5) or (6). Although the subject is different, the method is like that used for the harmonic oscillator in Section 3.9, and a review of that section will help with the following treatment.

Suppose that one has somehow found an eigenfunction of both L^2 and L_z with eigenvalues $\hbar^2\lambda$ and $\hbar m$. Since angular momentum has the units of $\hbar$, λ and m are dimensionless numbers. For the present these numbers can be used to label the eigenfunction.

$$L^2\psi_{\lambda m} = \hbar^2\lambda\psi_{\lambda m}, \tag{6.1-8}$$

$$L_z\psi_{\lambda m} = \hbar m\psi_{\lambda m}. \tag{6.1-9}$$

Now we shall construct an operator Υ that, when it operates on $\psi_{\lambda m}$, produces $\psi_{\lambda m+a}$, an eigenfunction with unchanged L^2 but with a different z component $\hbar(m + a)$; that is, Υ shifts L_z by an amount $\hbar a$.

$$\Upsilon\psi_{\lambda m} = \psi_{\lambda m+a},$$

where

$$L^2\psi_{\lambda m+a} = \hbar^2\lambda\psi_{\lambda m+a}$$

and

$$L_z\psi_{\lambda m+a} = \hbar(m + a)\psi_{\lambda m+a}.$$

Then

$$L_z\Upsilon\psi_{\lambda m} = \hbar(m + a)\Upsilon\psi_{\lambda m}.$$

Subtract from this equation (9) multiplied by Υ.

$$[L_z, \Upsilon]\psi_{\lambda m} = \hbar a\Upsilon\psi_{\lambda m}. \tag{6.1-10}$$

We want an Υ that works for any $\psi_{\lambda m}$, and therefore change (10) into the general operator equation

$$[L_z, \Upsilon] = \hbar a\Upsilon, \tag{6.1-11}$$

and then try to construct an Υ that satisfies it.

The operators that operate on angular momentum functions are L_x, L_y, and L_z and the identity. A term proportional to the identity or to L_z would give $\psi_{\lambda m}$ back again, and that would not do; $\Upsilon\psi_{\lambda m}$ must give $\psi_{\lambda m+a}$ with $a \neq 0$, $\psi_{\lambda m+a}$ is orthogonal to $\psi_{\lambda m}$, and therefore $\psi_{\lambda m+a}$ cannot contain any admixture proportional to $\psi_{\lambda m}$. How about a term in L_x^2? It also would not do for a similar reason. Since $L_x^2 = L^2 - L_y^2 - L_z^2$, it would, in general, introduce terms in the original $\psi_{\lambda m}$. One can argue in the same way that higher powers of any of the components of $\vec{L}$ should not enter, and that Υ will have to be a linear combination of L_x and L_y. We are not going to worry at this stage about the normalization of $\psi_{\lambda m+a} = \Upsilon\psi_{\lambda m}$, and therefore take the coefficient of L_x to be unity and focus on determining the coefficient b of L_y in

$$\Upsilon = L_x + bL_y. \tag{6.1-12}$$

This trial form substituted into (11) gives

$$[L_z, L_x + bL_y] = \hbar a(L_x + bL_y).$$

The basic commutation relations

$$[L_z, L_x] = i\hbar L_y \qquad \text{and} \qquad [L_z, L_y] = -i\hbar L_x$$

turn this equation into

$$i\hbar L_y - bi\hbar L_x = \hbar a L_x + \hbar ab L_y,$$

so that

$$i = ab \qquad \text{and} \qquad -ib = a.$$

Eliminating b yields $a^2 = 1$, and there are two possibilities;

$$\begin{aligned} a = +1: \quad b = +i, \qquad \Upsilon = L_x + iL_y \equiv L_+, \\ L_z(L_+\psi_{\lambda m}) = \hbar(m+1)(L_+\psi_{\lambda m}) \end{aligned} \tag{6.1-13}$$

$$\begin{aligned} a = -1: \quad b = -i, \qquad \Upsilon = L_x - iL_y \equiv L_-, \\ L_z(L_-\psi_{\lambda m}) = \hbar(m-1)(L_-\psi_{\lambda m}) \end{aligned} \tag{6.1-14}$$

The operator Υ can be one of two things, the *raising operator* L_+, which increases the z component by $\hbar$, or the *lowering operator* L_-, which decreases the z component by $\hbar$. Since L^2 commutes with all of its components, it is easy to check that the eigenvalue of L^2 is not changed by the application of $L_\pm$.

$$\begin{aligned} L^2\psi_{\lambda m} &= \hbar^2\lambda\psi_{\lambda m}, \\ L^2\psi_{\lambda m\pm 1} &= L^2 L_\pm \psi_{\lambda m} = L_\pm L^2\psi_{\lambda m} \\ &= \hbar^2\lambda L_\pm\psi_{\lambda m} = \hbar^2\lambda\psi_{\lambda m\pm 1}. \end{aligned}$$

Start with any $\psi_{\lambda m}$. By operating repeatedly with L_+, one can get $\psi_{\lambda m+1}$, $\psi_{\lambda m+2}$, ..., and by operating repeatedly with L_-, one can get $\psi_{\lambda m-1}$, $\psi_{\lambda m-2}$, ... If the process does not terminate, the magnitude of the z component of angular momentum could be made greater than the total, which would be nonsensical. This observation can be formalized like this:

$$\langle L^2\rangle = \langle L_x^2\rangle + \langle L_y^2\rangle + \langle L_z^2\rangle.$$

Since $\psi_{\lambda m}$ is an eigenstate of L^2 and L_z,

$$\hbar^2\lambda = \langle L_x^2\rangle + \langle L_y^2\rangle + m^2\hbar^2.$$

Since L_x and L_y are Hermitian, $\langle L_x^2\rangle \geq 0$, $\langle L_y^2\rangle \geq 0$, and

$$\lambda \geq m^2.$$

The operators $L_\pm$ take states up and down a ladder of L_z values, with rungs differing by $\hbar$. Since L^2 is fixed, the ladder must be finite; it has a highest and a lowest rung. Let the top rung have L_z eigenvalue $\hbar m_T$, and let the bottom rung have L_z eigenvalue $\hbar m_B$, where m_T^2 and m_B^2 are both $\leq \lambda$. Since m changes in integral steps,

$$m_T - m_B = n, \qquad n = 0, 1, 2, 3, \ldots. \tag{6.1-15}$$

Termination requires that

$$L_+ \psi_{\lambda m_T} = 0, \qquad L_- \psi_{\lambda m_B} = 0. \tag{6.1-16}$$

Note that

$$\begin{aligned} L_+ L_- &= (L_x + iL_y)(L_x - iL_y) = L_x^2 + L_y^2 - i[L_x, L_y] \\ &= L_x^2 + L_y^2 + \hbar L_z = L^2 - L_z^2 + \hbar L_z, \end{aligned}$$

$$L^2 = L_+ L_- + L_z^2 - \hbar L_z. \tag{6.1-17}$$

Similarly,

$$L^2 = L_- L_+ + L_z^2 + \hbar L_z. \tag{6.1-18}$$

With these two relations, and with the conditions (16), λ can be expressed in terms of m_T or of m_B.

$$\begin{aligned} L^2 \psi_{\lambda m_T} &= (L_- L_+ + L_z^2 + \hbar L_z)\psi_{\lambda m_T} \\ &= (0 + \hbar^2 m_T^2 + \hbar^2 m_T)\psi_{\lambda m_T}. \end{aligned}$$

Since also

$$\begin{aligned} L^2 \psi_{\lambda m_T} &= \hbar^2 \lambda \psi_{\lambda m_T}, \\ \lambda &= m_T^2 + m_T. \end{aligned} \tag{6.1-19}$$

In the same way,

$$\begin{aligned} L^2 \psi_{\lambda m_B} &= (L_+ L_- + L_z^2 - \hbar L_z)\psi_{\lambda m_B} \\ &= (0 + \hbar^2 m_B^2 - \hbar^2 m_B)\psi_{\lambda m_B}, \end{aligned}$$

$$\begin{aligned} L^2 \psi_{\lambda m_B} &= \hbar^2 \lambda \psi_{\lambda m_B}, \\ \lambda &= m_B^2 - m_B. \end{aligned} \tag{6.1-20}$$

The two equations (19) and (20) then require that

$$m_T^2 + m_T = m_B^2 - m_B,$$

which is solved by

$$m_T = -m_B \qquad \text{and} \qquad m_T = -1 + m_B.$$

The second solution contradicts (15). The result, $m_T = -m_B$, is really inevitable. Suppose that one transforms coordinates so that $z \to -z$. Then m_T and m_B exchange roles, and the essential physics has to be invariant. If I am standing on my head, then my m_B is your m_T, and our two points of view must be equally valid.

With

$$m_T = -m_B \equiv l, \qquad \lambda = l(l+1),$$

the results are summarized by

$$L^2\psi_{\lambda m} = \hbar^2 l(l+1)\psi_{\lambda m}, \tag{6.1-21}$$

$$L_z\psi_{\lambda m} = \hbar m\psi_{\lambda m}, \quad m = -l, -l+1, \ldots, l-1, l. \tag{6.1-22}$$

Since one gets from $-l$ to $+l$ in integer steps, $2l$ is an integer, and the angular momentum quantum number is either an integer or a half integer. The most familiar example of integral l is orbital angular momentum, although there are many particles that have integral intrinsic spin. Pions have zero spin, photons have unit spin, and the "graviton," the hypothetical quantum of the gravitational field, is believed to have spin two. Half-integral angular momentum occurs only as intrinsic spin. Electrons, protons, neutrons, neutrinos, and muons are examples of particles with spin $\frac{1}{2}$. An example of a spin $\frac{3}{2}$ particle is the Ω^-, a particle that has negative charge, a mass of about 1.78 times that of the proton, and a mean life of 1.3×10^{-10} sec.

The trickery with the construction and use of the $L_\pm$ has produced a set of discrete angular momentum states, but how can one be sure that we have caught all of the states? How does one know that there are not additional possibilities, perhaps rungs on the L_z ladder in between those that our arguments uncovered, that would permit $l = \frac{1}{4}$? It is indeed true that only integral and half-integral values of angular momentum can occur. A proof in standard mathematical language requires some group theory.[1] Here we will be satisfied with a physical argument, which is already contained in the remarks made above. Operations that change the projection of the angular momentum onto the z axis must be rotations. Any rotation can be generated by sums of powers of the components $\vec{L}$. However, as was argued in the construction of the trial form (12) for Υ, $L_x + bL_y$ is the most general thing that one can consider. This trial form led to $L_\pm$, and the conclusion that adjacent L_z eigenvalues must differ by $\hbar$. The magnitudes of the highest and lowest L_z eigenvalues must be the same. This result forces integral or half-integral values. These words are nothing but a recapitulation; they are simply an attempt to underline those parts of the argument that indicate that all possibilities are exhausted by $l = 0, \frac{1}{2}, 1, \frac{3}{2}, 2, \ldots$.

We got here by considering orbital angular momentum as a guide, but only the commutation relations (5) were used. Everything of importance is valid for

[1] Continuous unitary irreducible representations exist only for integral and half-integral values of angular momentum. See E. Merzbacher, *Quantum Mechanics*, 2d ed. (New York: John Wiley & Sons, Inc., 1968), Chap. 22.

any kind of angular momentum, orbital, spin, or a resultant of orbital and spin. It is usual and convenient to reserve the symbols L^2, $\vec{L}$, $L_\pm$, and l for orbital angular momentum, and to use J^2, $\vec{J}$, $J_\pm$, and j for the more general case. From here on, to avoid clumsy expressions like $\psi_{l(l+1)m}$, the eigenfunctions will be labeled with l, rather than $\lambda = l(l+1)$, or with j, as appropriate. Thus for specifically orbital angular momentum,

$$L^2\psi_{lm} = \hbar^2 l(l+1)\psi_{lm},$$

$$L_z\psi_{lm} = \hbar m\psi_{lm}$$

$$L_\pm\psi_{lm} \sim \psi_{lm\pm 1}$$

For the general case, where the angular momentum is not necessarily orbital,

$$J^2\psi_{jm} = \hbar^2 j(j+1)\psi_{jm}, \tag{6.1-23}$$

$$J_z\psi_{jm} = \hbar m\psi_{jm}, \tag{6.1-24}$$

$$J_\pm\psi_{jm} \sim \psi_{jm\pm 1}.$$

The question of normalization remains. Suppose ψ_{jm} is normalized. After operating on it with the raising operator J_+, the result has the quantum numbers j, $m+1$, but generally will not be normalized. Therefore one should write $J_+\psi_{jm} \sim \psi_{jm+1}$, and an equality obtains only if the right constant is introduced;

$$J_+\psi_{jm} = N^+_{jm}\psi_{jm+1}. \tag{6.1-25}$$

The constant N^+_{jm} must be selected so that, if ψ_{jm} is normalized, then ψ_{jm+1} is also normalized. Square the magnitude of (25) and integrate over all space.

$$|N^+_{jm}|^2 \int \psi^*_{jm+1}\psi_{jm+1} = \int (J_+\psi_{jm})^*(J_+\psi_{jm}).$$

Because $\psi_{jm+1} \equiv |j \quad m+1\rangle$ is normalized, the integral on the left,

$$\langle j \quad m+1 | j \quad m+1\rangle,$$

is unity.

$$\therefore \quad |N^+_{jm}|^2 = \int [(J_x + iJ_y)\psi_{jm}]^*(J_x + iJ_y)\psi_{jm}.$$

Since J_x and J_y are Hermitian,

$$\int (J_x\psi_{jm})^*\chi = \int \psi^*_{jm}J_x\chi,$$

$$\int (iJ_y\psi_{jm})^*\chi = -i\int (J_y\psi_{jm})^*\chi = -i\int \psi^*_{jm}J_y\chi,$$

where χ is any state function. Take χ to be $(J_x + iJ_y)\psi_{jm}$. Then

$$|N_{jm}^{+}|^2 = \int \psi_{jm}^{*}(J_x - iJ_y)(J_x + iJ_y)\psi_{jm}$$

$$= \int \psi_{jm}^{*}[J_x^2 + J_y^2 + i(J_xJ_y - J_yJ_x)]\psi_{jm}$$

$$= \int \psi_{jm}^{*}[J^2 - J_z^2 - \hbar J_z]\psi_{jm}$$

$$= \hbar^2[j(j+1) - m^2 - m] \int \psi_{jm}^{*}\psi_{jm}.$$

Since $|jm\rangle$ is normalized, the integral on the right is unity, and, within an arbitrary phase factor conventionally taken to be unity,

$$N_{jm}^{+} = \hbar\sqrt{j(j+1) - m(m+1)}. \qquad (6.1\text{-}26)$$

Similarly, N_{jm}^{-} is determined by the requirement that if $|jm\rangle$ and $|jm - 1\rangle$ are normalized,

$$J_{-}|jm\rangle = N_{jm}^{-}|j \quad m-1\rangle, \qquad (6.1\text{-}27)$$

and it is given by

$$N_{jm}^{-} = \hbar\sqrt{j(j+1) - m(m-1)}. \qquad (6.1\text{-}28)$$

PROBLEM 6.3

Verify (28).

The results of this section can be summarized in a simple geometric construction found in most introductory modern physics texts (Figure 6.1-1).[2] From (23), the square of the total angular momentum has eigenvalues $\hbar^2 j(j+1)$, where $j = 0$, or $\frac{1}{2}$, or 1, or $\frac{3}{2}$, or . . . . For any eigenstate of J^2, in a space with Cartesian axes J_x, J_y, and J_z, the tip of the angular momentum vector lies somewhere on a sphere of radius $\hbar\sqrt{j(j+1)}$. As seen in (24), an eigenstate of J^2 can also be an eigenstate of J_z with eigenvalue $\hbar m$, where $m = -j$, or $-j + 1$, . . . , or $+j - 1$, or $+j$. The tip of the angular momentum vector $\vec{J}$ must then also lie on the plane $J_z = m\hbar$. The intersection of the plane with the

[2] K. F. Ford, *Classical and Modern Physics*, (Lexington, Mass: Xerox College Publishing, 1972), Part VII.

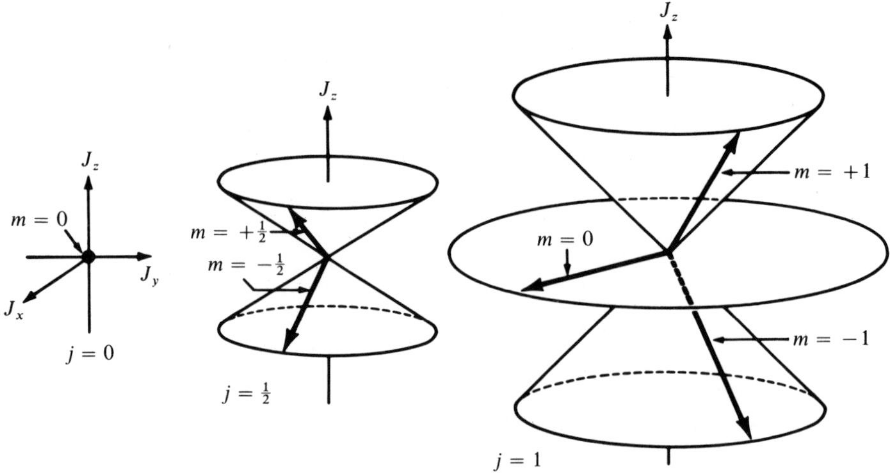

Fig. 6.1-1. Geometric representation of angular momentum for $j = 0, j = \frac{1}{2}$, and $j = 1$.

sphere is a circle of radius

$$J_x^2 + J_y^2 = J^2 - J_z^2.$$

One cannot know where on this circle $\vec{J}$ points because, if J_z is definite, J_x and J_y must be indefinite.

To study what can be known about J_x, it is convenient to write

$$J_x = \frac{J_x + iJ_y}{2} + \frac{J_x - iJ_y}{2} = \frac{J_+ + J_-}{2}. \qquad (6.1\text{-}29)$$

For any state $|jm\rangle$,

$$\langle J_x \rangle = \langle jm | J_x | jm \rangle = \langle jm | \frac{J_+ + J_-}{2} | jm \rangle$$

$$= \frac{N_{jm}^+}{2} \langle jm | j \quad m+1 \rangle + \frac{N_{jm}^-}{2} \langle jm | j \quad m-1 \rangle = 0$$

because states with different z components of angular momentum are orthogonal. Since $\langle J_x \rangle = 0$, $(\Delta J_x)^2 = \langle J_x^2 \rangle$. Note that

$$J_x^2 = \tfrac{1}{4}(J_+^2 + J_-^2 + J_+J_- + J_-J_+).$$

The first two terms contribute nothing because

$$\langle jm | J_\pm^2 | jm \rangle \sim \langle jm | j \quad m \pm 2 \rangle = 0.$$

The sum of (17) and (18) gives a useful expression for the other two terms,

$$J_+J_- + J_-J_+ = 2J^2 - 2J_z^2. \tag{6.1-30}$$

$$\begin{aligned}\therefore \quad (\Delta J_x)^2 &= \langle jm| \frac{J^2 - J_z^2}{2} |jm\rangle \\ &= \langle jm| \frac{\hbar^2 j(j+1) - \hbar^2 m^2}{2} |jm\rangle \\ &= \frac{\hbar^2}{2}[j(j+1) - m^2]. \end{aligned} \tag{6.1-31}$$

The uncertainty ΔJ_x is as small as it can be, $\hbar\sqrt{j/2}$, if $m = \pm j$, but the uncertainty cannot be zero unless $j = 0$. If it were, the uncertainty principle would be violated, because then J_x and J_z, two quantities that do not commute, would be known simultaneously.

PROBLEM 6.4

Explain the statement, "The uncertainty principle prevents $\vec{J}$ from pointing along the z axis." Suppose that the system is a macroscopic thing like a small gyroscope. Make a numerical estimate for the minimum angular uncertainty of $\vec{J}$.

PROBLEM 6.5

Use

$$J_y = \frac{J_+ - J_-}{2i} \tag{6.1-32}$$

to show that, for any $|jm\rangle$ that satisfies (23) and (24), $\langle J_y\rangle = 0$. Then obtain ΔJ_y.

One can understand (31) and the results of Problem 6.5 in terms of Figure 6.1-1. The tip of the angular momentum wanders on a circle of radius

$$J_x^2 + J_y^2 = J^2 - J_z^2 = \hbar^2[j(j+1) - m^2].$$

There is no preference for either the x or y direction, $\langle J_x^2\rangle = \langle J_y^2\rangle$, and the conclusions follow.

EXAMPLE

The state $|\frac{1}{2} \quad +\frac{1}{2}\rangle$ has $j = \frac{1}{2}$, magnitude of angular momentum $\hbar\sqrt{j(j+1)} = \hbar\sqrt{3}/2$, and z component of angular momentum $m\hbar = +\hbar/2$. The state $|\frac{1}{2} \quad -\frac{1}{2}\rangle$ has the same j, but has z component $-\hbar/2$.

a. Obtain a normalized angular momentum eigenstate with $j = \frac{1}{2}$ and x component of angular momentum equal to $-\hbar/2$.

b. With the system in the state found in **a**, J_z is measured. What are the possible results and their probabilities?

SOLUTION

a. The expansion postulate says that the desired state χ can be written as a superposition of states $|jm\rangle$. Any $j = \frac{1}{2}$ state is orthogonal to all $j \neq \frac{1}{2}$ states. Therefore χ can be written as a superposition of states $|\frac{1}{2}m\rangle$. There are only two, so it must be possible to write

$$\chi = c_+ |\tfrac{1}{2} \quad +\tfrac{1}{2}\rangle + c_- |\tfrac{1}{2} \quad -\tfrac{1}{2}\rangle.$$

Now impose the requirement that

$$J_x \chi = -\frac{\hbar}{2} \chi.$$

A useful trick is to write J_x in terms of the raising and lowering operators, as in (29).

$$J_x = \frac{J_+ + J_-}{2}.$$

With the normalization constants (26) and (28),

$$J_+ |\tfrac{1}{2} \quad -\tfrac{1}{2}\rangle = \hbar |\tfrac{1}{2} \quad +\tfrac{1}{2}\rangle, J_- |\tfrac{1}{2} \quad +\tfrac{1}{2}\rangle = \hbar |\tfrac{1}{2} \quad -\tfrac{1}{2}\rangle.$$

For $j = \frac{1}{2}$, $+\frac{1}{2}$ is the maximum possible m and $-\frac{1}{2}$ is the minimum possible m.

$$\therefore \quad J_+ |\tfrac{1}{2} \quad +\tfrac{1}{2}\rangle = 0, \qquad J_- |\tfrac{1}{2} \quad -\tfrac{1}{2}\rangle = 0.$$

$$\therefore \quad J_x \chi = \frac{J_+ + J_-}{2} (c_+ |\tfrac{1}{2} \quad +\tfrac{1}{2}\rangle + c_- |\tfrac{1}{2} \quad -\tfrac{1}{2}\rangle)$$

$$= \tfrac{1}{2}(0 + \hbar c_- |\tfrac{1}{2} \quad +\tfrac{1}{2}\rangle + \hbar c_+ |\tfrac{1}{2} \quad -\tfrac{1}{2}\rangle + 0).$$

If this result is to equal

$$-\frac{\hbar}{2} \chi = -\frac{\hbar}{2} (c_+ |\tfrac{1}{2} \quad +\tfrac{1}{2}\rangle + c_- |\tfrac{1}{2} \quad -\tfrac{1}{2}\rangle),$$

then $c_- = -c_+$. Since $|c_+|^2 + |c_-|^2 = 1$,

$$\chi = \frac{1}{\sqrt{2}} (|\tfrac{1}{2} \quad +\tfrac{1}{2}\rangle - |\tfrac{1}{2} \quad -\tfrac{1}{2}\rangle),$$

within an arbitrary phase factor. For this state a measurement of J_x definitely yields $-\hbar/2$, but the results of measurements of J_y and J_z cannot be predicted with certainty.

b. A measurement of J_z changes the state from χ to an eigenstate of J_z, that is, into either $|\frac{1}{2} \quad +\frac{1}{2}\rangle$ or into $|\frac{1}{2} \quad -\frac{1}{2}\rangle$. The result of the measurement must be one of the eigenvalues of J_z, that is, either $+\hbar/2$ or $-\hbar/2$. The probability of finding $+\hbar/2$ is $|c_+|^2 = \frac{1}{2}$, and the probability of finding $-\hbar/2$ is $|c_-|^2 = \frac{1}{2}$.

6.2 *Orbital Angular Momentum Eigenfunctions*

Section 6.1 is somewhat abstract. All results were obtained through manipulations of the commutation relations (6.1-5), and no explicit forms appeared for the $|jm\rangle$. In this section the Schrödinger equation is used to find functions of position that represent orbital angular momentum eigenstates.

Suppose that the potential energy, when written in terms of spherical polar coordinates, depends only on r, not on θ or φ. The problem might be that of an electron in a Coulomb field with $V = -Ze^2/4\pi\varepsilon_0 r$, or that of an atom in an isotropic harmonic oscillator potential $Kr^2/2$. Such Hamiltonians are invariant under all rotations, and commute with all components of $\vec{L}$ and with L^2. Section 3.7 shows that there must then be wave functions which are simultaneous eigenstates of energy, L^2 and one component of $\vec{L}$, say L_z.

To study problems of this kind, it is useful to write everything in terms of spherical polar coordinates (Figure 6.2-1). The transformation equations from and to Cartesians are

$$\begin{aligned} x &= r \sin\theta \cos\varphi, \\ y &= r \sin\theta \sin\varphi, \\ z &= r \cos\theta, \end{aligned} \tag{6.2-1}$$

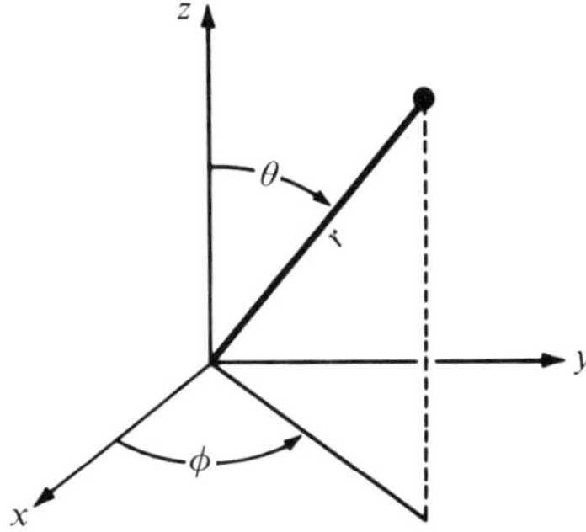

Fig. 6.2-1. Definition of the spherical polar coordinates.

and

$$r = \sqrt{x^2 + y^2 + z^2}, \qquad 0 \le r,$$
$$\theta = \cos^{-1}\left(\frac{z}{\sqrt{x^2 + y^2 + z^2}}\right), \qquad 0 \le \theta \le \pi, \tag{6.2-2}$$
$$\varphi = \tan^{-1}\left(\frac{y}{x}\right), \qquad 0 \le \varphi \le 2\pi.$$

The volume element $dx\, dy\, dz$ becomes $r^2\, dr \sin\theta\, d\theta\, d\varphi$. Derivatives are transformed by use of the chain rule,

$$\frac{\partial}{\partial x} = \frac{\partial r}{\partial x}\frac{\partial}{\partial r} + \frac{\partial \theta}{\partial x}\frac{\partial}{\partial \theta} + \frac{\partial \varphi}{\partial x}\frac{\partial}{\partial \varphi}$$
$$= \sin\theta \cos\varphi \frac{\partial}{\partial r} + \frac{\cos\theta \cos\varphi}{r}\frac{\partial}{\partial \theta} - \frac{\sin\varphi}{r \sin\theta}\frac{\partial}{\partial \varphi}.$$

The Laplacian becomes

$$\nabla^2\psi(r, \theta, \varphi) = \frac{1}{r^2}\frac{\partial}{\partial r}\left(r^2 \frac{\partial \psi}{\partial r}\right) + \frac{1}{r^2 \sin\theta}\frac{\partial}{\partial \theta}\left(\sin\theta \frac{\partial \psi}{\partial \theta}\right) + \frac{1}{r^2 \sin^2\theta}\frac{\partial^2 \psi}{\partial \varphi^2}. \tag{6.2-3}$$

Sometimes it is useful to change the form of the first term by the replacement

$$\frac{1}{r^2}\frac{\partial}{\partial r}\left(r^2 \frac{\partial \psi}{\partial r}\right) = \frac{1}{r}\frac{\partial^2}{\partial r^2}(r\psi). \tag{6.2-4}$$

Review and check these geometric and calculus results; they are essential for what follows.

The stationary-state Schrödinger equation, with $V = V(r)$ only, is

$$-\frac{\hbar^2}{2\mu}\nabla^2\psi + V(r)\psi = E\psi. \tag{6.2-5}$$

The mass is called μ here because many problems of this kind deal with two particles that are moving around their common center of mass. For them μ is the reduced mass given by (4.1-15), and the coordinates r, θ, φ are the relative coordinates that locate particle 2 with respect to particle 1.

Write ∇^2 as in (3) and look for a separable solution of the form

$$\psi(r, \theta, \varphi) = R(r)Y(\theta, \varphi). \tag{6.2-6}$$

Substitute and divide by $-\hbar^2 RY/2\mu r^2$.

$$\frac{1}{R}\frac{d}{dr}\left(r^2\frac{dR}{dr}\right) + \frac{2\mu r^2}{\hbar^2}[E - V(r)]$$

$$= -\frac{1}{Y}\left[\frac{1}{\sin\theta}\frac{\partial}{\partial\theta}\left(\sin\theta\frac{\partial Y}{\partial\theta}\right) + \frac{1}{\sin^2\theta}\frac{\partial^2 Y}{\partial\varphi^2}\right] \equiv \lambda, \qquad (6.2\text{-}7)$$

a constant, because one side depends on only r, the other side depends on only θ and φ, and the coordinates can be varied independently. The separation has worked and there are now two equations rather than one. The ordinary differential equation for $R(r)$ cannot be solved until a specific choice is made for $V(r)$. The equation for $Y(\theta, \varphi)$, however, is applicable to any problem that involves a spherically symmetric $V(r)$, and it is the equation that determines the angular momentum eigenfunctions.

With the transformation equations (1) and (2) and chain-rule replacements of the derivatives, the three Cartesian components of $\vec{L}$ can be expressed in terms of r, θ, φ and derivatives with respect to these coordinates.

$$L_x = \frac{\hbar}{i}\left(y\frac{\partial}{\partial z} - z\frac{\partial}{\partial y}\right) = \frac{-\hbar}{i}\left(\sin\varphi\frac{\partial}{\partial\theta} + \cot\theta\cos\varphi\frac{\partial}{\partial\varphi}\right), \qquad (6.2\text{-}8)$$

$$L_y = \frac{\hbar}{i}\left(z\frac{\partial}{\partial x} - x\frac{\partial}{\partial z}\right) = \frac{\hbar}{i}\left(\cos\varphi\frac{\partial}{\partial\theta} - \cot\theta\sin\varphi\frac{\partial}{\partial\varphi}\right), \qquad (6.2\text{-}9)$$

$$L_z = \frac{\hbar}{i}\left(x\frac{\partial}{\partial y} - y\frac{\partial}{\partial x}\right) = \frac{\hbar}{i}\frac{\partial}{\partial\varphi}. \qquad (6.2\text{-}10)$$

PROBLEM 6.6

Verify (10).

Although these expressions are written in terms of spherical polar coordinates, they are still the x, y, and z components of $\vec{L}$. Squaring and adding these three operators gives L^2,

$$L^2 = L_x^2 + L_y^2 + L_z^2 = -\hbar^2\left[\frac{1}{\sin\theta}\frac{\partial}{\partial\theta}\left(\sin\theta\frac{\partial}{\partial\theta}\right) + \frac{1}{\sin^2\theta}\frac{\partial^2}{\partial\varphi^2}\right]. \qquad (6.2\text{-}11)$$

Therefore, *mirabile dictu*, the $Y(\theta, \varphi)$ equation contained in (7) is actually

$$L^2 Y(\theta, \varphi) = \hbar^2\lambda Y(\theta, \varphi), \qquad (6.2\text{-}12)$$

and the energy operator in the Schrödinger equation is

$$\left(\frac{p^2}{2\mu} + V\right)\psi = \left(\frac{-\hbar^2}{2\mu}\nabla^2 + V\right)\psi$$
$$= \frac{-\hbar^2}{2\mu}\left[\frac{1}{r^2}\frac{\partial}{\partial r}\left(r^2\frac{\partial\psi}{\partial r}\right)\right] + \left(\frac{L^2}{2\mu r^2} + V\right)\psi. \tag{6.2-13}$$

The $L^2/2\mu r^2$ term acts like an additional potential that pushes the particle away from the origin; in form and behavior it is just like the "centrifugal potential" in classical mechanics that arises in the equation of motion for a particle moving moving around a force center.[3]

We already know from (6.1-21) that λ, the eigenvalue of L^2, is $l(l + 1)$, and it was with forethought that the name λ was given to the separation constant in (7). The result $\lambda = l(l + 1)$ will emerge again as a by-product of treating (12) as a differential equation problem.

It is easy to separate the partial differential equation (12) into two ordinary differential equations. Substitute

$$Y(\theta, \varphi) = \Theta(\theta)\Phi(\varphi), \tag{6.2-14}$$

divide by $-\Theta\Phi/\sin^2\theta$, and rearrange.

$$\frac{\sin\theta}{\Theta}\frac{d}{d\theta}\left(\sin\theta\frac{d\Theta}{d\theta}\right) + \lambda\sin^2\theta = -\frac{1}{\Phi}\frac{d^2\Phi}{d\varphi^2} \equiv m^2, \tag{6.2-15}$$

a constant, because the left side depends only on θ, the right side depends only on φ, and these coordinates can be varied independently. The two separated equations are

$$\frac{1}{\sin\theta}\frac{d}{d\theta}\left(\sin\theta\frac{d\Theta}{d\theta}\right) + \left(\lambda - \frac{m^2}{\sin^2\theta}\right)\Theta = 0, \tag{6.2-16}$$

$$\frac{d^2\Phi}{d\varphi^2} + m^2\Phi = 0. \tag{6.2-17}$$

The useful solutions are those that are eigenfunctions of L^2 and of L_z. The general solution of the φ equation (17) is

$$\begin{aligned} \Phi &= Ae^{im\varphi} + Be^{-im\varphi}, \qquad m \neq 0 \\ &= C + D\varphi, \qquad m = 0, \end{aligned} \tag{6.2-18}$$

but there is also the requirement that Φ must satisfy the L_z eigenvalue equation (6.1-9). With (10) for the L_z operator,

$$\frac{\hbar}{i}\frac{d}{d\varphi}\Phi = \hbar m\Phi,$$

[3] K. R. Symon, *Mechanics*, 3d ed. (Reading, Mass: Addison-Wesley Publishing Co., Inc., 1971), Sec. 3-13.

where $\hbar m$ is the necessarily real eigenvalue. The eigenfunctions are

$$\Phi(\varphi) = \frac{e^{im\varphi}}{\sqrt{2\pi}}, \tag{6.2-19}$$

with m positive, negative, or zero; that is, in (18), for $m \neq 0$, either $e^{im\varphi}$ or $e^{-im\varphi}$ is retained, and for $m = 0$, the constant D is zero. That D must be zero is also clear on other grounds. The probability density is proportional to $|\Phi|^2$, which changes from $|C|^2$ to $|C + 2\pi D|^2$ under a rotation from $\varphi = 0$ to $\varphi = 2\pi$. A rotation through 2π should not change any measurable, and D must therefore be zero. The constant $1/\sqrt{2\pi}$ that multiplies $e^{im\varphi}$ is arbitrary, but is a convenient choice because it normalizes Φ in the sense that

$$\int_0^{2\pi} \Phi^* \Phi \, d\varphi = 1.$$

The next step is to look at the θ equation. With $w \equiv \cos\theta$, $P(w) \equiv \Theta(\theta)$, (16) becomes

$$\frac{d}{dw}\left[(1 - w^2)\frac{dP}{dw}\right] + \left(\lambda - \frac{m^2}{1 - w^2}\right)P = 0. \tag{6.2-20}$$

First, study (20) for the lowest possible value of m^2. Section 6.1 shows that the angular momentum quantum numbers l and m must be either integral or half-integral, and that $m = -l, -l + 1, \ldots, +l - 1, +l$; see (6.1-22). If l is half-integral, the lowest value of m^2 is $\frac{1}{4}$, while if l is integral, the lowest value of m^2 is zero. If $m^2 = \frac{1}{4}$, or any other number that is not the square of an integer, then the solutions to (20) are special functions known as hypergeometric functions. These functions are useful for a variety of applications,[4] but they lead to singular and unphysical behavior at $w = \pm 1$, and cannot describe the desired angular momentum eigenfunctions.[5] It follows that, as anticipated in Section 6.1, orbital angular momentum can be described only by integral, not half-integral, quantum numbers.

Since m is an integer, the L_z eigenfunctions (18) are single valued in the sense that $\Phi(\varphi + 2\pi) = \Phi(\varphi)$,

$$e^{im(\varphi + 2\pi)} = e^{2\pi i m} e^{im\varphi} = e^{im\varphi}.$$

This single-valuedness could not have been required as a condition at the beginning of the argument, even though it would have produced the correct conclusion that m must be an integer for orbital angular momentum. The

[4] J. D. Jackson, *Classical Electrodynamics*, 2d ed. (New York: John Wiley & Sons, Inc., 1975), Sec. 3.4, and references there.

[5] M. L. Whipman, *Am. J. Phys.* **34**, 656 (1966). One can also show by general operator methods that half-integral values for orbital angular momentum lead to an inconsistency; see H. Buchdahl, *Am. J. Phys.* **30**, 829 (1962) and E. Merzbacher, *Am. J. Phys.* **31**, 549 (1963).

measurable quantities are expectation values $\int \psi^*\Omega\psi$, and with $\psi \sim e^{im\varphi}$, they are invariant under rotations through 2π for any m. In fact, for half-integral spin, ψ changes sign under rotation through 2π. This matter is discussed further in Chapter 8.

For integral m, (20) is called the *associated Legendre equation*, and the $m = 0$ form,

$$(1 - w^2)\frac{d^2P}{dw^2} - 2w\frac{dP}{dw} + \lambda P = 0, \tag{6.2-21}$$

is called the *Legendre equation.*

The Legendre equation can be solved by an approach like that used to find the Hermite polynomials in Section 2.7. Look for a series solution

$$P(w) = \sum_{\nu=0}^{\infty} a_\nu w^{s+\nu}$$

with $a_0 \neq 0$, so that s is the lowest power of w that occurs. Substitute this series into (21).

$$\sum_{\nu=0}^{\infty} \{(\nu + s)(\nu + s - 1)a_\nu w^{\nu+s-2} - [(\nu + s)(\nu + s + 1) - \lambda]a_\nu w^{\nu+s}\} = 0. \tag{6.2-22}$$

The coefficient of each power of w must be zero. Since $a_0 \neq 0$, the requirement that the coefficient of w^{s-2} be zero provides the *indicial equation*

$$s(s - 1) = 0, \qquad s = 0 \text{ or } 1. \tag{6.2-23}$$

The requirement that the coefficient of w^{s-1} be zero demands

$$(s + 1)sa_1 = 0. \tag{6.2-24}$$

For all higher powers, setting the coefficient of $w^{s+\nu}$ equal to zero yields the *recursion relation*

$$a_{\nu+2} = \frac{(\nu + s)(\nu + s + 1) - \lambda}{(\nu + s + 2)(\nu + s + 1)}a_\nu. \tag{6.2-25}$$

As $\nu \to \infty$, $(a_{\nu+2}/a_\nu) \to 1$. Therefore the series converges for $|w| < 1$ and diverges for $|w| = 1$. Since $w = \cos\theta$, $|w| = 1$ occurs at $\theta = 0$ and π, and solutions that diverge there are not acceptable. The series must be cut off so that it becomes a polynomial, with a finite highest power of w. Call this highest power l, and call the corresponding polynomial $P_l(w)$. If the last nonzero coefficient is $a_{\nu\,\max}$, then

$$l = \nu_{\max} + s, \tag{6.2-26}$$

and the series is cut off there by the choice

$$\lambda = l(l + 1). \tag{6.2-27}$$

The eigenvalue has in this approach been determined by the requirement that the wave function must be finite at $\theta = 0$ and π. With a λ chosen to cut off the series that is, through the recursion relation (25), proportional to a_0, it is not possible to cut off the series that is proportional to a_1. One must therefore choose $a_1 = 0$ so that the $a_\nu \neq 0$ only for even ν. Equation (24) is then no restriction on s.

Except for an arbitrary multiplying constant, the $P_l(w)$ are now determined, and can be calculated explicitly as follows. For even l, choose $s = 0$ and begin the series with a constant a_0. For odd l, choose $s = 1$ and begin the series with $a_0 w$. Use the recursion relation (25), with $s = 0$ or 1 and $\lambda = l(l + 1)$, to calculate successively higher coefficients until the process terminates by giving zero.

It is a convenient convention to choose a_0 so that all $P_l(w)$ are unity at $w = 1$, that is, at $\theta = 0$. With this convention the first six $P_l(w)$ are

$$\begin{aligned} P_0 &= 1 \\ P_1 &= w \\ P_2 &= \tfrac{1}{2}(3w^2 - 1) \\ P_3 &= \tfrac{1}{2}(5w^3 - 3w) \\ P_4 &= \tfrac{1}{8}(35w^4 - 30w^2 + 3) \\ P_5 &= \tfrac{1}{8}(63w^5 - 70w^3 + 15w) \end{aligned} \tag{6.2-28}$$

PROBLEM 6.7

a. By using the recursion relations as outlined above, verify (28) for P_0, P_1, and P_2.

b. The $P_l(w)$ are *orthogonal functions* that satisfy

$$\int_{-1}^{+1} P_{l'}(w)P_l(w)dw = \frac{2\delta_{l'l}}{2l + 1}. \tag{6.2-29}$$

Verify explicitly that (29) is right if l and l' are 0, 1, or 2. Can you argue from Section 6.1 that the P_l must be orthogonal functions?

c. It can be shown directly from the Legendre equation that the P_l can also be calculated from a *Rodrigues formula*

$$P_l(w) = \frac{1}{2^l l!}\frac{d^l}{dw^l}(w^2 - 1)^l. \tag{6.2-30}$$

Verify that this expression gives P_0, P_1, and P_2 correctly. These and other properties of the $P_l(w)$ are derived in many mathematics and mathematical physics texts.[6]

---

[6] H. Margenau and G. M. Murphy, *The Mathematics of Physics and Chemistry*, 2d ed. (Princeton: D. Van Nostrand Co, Inc., 1956), Secs. 2.11, 3.3–3.7; J. Mathews and R. L. Walker, *Mathematical Methods of Physics*, 2d ed. (New York: W. A. Benjamin, Inc., 1970), Sec. 7.1.

The solution for the θ dependence of the wave function is now complete for the special case $m = 0$, and is given by the $P_l(\cos\theta)$. Problem 6.8 shows how to get the $m \neq 0$ solutions.

PROBLEM 6.8

Suppose that $P_l(w)$ satisfies the Legendre equation (21). Show that

$$P_{lm}(w) = (1 - w^2)^{|m|/2} \frac{d^{|m|}}{dw^{|m|}} P_l(w) \tag{6.2-31}$$

satisfies the associated Legendre equation (20). Then use this result to calculate P_{11}, P_{21}, and P_{22}. Notice that $P_{lm} = P_{l-m}$.

The function $Y(\theta, \varphi)$ is the product of the results (19) and (31), and will be labeled by the indices l, m.

$$Y_{lm}(\theta, \varphi) \sim \Phi(\varphi)\Theta(\theta) \sim e^{im\varphi} P_{lm}(\cos\theta). \tag{6.2-32}$$

It is useful to normalize the Y_{lm} in the sense that

$$\int_{\theta=0}^{\pi} \int_{\varphi=0}^{2\pi} | Y_{lm}(\theta, \varphi)|^2 \sin\theta \, d\theta \, d\varphi = 1. \tag{6.2-33}$$

A few of the Y_{lm} are listed below.

$$\begin{aligned}
Y_{00} &= \frac{1}{\sqrt{4\pi}} \\
Y_{10} &= \sqrt{\frac{3}{4\pi}} \cos\theta \\
Y_{1\pm1} &= \mp\sqrt{\frac{3}{8\pi}} \sin\theta e^{\pm i\varphi} \\
Y_{20} &= \sqrt{\frac{5}{16\pi}} (3\cos^2\theta - 1) \\
Y_{2\pm1} &= \mp\sqrt{\frac{15}{8\pi}} \cos\theta \sin\theta \, e^{\pm i\varphi} \\
Y_{2\pm2} &= \sqrt{\frac{15}{32\pi}} \sin^2\theta \, e^{\pm 2i\varphi}
\end{aligned} \tag{6.2-34}$$

PROBLEM 6.9

Use the results of Problem 6.8 and (32) and (33) to verify (34) for Y_{00}, Y_{10}, and $Y_{1\,-1}$.

It can be shown that, for all l and m,

$$Y_{lm}(\theta, \varphi) = (-1)^{(m+|m|)/2} \sqrt{\frac{2l+1}{4\pi} \frac{(l-m)!}{(l+m)!}}\, P_{lm}(\cos\theta) e^{im\varphi}. \tag{6.2-35}$$

These functions are the normalized *spherical harmonics*. The general normalization constant and various properties of the Y_{lm} are discussed in many texts.[7] Notice the $(-1)^{(m+|m|)/2}$ that appears in the examples (34) and the general expression (35). This factor is, of course, not relevant to normalization. It is included so that the different Y_{lm} with the same l are consistent with the way the raising and lowering operators $L_\pm$ work according to (6.1-25) and (6.1-27).

The Y_{lm} are the eigenfunctions of L^2 and L_z.

$$L^2 Y_{lm}(\theta, \varphi) = \hbar^2 l(l+1) Y_{lm}(\theta, \varphi) \tag{6.2-36}$$

$$L_z Y_{lm}(\theta, \varphi) = \hbar m Y_{lm}(\theta, \varphi) \tag{6.2-37}$$

Since different Y_{lm} have different eigenvalues of L^2 or L_z or both, they are orthogonal. Since normalization has been imposed, they form an orthonormal set,

$$\int_0^{2\pi} \int_0^{\pi} Y_{lm}(\theta, \varphi)^* Y_{l'm'}(\theta, \varphi) \sin\theta \, d\theta \, d\varphi = \delta_{ll'} \delta_{mm'}. \tag{6.2-38}$$

One way of getting the Y_{lm} is now complete. The spherical harmonics are so important that in a more leisurely world one would derive them a dozen different ways. We will compromise and only sketch a second approach[8] that uses the raising and lowering operators $L_\pm$ developed in Section 6.1.

From (8) and (9),

$$L_\pm = L_x \pm iL_y = \pm\hbar e^{\pm i\varphi}\left(\frac{\partial}{\partial\theta} \pm i \cot\theta \frac{\partial}{\partial\varphi}\right). \tag{6.2-39}$$

[7] For example, D. Park, *Introduction to the Quantum Theory*, 2d ed. (New York: McGraw-Hill Book Company, 1974), Appendix 4, gives, among other useful information, a comparison between different notations for the P_{lm}, which are often written P_l^m, and the Y_{lm}, which are sometimes written Y_l^m. The spherical harmonics also occur in electromagnetism, acoustics, fluid dynamics, and many other fields that are concerned with equations for the form $\nabla^2\psi + F(r)\psi = 0$.

[8] D. S. Saxon, *Elementary Quantum Mechanics* (San Francisco: Holden-Day, Inc., 1968), Chap. X; S. Gasiorowicz, *Quantum Physics* (New York: John Wiley & Sons. Inc., 1974), Chap. 10.

From (14) and (19) it follows that

$$Y_{ll}(\theta, \varphi) = \frac{e^{il\varphi}}{\sqrt{2\pi}} \Theta_{ll}(\theta). \tag{6.2-40}$$

Since l is the highest possible value of m, operating on Y_{ll} with the raising operator must give zero.

$$L_+ Y_{ll} = \hbar e^{i\varphi}\left(\frac{\partial}{\partial\theta} + i \cot\theta \frac{\partial}{\partial\varphi}\right)\frac{e^{il\varphi}}{\sqrt{2\pi}} \Theta_{ll}(\theta) = 0,$$

$$\left(\frac{d}{d\theta} - l \cot\theta\right)\Theta_{ll}(\theta) = 0,$$

$$\Theta_{ll} = A \sin^l \theta.$$

The constant A is determined by the requirement

$$1 = \int_0^\pi |A \sin^l \theta|^2 \sin\theta \, d\theta = \frac{2^{2l+1}(l!)^2}{(2l+1)!} |A|^2.$$

$$\therefore \quad Y_{ll}(\theta, \varphi) = \frac{1}{2^l l!} \sqrt{\frac{(2l+1)!}{4\pi}} e^{il\varphi} \sin^l \theta.$$

Now all of the normalized Y_{lm} for a given l can be generated by using (6.1-27), with the lowering operator L_- expressed as in (39).

$$Y_{ll-1} = \frac{1}{N_{ll}^-} L_- Y_{ll}, \; Y_{ll-2} = \frac{1}{N_{ll-1}^-} L_- Y_{ll-1}, \ldots .$$

As observed above, everything in this section applies to any problem in which $V = V(r)$ only. The radial equation contained in (7) remains to be solved, and the procedure and results depend on the actual $V(r)$ that one selects. However, the angular dependence is always described by the $Y_{lm}(\theta, \varphi)$. A solution

$$\psi(r, \theta, \varphi) = R(r) Y_{lm}(\theta, \varphi) \tag{6.2-41}$$

is automatically an eigenfunction of L^2 with eigenvalue $\hbar^2 l(l+1)$ and of L_z with eigenvalue $\hbar m$. Since the Y_{lm} are normalized on a sphere of unit radius, ψ itself is normalized by the requirement

$$\int_0^\infty |R(r)|^2 r^2 \, dr = 1. \tag{6.2-42}$$

The parity of $\psi(r, \theta, \varphi)$ is determined entirely by the orbital angular momentum quantum number l. A reflection

$$x \to -x, \qquad y \to -y, \qquad z \to -z$$

gives

$$\theta \to \pi - \theta, \qquad \varphi \to \varphi + \pi,$$

and leaves r unchanged, so that $R(r)$ is unaffected. With the parity operator $\mathscr{P}$ introduced in Section 3.6,

$$\mathscr{P}\Psi(\vec{r}, t) = \Psi(-\vec{r}, t), \tag{6.2-43}$$

$$\mathscr{P}Y_{lm}(\theta, \varphi) = Y_{lm}(\pi - \theta, \varphi + \pi) = (-1)^l Y_{lm}(\theta, \varphi). \tag{6.2-44}$$

PROBLEM 6.10

a. Check explicitly that (44) is right for Y_{00}, Y_{1-1} and Y_{2+1}.

b. Use (31), (32), and the structure of the $P_l(\cos\theta)$ to prove (44).

PROBLEM 6.11

a. An angular momentum eigenstate with $L^2 = 2\hbar^2$ and $L_z = 0$ is described by $Y_{10}(\theta, \varphi)$. Obtain a normalized angular momentum eigenstate in which L^2 is the same, that is, in which l is still 1, but which has $L_x = 0$. (*Suggestion*: It must be possible to express the desired eigenfunction of L_x as a linear combination of Y_{1-1}, Y_{10}, and Y_{1+1}, the three eigenfunctions of L_z; why? Write such a linear combination and require that L_x operating on it gives zero.)

b. With the system in the state found in (**a**), L_z is measured. What are the possible results and their probabilities?

SLAC

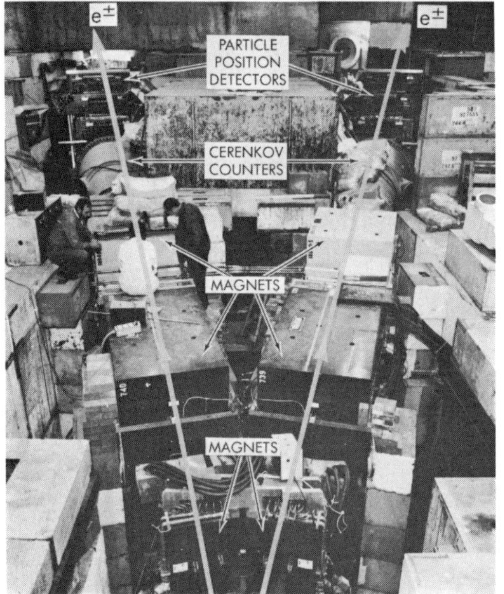

BNL

The discovery of the J/ψ, a particle understood as a bound state of a quark and its antiparticle; see Section 7.4. Top: At the Stanford Linear Accelerator Center, a counter array surrounds a region where electrons and positrons collide. B. Richter and his collaborators found, for combined e^- and e^+ energy near 3.1 GeV, that there was a large yield of decay particles. Bottom: At Brookhaven National Laboratory, high energy protons struck a target below the picture, and the simultaneous appearance of electrons and positrons was analyzed by the magnets and counters shown. C. C. Ting and his collaborators found that there is a high probability for the e^- and e^+ to have 3.1 GeV total energy in their own center of mass system. In both the Stanford and Brookhaven experiments, the J/ψ is formed in the collision and then decays into the products that are detected.

7

Central Force Bound States

Chapter 6 shows that if $V = V(r)$ only, then the Schrödinger equation can be separated in spherical polar coordinates. The θ, φ dependence is given by the orbital angular momentum eigenfunctions $Y_{lm}(\theta, \varphi)$ for any $V(r)$. The r dependence is determined by the left half of (6.2-7) and does depend on the form of $V(r)$. This chapter examines some mathematically tractable forms of $V(r)$ that describe physically interesting systems.

7.1 *Generalities and the Sphere of Constant Potential Energy*

With $\lambda = l(l + 1)$, and with the radial derivative expressed as in (6.2-4), the left half of (6.2-7) shows that $R(r)$ obeys

$$-\frac{\hbar^2}{2\mu}\frac{d^2(rR)}{dr^2} + \left[V(r) + \frac{\hbar^2 l(l+1)}{2\mu r^2}\right]rR = ErR, \tag{7.1-1}$$

or, with

$$u(r) = rR(r), \tag{7.1-2}$$

$$-\frac{\hbar^2}{2\mu}\frac{d^2u}{dr^2} + \left[V(r) + \frac{\hbar^2 l(l+1)}{2\mu r^2}\right]u = Eu. \tag{7.1-3}$$

Equation (3) is like the one-dimensional Schrödinger equation except for two things. First, there is an effective potential equal to the actual potential plus the "centrifugal potential" $L^2/2\mu r^2$; this term was discussed in Section 6.2 after it appeared in (6.2-13). Second, for $R(r)$ to be finite at the origin, $u = rR$ must be zero at the origin, and so $u(r)$ obeys a boundary condition which is different from that applicable to the one-dimensional $\psi(x)$.

To study the behavior of u near the origin, substitute $u \sim r^s$ into (3).

$$-\frac{\hbar^2}{2\mu} s(s-1)r^{s-2} + V(r)r^s + \frac{\hbar^2 l(l+1)}{2\mu} r^{s-2} = Er^s. \tag{7.1-4}$$

If, for small r, $V(r)$ behaves like r^q with $q > -2$, then the lowest power of r in (4) is $s - 2$. The vanishing of its coefficient requires that $s(s-1) = l(l+1)$, so that $s = -l$ or $s = l + 1$. The $-l$ solution must be rejected because $l \geq 0$, and $u \sim r^{-l}$ violates the $u(0) = 0$ boundary condition.

$$\therefore \quad s = l + 1, \qquad u \sim r^{l+1}, \qquad R \sim r^l \text{ for small } r. \tag{7.1-5}$$

This result is not applicable to all potentials; both nuclear and molecular calculations use models in which strongly repulsive and highly singular forms of $V(r)$ keep particles away from each other. However, the condition $V \sim r^q$, $q > -2$ is satisfied by many interesting forms of V, including the Coulomb potential ($q = -1$), a sphere of constant potential ($q = 0$), the harmonic oscillator ($q = +2$) and linear potentials ($q = +1$). For such forms (5) is the indicial equation that determines the lowest power of r in an expansion of the solution. It is also a physically sensible result that describes how particles with nonzero angular momentum are unlikely to be near the origin.

The possible energies are determined by the eigenvalue equation (3) that depends on $V(r)$ and on the square of the orbital angular momentum $\hbar^2 l(l+1)$, but not on $\hbar m$, the z component of angular momentum. For a given l different values of m describe different orientations of an angular momentum of given magnitude. Since the Hamiltonian is unchanged by rotations, the energy cannot be affected by orientation. The range of m, $-l$ to $+l$, contains $2l + 1$ possibilities, so that the rotational symmetry produces a $2l + 1$ degeneracy of states labeled with definite energy and l.

A simplified picture of a neutron inside a nucleus provides a good application. The neutron is relatively free to move about in the interior of the nucleus, where it is pulled roughly equally in all directions, but it is pulled back strongly if it tries to pass beyond the edge at $r =$ a few $\times\ 10^{-13}$ cm. Such a force is described by the potential energy curve shown in Figure 7.1-1. The flat part at $-V_0$ describes the few $\times\ 10^{-13}$ cm where the neutron thinks it is safely inside. The region near $r = b$, where V changes from $-V_0$ to zero, is determined by the decreasing density of nucleons at the edge of the nucleus, and by the range of the nucleon-nucleon force.

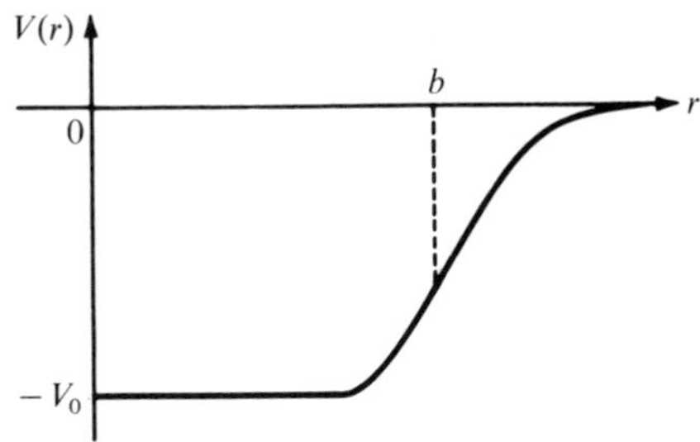

Fig. 7.1-1. A model of the potential energy seen by a neutron in a nucleus.

Obviously, such a smoothed-out description of the force is only an approximation, but it works well for many purposes. It is similar to the free electron model of solids, Section 5.1. Why do both of these stories work rather well for particles that live in a complicated environment and are subject to frequent collisions? Both nucleons and electrons obey the Pauli exclusion principle. If they try to scatter with only a modest change in energy, they will probably find that they are trying to scatter into a state that is already occupied. The exclusion principle therefore suppresses much of the scattering that would otherwise be possible, and lengthens the mean free path of the particles. A consequence is that a nucleon bound inside a nucleus can have a mean free path greater than the circumference of the nucleus, and be described with surprising success by a smooth potential. However, a nucleon would not have much chance of getting through the surface of the nucleus without collisions if it came in with a positive energy of several MeV.

PROBLEM 7.1

The success of models with smooth potentials suggests long mean free paths for nucleons in nuclei. However, a nucleon incident with $E = +$ a few MeV has negligible chance of even penetrating the surface without a strong interaction. Is there a discrepancy here?

To make the mathematics easier, pretend that $V(r)$ increases from $-V_0$ to 0 over a negligible distance, so that we have a sphere of constant potential energy:

$$\begin{aligned} V(r) &= -V_0, \qquad 0 \le r \le b, \\ &= 0, \qquad r > b. \end{aligned} \tag{7.1-6}$$

The solutions describe standing waves in a spherical region. If V_0 is very large, so that the waves go very nearly to zero at $r = b$, the problem is the same as that of resonant sound waves in a spherical cavity.[1]

For the special case $l = 0$, where the "centrifugal potential" term is absent, the problem is particularly simple. With the binding energy $B \equiv -E$, (3) becomes

$$-\frac{\hbar^2}{2\mu}\frac{d^2u}{dr^2} - V_0 u = -Bu, \qquad r \leq b,$$

$$-\frac{\hbar^2}{2\mu}\frac{d^2u}{dr^2} = -Bu, \qquad r > b.$$

The $r \leq b$ solution is

$$u = A \sin Kr + D \cos Kr, \qquad \hbar K = \sqrt{2\mu(V_0 - B)}.$$

Now D must be zero because $u(0) = 0$. It is this fact that makes the calculation different from that for the one-dimensional case with $V(x)$ given by (2.5-1). If, however, the wave function is forced to be zero at the origin as in Problem 2.11, then the two calculations look the same. The $r > b$ solution is

$$u = Ce^{-\alpha r} + Fe^{+\alpha r}, \qquad \hbar\alpha = \sqrt{2\mu B},$$

where F must be zero to make the solution decrease at large r. Continuity at b requires

$$A \sin Kb = Ce^{-\alpha b},$$

continuity of derivative requires

$$KA \cos Kb = -\alpha Ce^{-\alpha b},$$

and dividing the second by the first condition yields

$$K \cot Kb = -\alpha. \tag{7.1-7}$$

With $b = a/2$ this equation is the same as (2.5-9), which determines the *odd* solutions in a one-dimensional well; the condition that looks like the one for the one-dimensional even solutions is absent because of the requirement that $u(0) = 0$.

It follows that a three-dimensional well can be too shallow to have any bound states, while the one-dimensional well (2.5-1) must have at least one bound state. In three dimensions $u(0) = 0$, and the interior solution must be a sine, not a cosine, function. As shown in Figure 7.1-2, the argument must advance beyond

[1] J. W. Strutt, Baron Rayleigh, *The Theory of Sound*, 2d ed. (1894) reprinted (New York: Dover Publications, Inc., 1945), Secs. 331–332. Many calculations in Lord Rayleigh's nineteenth century book can be translated directly into quantum mechanics.

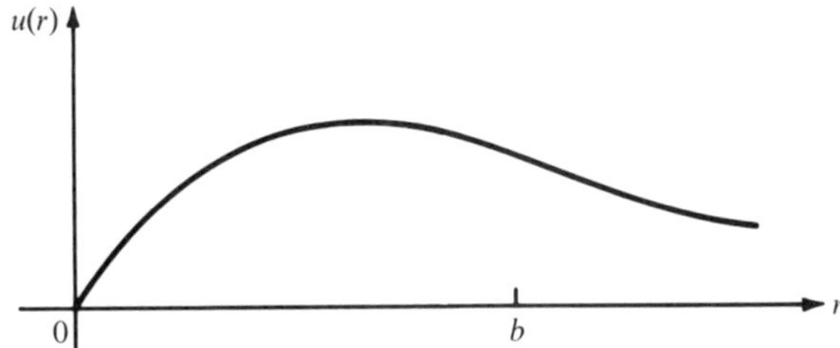

Fig. 7.1-2. The ground state in the potential given by (7.1-1) if $V_0 b^2$ is slightly more than the minimum required for binding.

$\pi/2$ in the $r < b$ region so that u has a negative slope at b; otherwise it is impossible to hook on a decreasing exponential. That is,

$$\frac{\pi^2}{4} < K^2 b^2 = \frac{2\mu}{\hbar^2}(V_0 - B)b^2 < \frac{2\mu}{\hbar^2} V_0 b^2. \tag{7.1-8}$$

PROBLEM 7.2

Find the condition, similar to (8), for the existence of *two* bound states with $l = 0$.

It is interesting to sketch a rough picture of the deuteron by choosing appropriate parameters in the potential energy given by (6). The neutron-proton force is not actually central, that is, it cannot be described entirely by a $V(r)$. There are noncentral forces which mix the orbital angular momentum states of the deuteron so that it is about 95% in an $l = 0$, or S, state, and about 5% in an $l = 2$, or D, state (see Section 9.3). The *total* angular momentum is always unity, and is the resultant of the two half-integral spins of the nucleons plus the orbital motion, like this:

$$0.95\begin{pmatrix}\uparrow \text{ spin of } n \\ \uparrow \text{ spin of } p \\ \cdot\, l = 0\end{pmatrix} + 0.05\begin{pmatrix}\Big\uparrow & \downarrow \text{ spin of } n \\ \Big| & \downarrow \text{ spin of } p \\ l = 2 & \end{pmatrix}$$

Some spin formalism is needed to deal with this situation properly. Here we shall sweep under the rug the few percent of D state and pretend that the deuteron is in the lowest S state with V described by (6).

The reduced mass of the neutron-proton system is, for neutron and proton masses M_n and M_p,

$$\left(\frac{1}{M_p} + \frac{1}{M_n}\right)^{-1} = \left(\frac{1}{M_p} + \frac{1}{1.00138 M_p}\right)^{-1} = \frac{469.5 \text{ MeV}}{c^2},$$

just about half the proton or neutron mass. The range of the force is, according to the argument in Section 1.3, about $\hbar/M_\pi c = 1.4 \times 10^{-13}$ cm, a reasonable choice for the radius b of the well. The binding energy B equals the energy of the gamma ray seen from the reaction $n + p \to d + \gamma$ if the incident particles have negligible kinetic energy; the result is $B = 2.225$ MeV. With these values (7) becomes

$$\tan\left(0.327\sqrt{\frac{V_0}{B} - 1}\right) = -\sqrt{\frac{V_0}{B} - 1}.$$

Numerical solution yields

$$0.327\sqrt{\frac{V_0}{B} - 1} = Kb = 1.76 \text{ radians} = 101^\circ, \qquad V_0 = 30.B = 67.\text{ MeV}.$$

Because of the limits of the model, the precise numbers are not particularly significant, but some qualitative features are important. The results reinforce the observations in Section 1.3 that the deuteron is just barely bound. The binding energy is much less than the depth of the well, and the interior wave function, $\sin Kr$, just manages to turn over before the edge of the well, as shown by $Kb = 90^\circ + 11^\circ$.

Once you are convinced that the deuteron is almost unbound, you can estimate the depth of the well without solving the transcendental equation (7). From the inequality (8),

$$V_0 > \frac{\hbar^2}{2\mu b^2}\frac{\pi^2}{4},$$

somewhat. With

$$b = \frac{\hbar}{M_\pi c} = \frac{\hbar c}{140 \text{ MeV}} \qquad \text{and} \qquad 2\mu = \frac{940 \text{ MeV}}{c^2},$$

$$V_0 > \frac{\dfrac{\hbar^2\pi^2}{4}}{\left(\dfrac{940 \text{ MeV}}{c^2}\right)\left(\dfrac{\hbar c}{140 \text{ MeV}}\right)^2} = 50 \text{ MeV},$$

somewhat.

PROBLEM 7.3

Put together the normalized deuteron wave function for the model of this section. Then calculate the probability that the neutron and proton are outside each other's force range.

Let us leave the deuteron for the present and look briefly at the general problem of solutions in the $V(r)$ given by (6) for orbital angular momentum quantum number $l > 0$. First, we need the $r < b$ solutions to (1), with $V = -V_0$, $E = -B < 0$. In terms of the variable $\rho = Kr$, the solutions $R(\rho)$, for $r < b$, are the *spherical Bessel functions* $j_l(\rho)$ and the *spherical Neumann functions* $n_l(\rho)$. Here are these functions for $l = 0, 1, 2$ and for general l:

$$
\begin{aligned}
j_0 &= \frac{\sin\rho}{\rho} & n_0 &= -\frac{\cos\rho}{\rho} \\
j_1 &= \frac{\sin\rho}{\rho^2} - \frac{\cos\rho}{\rho} & n_1 &= -\frac{\cos\rho}{\rho^2} - \frac{\sin\rho}{\rho} \\
j_2 &= \left(\frac{3}{\rho^3} - \frac{1}{\rho}\right)\sin\rho - \frac{3}{\rho^2}\cos\rho & n_2 &= -\left(\frac{3}{\rho^3} - \frac{1}{\rho}\right)\cos\rho - \frac{3}{\rho^2}\sin\rho \qquad (7.1\text{-}9) \\
j_l &= \rho^l\left(\frac{-1}{\rho}\frac{d}{d\rho}\right)^l \frac{\sin\rho}{\rho} & n_l &= -\rho^l\left(\frac{-1}{\rho}\frac{d}{d\rho}\right)^l \frac{\cos\rho}{\rho}
\end{aligned}
$$

The exterior solutions that drop off exponentially are the *spherical Hankel functions* $h_l^{(1)}(i\eta) \equiv j_l(i\eta) + in_l(i\eta)$, with $\eta \equiv \alpha r$:

$$
\begin{aligned}
h_0^{(1)}(i\eta) &= -\frac{e^{-\eta}}{\eta} \\
h_1^{(1)}(i\eta) &= i\left(\frac{1}{\eta} + \frac{1}{\eta^2}\right)e^{-\eta} \qquad (7.1\text{-}10) \\
h_2^{(1)}(i\eta) &= \left(\frac{1}{\eta} + \frac{3}{\eta^2} + \frac{3}{\eta^3}\right)e^{-\eta}.
\end{aligned}
$$

The solutions that increase exponentially at large r are $h_l^{(2)}(i\eta) \equiv j_l(i\eta) - in_l(i\eta)$. As with other special functions, it is sufficient for our needs to observe that, by substitution, one can check that these functions are the solutions, and that their general properties can be looked up.[2]

The $n_l(\rho)$ become infinite as $\rho \to 0$. Therefore only the $j_l(\rho)$ can be retained for the interior solutions. The energy eigenvalues are fixed by the requirement that the values and derivatives of the j_l equal those of the corresponding $h_l^{(1)}(i\eta)$ at $r = b$.

PROBLEM 7.4

a. Check that for the $V(r)$ of (6), the $l = 1$ solution is given correctly by expressions (9) and (10).

[2] L. I. Schiff, *Quantum Mechanics*, 3d ed. (New York: McGraw-Hill Book Company, 1968), Sec. 15, and references there.

b. Obtain a transcendental equation, analogous to (7) for the $l = 0$ case, that determines the energy of the lowest $l = 1$ state.

c. Find the condition, similar to (8) for the $l = 0$ case, that V_0 and b must satisfy if there is to be at least one $l = 1$ state.

7.2 *The Isotropic Harmonic Oscillator*

Once again we return to the harmonic oscillator. It is directly relevant to the study of atoms that are oscillating about equilibrium positions in solids and in molecules, as is mentioned in Section 2.7. In solids the atoms can, for some purposes, be treated as if they were in a fixed potential. In diatomic molecules such as H_2, O_2, and HCl, the center of mass is unaccelerated, and the problem is that of a single particle with reduced mass in the relative coordinates. The harmonic oscillator wave functions are also used to describe systems, such as nucleons in nuclei, for which $V(r)$ is not represented very well by $Kr^2/2$. The wave functions have functional forms that are easy to use, they can provide a fair first approximation, and, since they are complete, they can be superposed to construct accurate solutions.

Section 2.7 examines the three-dimensional harmonic oscillator in Cartesian coordinates. The problem separates into three one-dimensional problems (2.1-7) because

$$V(x, y, z) = \frac{K_x x^2}{2} + \frac{K_y y^2}{2} + \frac{K_z z^2}{2}$$

has the form

$$V_x(x) + V_y(y) + V_z(z).$$

If the oscillator is isotropic in the sense that $K_x = K_y = K_z \equiv K$, then the potential can equally well be viewed as

$$V(r) = \frac{Kr^2}{2},$$

and all the results of Chapter 6 apply. We are going to find a few solutions in this spirit, and then look at the relations between them and the $\psi(x, y, z)$ of Section 2.7.

Equation (7.1-3) for this example is

$$-\frac{\hbar^2}{2\mu}\frac{d^2u}{dr^2} + \left[\frac{Kr^2}{2} + \frac{\hbar^2 l(l+1)}{2\mu r^2}\right]u = Eu. \tag{7.2-1}$$

For $l = 0$, just as for the rectangular well, everything looks like the corresponding one-dimensional case except for the condition $u(0) = 0$. The solutions are therefore those involving the *odd* Hermite polynomials H_1, H_3, ... The ground state is described by

$$u(r) \sim H_1(\beta r)e^{-\beta^2 r^2/2}, \qquad \beta \equiv \sqrt{\frac{\mu\omega}{\hbar}}, \qquad \omega \equiv \sqrt{\frac{K}{\mu}}.$$

Since $H_1(\beta r) = 2\beta r$, and since

$$\psi(r, \theta, \varphi) = Y_{00}(\theta, \varphi)R(r) = \frac{1}{\sqrt{4\pi}}\frac{u(r)}{r} \sim e^{-\beta^2 r^2/2},$$

the wave function is the same as

$$\psi(x, y, z) \sim e^{-\beta^2(x^2+y^2+z^2)/2}.$$

The energy as seen from the r, θ, φ, point of view is $(1 + \frac{1}{2})\hbar\omega$ because the ground state looks like the first excited state of the one-dimensional problem. The energy as seen from the x, y, z point of view is $(\frac{1}{2} + \frac{1}{2} + \frac{1}{2})\hbar\omega$ because the three one-dimensional motions each contribute $\hbar\omega/2$. This exact equivalence is inevitable because the ground state is not degenerate; there is only one wave function, and one must find it via any approach.

The equivalence is slightly more subtle for the excited states because they are degenerate. From the x, y, z point of view, the energy $5\hbar\omega/2$ can be reached by exciting the x motion and leaving the y and z motions in the ground state, or by exciting only the y motion, or by exciting only the z motion. There are then three degenerate wave functions

$$\begin{aligned}\psi_{100}(x, y, z) &\sim xe^{-\beta^2(x^2+y^2+z^2)/2},\\ \psi_{010}(x, y, z) &\sim ye^{-\beta^2(x^2+y^2+z^2)/2},\\ \psi_{001}(x, y, z) &\sim ze^{-\beta^2(x^2+y^2+z^2)/2}.\end{aligned}$$

The function ψ_{001} can be written directly in terms of r, θ, and φ as an angular momentum eigenstate,

$$\psi_{001}(x, y, z) \sim r\cos\theta e^{-\beta^2 r^2/2} \sim Y_{10}(\theta, \varphi)re^{-\beta^2 r^2/2};$$

it is a state with $l = 1$, $m = 0$. Check that it does indeed satisfy (1) with $l = 1$ and $E = 5\hbar\omega/2$. The other two functions, $\psi_{100}(x, y, z)$ and $\psi_{010}(x, y, z)$, cannot be translated into angular momentum eigenfunctions directly, but linear combinations of the two can:

$$\begin{aligned}\psi_{100} \pm i\psi_{010} &\sim (x \pm iy)e^{-\beta^2 r^2/2}\\ &= (r\sin\theta\cos\varphi \pm ir\sin\theta\sin\varphi)e^{-\beta^2 r^2/2}\\ &= \sin\theta e^{\pm i\varphi} re^{-\beta^2 r^2/2}\\ &\sim Y_{1\pm1}(\theta, \varphi)re^{-\beta^2 r^2/2}.\end{aligned}$$

From either point of view there are three degenerate states at $E = 5\hbar\omega/2$. In the x, y, z description, the states are the $\psi_{100}, \psi_{010}, \psi_{001}$ listed above. In the r, θ, φ description, they are states with $l = 1$, and with the three orientations $m = 0, \pm 1$. The x, y, z states and the angular momentum eigenstates are not simply equal, but each member of one set can be written as a linear combination of the members of the other set.

More highly excited states involve larger sets of degenerate wave functions, but the principle is the same. As shown in Problem 2.21, the number of states with energy $(n + \frac{3}{2})\hbar\omega$ is $(n + 1)(n + 2)/2$, so that there are six linearly independent states with energy $7\hbar\omega/2$. In the language of Section 2.7, these states have (n_x, n_y, n_z) equal to (0, 1, 1), (1, 0, 1), (1, 1, 0), (2, 0, 0), (0, 2, 0), or (0, 0, 2). The corresponding set of angular momentum eigenstates contains one $l = 0$ state and five $l = 2$ states with $m = -2, -1, 0, +1,$ and $+2$.

PROBLEM 7.5

Find the $l = 0$, $E = 7\hbar\omega/2$ state by trying

$$\frac{u(r)}{r} = R(r) = (a + br^2)e^{-\beta^2 r^2/2}$$

in (1). Can Section 2.7 be used to check the result?

PROBLEM 7.6

Write the (0, 0, 2) solution for the isotropic harmonic oscillator in Cartesian coordinates. Then express this solution as a linear combination of angular momentum eigenfunctions.

Equation (1) can be solved in a systematic and general way by techniques much like those used in Section 2.7.[3] Substitution of

$$\frac{u}{r} = R(r) = v(r)e^{-\beta^2 r^2/2}$$

turns (1) into an equation for $v(r)$. A power series trial solution yields a recursion relation. The series generated by the recursion relation diverges unless it is cut off and turned into a polynomial. The required termination fixes the energy eigenvalues. The polynomials in r that are determined by this procedure are called *Laguerre polynomials.*

EXAMPLE

Calculate $\langle r \rangle$ for the ground state and for the three degenerate first excited states of the isotropic harmonic oscillator.

[3] Details of the procedure and properties of the functions involved are given by J. L. Powell and B. Crasemann, *Quantum Mechanics* (Reading, Mass.: Addison-Wesley Publishing Co., Inc., 1961), Secs. 7.4 and 7.5. A different approach is used by R. H. Dicke and J. P. Wittke, *Introduction to Quantum Mechanics* (Reading, Mass.: Addison-Wesley Publishing Co., Inc., 1960), Sec. 10–3. Raising and lowering operators are developed for the three quantities energy, total angular momentum, and z component of angular momentum. All states can then be generated by repeated application of these operators to the ground state wave function.

SOLUTION

The ground state wave function is $CY_{00}(\theta, \varphi)e^{-\beta^2 r^2/2}$. Normalization requires, for $d\Omega \equiv \sin\theta\, d\theta\, d\varphi$,

$$|C|^2 \iint |Y_{00}|^2 \, d\Omega \int_0^\infty e^{-\beta^2 r^2} r^2 \, dr = |C|^2 \times 1 \times \frac{\sqrt{\pi}}{4\beta^3} = 1.$$

Therefore for the ground state,

$$\langle r \rangle_0 \equiv \iiint |\psi|^2 r \, d\Omega r^2 \, dr$$

$$= \frac{4\beta^3}{\sqrt{\pi}} \iint |Y_{00}|^2 \, d\Omega \int_0^\infty e^{-\beta^2 r^2} r^3 \, dr = \frac{4\beta^3}{\sqrt{\pi}} \times 1 \times \frac{1}{2\beta^4} = \frac{2}{\beta\sqrt{\pi}}.$$

The three degenerate first excited states are $C' Y_{1m}(\theta, \varphi) r e^{-\beta^2 r^2/2}$, $m = -1$, 0, $+1$. Normalization requires

$$|C'|^2 \iint |Y_{1m}|^2 \, d\Omega \int_0^\infty r^2 e^{-\beta^2 r^2} r^2 \, dr = |C'|^2 \times 1 \times \frac{3\sqrt{\pi}}{8\beta^5} = 1.$$

Therefore for any of the three states,

$$\langle r \rangle_1 = \frac{8\beta^5}{3\sqrt{\pi}} \iint |Y_{1m}|^2 \, d\Omega \int_0^\infty e^{-\beta^2 r^2} r^5 \, dr$$

$$= \frac{8\beta^5}{3\sqrt{\pi}} \times 1 \times \frac{1}{\beta^6} = \frac{8}{3\beta\sqrt{\pi}}.$$

Note that $\langle r \rangle_1$ is somewhat larger than $\langle r \rangle_0$. This result can be viewed as a consequence of the outward push of orbital angular momentum: $\langle r \rangle_0$ refers to the lowest $l = 0$ state while $\langle r \rangle_1$ refers to the lowest $l = 1$ state. The effect is not very large because $V = Kr^2/2$ rises rapidly to resist this outward push.

It is interesting to make some numerical estimates for an atom in a molecule or a solid. Typical chemical energies, of the order of an eV, are required to displace an atom by one or two Å.

$$\therefore \quad 1 \text{ eV} \simeq K \frac{(1 \text{ Å})^2}{2}, \qquad K \simeq \frac{2 \times 10^{16} \text{ eV}}{\text{cm}^2}.$$

The mass is of the order of ten nucleon masses: $\mu \simeq 10^{10}$ eV/c^2. The classical frequency is therefore

$$\omega = \sqrt{\frac{K}{\mu}} \simeq \frac{4 \times 10^{13}}{\text{sec}}.$$

The energy difference between adjacent states, $\hbar\omega$, is 0.03 eV, and the radiation emitted in transitions is in the infrared region. Because thermal energy at 300°K is of the same magnitude, the atom is likely to be in one of the lowest few vibrational states at room temperature.

$$\beta = \sqrt{\frac{\mu\omega}{\hbar}} \simeq \frac{8 \times 10^8}{\text{cm}}.$$

$$\therefore \quad \langle r \rangle_0 \simeq \frac{2}{\left(\dfrac{8 \times 10^8}{\text{cm}}\right)\sqrt{\pi}} = 0.14 \text{ Å}.$$

Even at low temperature, with only the ground state occupied, the mean distance of an atom from its equilibrium position is about a tenth of the interatomic separation.

7.3 *The Coulomb Potential*

Many important systems are described by bound states in a Coulomb potential. There is the ordinary hydrogen atom, with $V(r) = -e^2/4\pi\varepsilon_0 r$ and $\mu = M_e(1836.15/1837.15)$. This system was from the beginning the most important testing ground for quantum theory. There are the ions with one remaining electron, He^+, Li^{++}, Be^{+++}, ..., with potential energies $-Ze^2/4\pi\varepsilon_0 r$. *Positronium*, a bound state of an electron and a positron, has $\mu = M_e/2$. There is a great variety of so-called exotic atoms, formed when a negative particle such as μ^-, π^-, K^-, $\bar{p}$, Σ^-, Ξ^-, or Ω^- is captured in the Coulomb field of a nucleus. For all these systems, $-Ze^2/4\pi\varepsilon_0 r$ is an approximation, but for many it is a good approximation. The complicating factors include the effect of spin and the associated magnetic moment interactions, the fact that the nucleus is not a point charge, the strong specifically nuclear interaction that acts in, for example, the π^--nucleus system, relativistic effects, and vacuum polarization.[4]

[4] As is mentioned in Section 1.3, the Coulomb interaction can be viewed as the consequence of virtual photons that are exchanged between the interacting charges. The virtual photons can, in turn, create virtual electrons and positrons. The region between charges is therefore a busy complicated environment. The virtual electrons and positrons are pushed around by the electric field so as to shield partially the interacting charges. When the charges come very close to each other, the effectiveness of this shielding is reduced. One must use a quantum field theory treatment of the electromagnetic interaction to compute the effect of this "vacuum polarization," and a quantitative treatment is beyond the scope of this book. J. J. Sakurai, *Advanced Quantum Mechanics* (Reading, Mass.: Addison-Wesley Publishing Co., Inc., 1967) provides a good introduction to these matters, and presents on pp. 138–139 a clear qualitative picture of vacuum polarization.

Equation (7.1-3) is

$$-\frac{\hbar^2}{2\mu}\frac{d^2u}{dr^2} + \left[\frac{-Ze^2}{4\pi\varepsilon_0 r} + \frac{\hbar^2 l(l+1)}{2\mu r^2}\right]u = Eu. \tag{7.3-1}$$

It is useful to express lengths and energies in units that are natural to the problem. A characteristic length is the Bohr radius $4\pi\varepsilon_0\hbar^2/\mu e^2$ which, for $\mu = M_e$, is 0.529177×10^{-8} cm. Define the dimensionless quantity ρ by

$$r = \frac{4\pi\varepsilon_0\hbar^2}{\mu e^2}\rho.$$

A characteristic energy is $\mu e^4/2\hbar^2(4\pi\varepsilon_0)^2$, the *Rydberg* energy, which, for $\mu = M_e$, is 13.6058 eV. Notice that this is half of the potential energy of two particles of charge e separated by one Bohr radius. We are here interested in bound states, with $E < 0$, and introduce the dimensionless quantity b through

$$E = \frac{-\mu e^4}{2\hbar^2(4\pi\varepsilon_0)^2}b^2.$$

These substitutions turn (1) into

$$\frac{d^2u}{d\rho^2} - \left[-\frac{2Z}{\rho} + \frac{l(l+1)}{\rho^2} + b^2\right]u = 0. \tag{7.3-2}$$

As $\rho \to \infty$, the Coulomb and centrifugal terms become small compared with b^2, the equation behaves like

$$\frac{d^2u}{d\rho^2} \simeq b^2u, \tag{7.3-3}$$

and $u \sim e^{\pm b\rho}$. The $e^{+b\rho}$ solution is unacceptable, and it follows that the large ρ behavior is dominated by $e^{-b\rho}$. With the substitution

$$u(\rho) = v(\rho)e^{-b\rho}, \tag{7.3-4}$$

(2) becomes

$$\frac{d^2v}{d\rho^2} - 2b\frac{dv}{d\rho} - \left[\frac{l(l+1)}{\rho^2} - \frac{2Z}{\rho}\right]v = 0. \tag{7.3-5}$$

We are going to solve (5) by substituting a series for v. It will turn out that the series will have to be cut off to yield a wave function that decreases at large distances, so that the function u is a decreasing exponential times a polynomial. This preview is provided to point out how much the method is like that used for the harmonic oscillator in Section 2.7. There the asymptotic behavior of the solution is a Gaussian, and ψ is the Gaussian times the Hermite polynomials. Compare the two solutions step by step.

The discussion at the beginning of this chapter shows that the lowest power in a series for $v(\rho)$ is ρ^{l+1}; see (7.1-5). A power series for $v(\rho)$ therefore has the form

$$v(\rho) = \sum_{\nu=0}^{\infty} C_\nu \rho^{\nu+l+1}, \qquad C_0 \neq 0. \tag{7.3-6}$$

Substitute (6) into (5).

$$\sum_{\nu=0}^{\infty} [C_\nu(\nu + l + 1)(\nu + l)\rho^{\nu+l-1} - 2bC_\nu(\nu + l + 1)\rho^{\nu+l} - l(l + 1)C_\nu\rho^{\nu+l-1} + 2ZC_\nu\rho^{\nu+l}] = 0.$$

Set the coefficient of $\rho^{\nu+l}$ equal to zero.

$$C_{\nu+1} = \frac{2b(\nu + l + 1) - 2Z}{(\nu + 1)(\nu + 2l + 2)} C_\nu. \tag{7.3-7}$$

We are well on the way to success because a recursion relation has emerged. Given l, Z, b, and C_0, the function $v(\rho)$ is determined. Since (5) is a second-order differential equation, there are two linearly independent solutions, but one of them is eliminated by the condition $v(0) = 0$. The series generated by the recursion relation (7) is therefore *the* solution of interest.

For large ν, $\nu \gg l$ and $\nu \gg Z$,

$$\frac{C_{\nu+1}}{C_\nu} \simeq \frac{2b}{\nu},$$

the same ratio as for the power series expansion of $e^{+2b\rho}$. If the series is not cut off, $u \sim e^{-b\rho}e^{+2b\rho} = e^{+b\rho}$, and the wave function would increase as the radius increases. There must be a cutoff, that is, a first coefficient, $C_{n'}$ that equals zero and assures, because of (7), that all higher coefficients are also zero. If C_0 were zero, the entire wave function would be zero. The cutoff must be produced by the condition

$$b(l + n') = Z, \qquad n' = 1, 2, 3, \ldots, \tag{7.3-8}$$

for which (7) gives $C_{n'} = 0$, as desired. Since it determines the r dependence of the wave function, n' is called the *radial quantum number*. Let $l + n' = n$. Since $l = 0, 1, 2, \ldots$, n can be $1, 2, 3, \ldots$. With $b = Z/n$,

$$E = \frac{-\mu e^4 Z^2}{2\hbar^2(4\pi\varepsilon_0)^2} \frac{1}{n^2}, \tag{7.3-9}$$

and the familiar $1/n^2$ form of the hydrogen spectrum has emerged. Since n determines the energy, it is called the *principal quantum number*.

The very famous expression (9) describes, to good approximation, the spectrum of hydrogen. This result was developed empirically in the nineteenth century, beginning with Balmer's work in 1885. It became clear before the advent of quantum theory that, in our notation, the frequencies of the radiations from hydrogen satisfy

$$\begin{aligned}\nu_{if} &= \frac{E_i - E_f}{h} = \frac{1}{h}\frac{\mu e^4}{2(4\pi\varepsilon_0)^2\hbar^2}\left(\frac{1}{n_f^2} - \frac{1}{n_i^2}\right) \\ &= \frac{13.6\text{ eV}}{h}\left(\frac{1}{n_f^2} - \frac{1}{n_i^2}\right) = \frac{3.29 \times 10^{15}}{\text{sec}}\left(\frac{1}{n_f^2} - \frac{1}{n_i^2}\right),\end{aligned} \tag{7.3-10}$$

where the subscripts i and f refer to the initial and final states in a transition.

There is a high degree of degeneracy. The energy is determined by the sum $n' + l = n$, so that $(n' = n, l = 0)$, $(n' = n - 1, l = 1), \ldots, (n' = 1, l = n - 1)$ all give the same E. This feature is a special property of the Coulomb field. As for any spherically symmetric Hamiltonian, each value of l allows $2l + 1$ different possibilities for m that must have the same E. The degeneracy of an energy eigenstate with principal quantum number n is therefore

$$\sum_{l=0}^{n-1} (2l + 1) = n^2 \tag{7.3-11}$$

if spin is not considered.

The energy levels are shown in Figure 7.3-1. Each of the levels there represents $4(2l + 1)$ states if the system is the ordinary hydrogen atom because

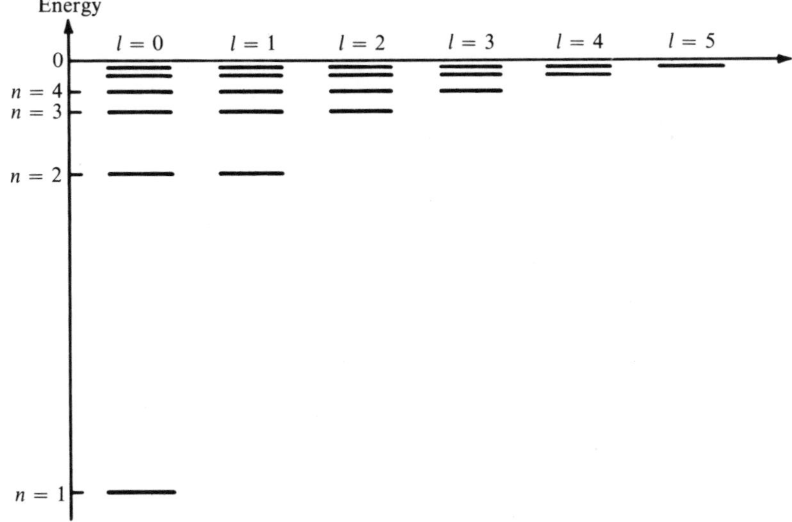

Fig. 7.3-1. The hydrogen-like spectrum. For hydrogen the $n = 1$ state is at -13.6 eV.

m can have $2l + 1$ values, and in addition there are two possible orientations of the electron spin and of the proton spin.

For any choice of n' and l, the recursion relation (7) becomes

$$C_{\nu+1} = \frac{2Z}{n} \frac{\nu + 1 - n'}{(\nu + 1)(\nu + 2l + 2)} C_\nu. \tag{7.3-12}$$

The method used to generate any desired $v(\rho)$ is the same as that used for the Hermite and Legendre polynomials. The lowest constant C_0 can be fixed by normalization. Successively higher C_ν are given as multiples of C_0 until the process terminates because $\nu + 1 = n'$. The $v(\rho)$ so obtained are called *associated Laguerre polynomials*, and their general properties are discussed in various texts.[5]

The complete wave functions can now be assembled. For example, the ground state is given by $n = 1$, $n' = 1$, and $l = 0$. The polynomial $v(\rho)$ is then just $C_0 \rho$, $u(\rho) = C_0 \rho e^{-Z\rho}$, and $R(\rho) = C_0 e^{-Z\rho}$. The wave function is $R(r)Y_{00} = R(r)/\sqrt{4\pi}$. The equations and graphs of hydrogen atom wave functions are given in great detail by Pauling and Wilson.[6]

PROBLEM 7.7

a. Put together the normalized ground state wave function for hydrogen. Use the symbol

$$a = 4\pi\varepsilon_0 \frac{\hbar^2}{\mu e^2} \tag{7.3-13}$$

for the Bohr radius.

b. Plot versus r the probability of finding the electron at r. Be sure to remember that the volume element is proportional to r^2. Compute the most probable r.

c. Compute the expectation value of r.

PROBLEM 7.8

Obtain the four normalized wave functions $\psi_{nlm}(r, \theta, \varphi)$ that correspond to the first possible energy above the ground state.

PROBLEM 7.9

There is evidence for an example of a Coulomb bound state of two unstable particles, a pion and muon of opposite charge.[7] Obtain the energy levels

[5] E. Merzbacher, *Quantum Mechanics*, 2d ed. (New York: John Wiley & Sons, Inc., 1970), Chap. 10; L. I. Schiff, *Quantum Mechanics*, 3d ed. (New York: McGraw-Hill Book Company, 1968), Sec. 16.

[6] L. Pauling and E. B. Wilson, *Introduction to Quantum Mechanics* (New York: McGraw-Hill Book Company, 1935), Chap. V. The older quantum texts should not be forgotten. They provide explicit details that have been crowded out of more recent books.

[7] R. Coombes et al., *Phys. Rev. Lett.* **37**, 249 (1976).

for this system. Assume that for very high-lying states, the natural width is determined primarily by the mean life of the pion. How large must n be if the natural width equals the spacing between levels?

One should know some of the jargon used to describe these states. Electrons with $n = 1$ are said to be in the K shell, those with $n = 2$ in the L shell, those with $n = 3$ in the M shell, and so on through the alphabet. The $l = 0$ states are called S states; the $l = 1$, P states; the $l = 2$, D states; the $l = 3$, F states; the $l = 4$, G states; and so on. These peculiar designations come from the old days of spectroscopy, when the various series of transitions were labeled *S*harp, *P*rincipal, *D*iffuse, and *F*undamental. For higher l one continues through the alphabet, $G, H, I, \ldots$, with J omitted because it is doing too many other things.

Try to get some feeling for the behavior and form of these wave functions. For example, there is plenty of good physics in the observation that $R(r) \sim r^l$ near the origin. For which state is the finite size of the nucleus most important? If a π^- is captured by the Coulomb field of a nucleus, is the specifically nuclear interaction important in S states? In P states? In D states?

PROBLEM 7.10

Obtain the expectation value of the potential energy for the ground state in a Coulomb field. What then is the expectation value of the kinetic energy? Show that

$$\sqrt{\langle \text{speed}^2 \rangle} = \frac{Zc}{137.04}, \tag{7.3-14}$$

where c is the speed of light.

7.4 *Linear Potentials and Quarks*

For a potential that varies linearly with radius,

$$V(r) = gr - V_0, \tag{7.4-1}$$

(7.1-3) is

$$-\frac{\hbar^2}{2\mu}\frac{d^2u}{dr^2} + \left[gr - V_0 + \frac{\hbar^2 l(l+1)}{2\mu r^2}\right]u = Eu. \tag{7.4-2}$$

For $l > 0$ this equation cannot be solved in terms of standard tabulated functions, and approximation methods must be used. However, for $l = 0$ most of the job has already been done in Section 2.9. With

$$\xi \equiv \left(r - \frac{E + V_0}{g}\right)\left(\frac{2\mu g}{\hbar^2}\right)^{1/3}, \tag{7.4-3}$$

$$\eta(\xi) \equiv u(r),$$

the $l = 0$ equation becomes

$$\frac{d^2\eta}{d\xi^2} - \xi\eta = 0, \tag{7.4-4}$$

the same as (2.9-3). Section (2.9) examines one-dimensional wave functions that are forced to be zero at $x = 0$ by a $V(x)$ that is infinite for $x < 0$. Here mathematically equivalent behavior results from the boundary condition $u(0) = 0$. The same relation between one-dimensional problems and three-dimensional $l = 0$ problems is mentioned in Sections 7.1 and 7.2. Of course, in these three-dimensional problems, it is $u(r)$ that goes to zero at $r = 0$, while the actual wave function is proportional to $u(r)/r$ and is different from zero at $r = 0$.

The solution of (4) is the Airy integral defined by (2.9-4) and graphed in Figure 2.9-1. Since $u(0) = 0$, $\xi(r = 0)$ must be one of the ξ_n, the zeros of the Airy integral. As in Problem 2.27, the energy eigenvalues E_n are therefore determined by the condition

$$\xi_n = \left(0 - \frac{E_n + V_0}{g}\right)\left(\frac{2\mu g}{\hbar^2}\right)^{1/3},$$

or

$$E_n = |\xi_n|\left(\frac{g^2\hbar^2}{2\mu}\right)^{1/3} - V_0. \tag{7.4-5}$$

The result is (2.9-12) except for the appearance of the reduced mass μ rather than the mass M, and the shift in the origin of energy by V_0.

As is mentioned in Section 2.9, the quark model of elementary particles provides one of the motivations for looking at linear potentials. This subject can be discussed a little more in terms of the three-dimensional formalism of this chapter.

During recent years the number of known "elementary particles" has grown to roughly a hundred. This number exceeds substantially that of the elements known in 1869 when D. I. Mendeléef and L. Meyer published tables that emphasized regularities and periodicities in chemical behavior. We have again reached a time when the number and variety of objects previously accepted as unanalyzable is distressingly large, and so we begin again a search for an underlying simpler structure.

Regularities in the properties of particles are now well established.[8] It is possible to arrange the particles in groups that show similar behavior and that reveal symmetries in their properties. Speculations regarding underlying structures that might explain these properties are beginning to show some success. A promising theory is that most particles are made of basic constituents that have been named *quarks*, objects that have spin $\frac{1}{2}$ and, possibly, fractional charges. Pions, kaons, and other *mesons*, that is, particles with integral spin (0, 1, ...) that interact strongly with nuclei, are viewed as structures that contain a quark and an antiquark. Protons, neutrons, and other *baryons*, that is, particles with half-integral spin ($\frac{1}{2}$, $\frac{3}{2}$, ...) that interact strongly with nuclei, are viewed as structures that contain three quarks. The mesons and baryons together include all the particles that interact strongly and are called hadrons. The quark models describe the properties of the known hadrons so well that they must be taken seriously.

For some time there has been a vigorous and apparently unsuccessful search for isolated quarks. One experiment may have found evidence for fractional charges, but the situation is not clear.[9] It is as if, while all attempts to detect isolated electrons ended in failure, we have a successful electron theory of atoms that explains spectra and chemical reactions very well. One possible resolution is that there is something about the force between quarks that prevents their being isolated.

A suggestion that makes sense for a variety of reasons is that the force between quarks is constant, independent of their separation, for a relatively large distance.[8] Such a force is described by a linear potential of the form (1), which shows that it costs more and more energy as one of the quarks is pulled further and further away from the other quarks that constitute a hadron. When the energy reached is more than twice the rest energy of a quark, a quark-antiquark pair can be produced, just as an electron-positron pair can be produced if there is enough electromagnetic energy. The newly created quark can fall back to replace the quark that has been pulled out, while the newly created antiquark joins the quark that is being removed to form a meson. A lot of energy has been expended, but the outcome is the creation of a meson rather than the isolation of a quark. If the potential energy has the form (1) for all r, then isolated quarks can never be seen. If the form (1) is a good approximation out to a rather large radius beyond which $V(r)$ decreases to zero, then isolated quarks might occur, but with very small probability.

[8] S. D. Drell, "Electron-Positron Annihilation and the New Particles," *Scientific American* **232** (6), 50 (1975); S. L. Glashow, "Quarks with Color and Flavor," *Scientific American* **233** (4), 38 (1975); Y. Nambu, "The Confinement of Quarks," *Scientific American* **235** (5), 48 (1976); S. D. Drell, *Am. J. Phys.* **46**, 597 (1978).

[9] G. S. LaRue, W. M. Fairbank, and A.F. Hebard, *Phys. Rev. Lett.* **38**, 1011 (1977); L. W. Jones, *Rev. Mod. Phys.* **49**, 717 (1977); R. N. Boyd, D. Elmore, A. C. Melissinos, and E. Sugarbaker, *Phys. Rev. Lett.* **40**, 216 (1978); J. P. Schiffer, T. R. Renner, D. S. Gemmell, and F. P. Mooring, *Phys. Rev. D* **17**, 2241 (1978); and references there.

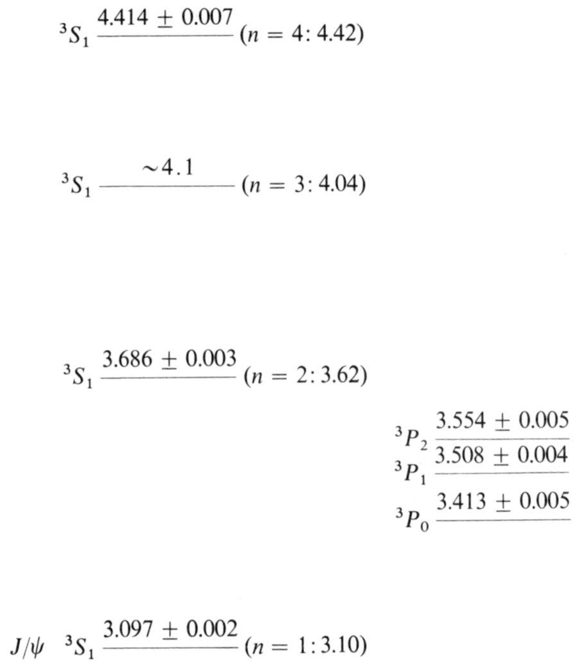

$$^3S_1 \ \underline{\ 4.414 \pm 0.007\ } \ (n = 4{:}\ 4.42)$$

$$^3S_1 \ \underline{\ \sim 4.1\ } \ (n = 3{:}\ 4.04)$$

$$^3S_1 \ \underline{\ 3.686 \pm 0.003\ } \ (n = 2{:}\ 3.62)$$

$$^3P_2 \ \underline{\ 3.554 \pm 0.005\ }$$

$$^3P_1 \ \underline{\ 3.508 \pm 0.004\ }$$

$$^3P_0 \ \underline{\ 3.413 \pm 0.005\ }$$

$$J/\psi \ \ ^3S_1 \ \underline{\ 3.097 \pm 0.002\ } \ (n = 1{:}\ 3.10)$$

Fig. 7.4-1. Some states of the quark-antiquark system called "charmonium." The angular momentum assignments are shown to the left of the level, and the experimental Mc^2 of each meson is given in GeV $\equiv 10^9$ eV on the line that indicates the level. For the S states, the energies calculated from (6) are shown in parentheses to the right of the level.

If (1) is approximately right, then it should be possible to find two-quark systems with energy levels given approximately by (5). It appears that such systems do indeed exist, and that one is a combination of a heavy quark, called the *charmed quark*, and its antiparticle. These two objects move around their common center of mass, much as an electron and a positron do when they combine to form positronium. In the tradition of whimsy that dominates the assignment of names in this field, the system is now known as *charmonium.* Our emphasis here is not on the general structure and nomenclature of quark theory, which are described in the References 8, but on the dynamics of this one system.

In 1974 experiments by C. C. Ting and collaborators at Brookhaven National Laboratory, and by B. Richter and collaborators at the Stanford Linear Accelerator Center, led to the discovery[10] of a heavy meson now called

[10] For this discovery, C. C. Ting and B. Richter received the Nobel Prize in Physics in 1976. Ting wanted to call the new particle J while Richter wanted to call it ψ. Their Nobel lectures, published in *Rev. Mod. Phys.* **49**, 235 and 251 (1977), contain interesting background and reminiscences.

the J/ψ. Subsequent experiments showed that the J/ψ is a member of a family of mesons[11] shown in Figure 7.4-1, and that these mesons can be interpreted as the energy levels of charmonium. Figure 7.4-1 is organized somewhat like Figure 7.3-1, which shows the hydrogen-like energy levels. The angular momenta of the mesons can be inferred from their production and decay. The $l = 0$, or S, states are arranged in order of increasing energy. Three $l = 1$, or P, states are also shown.[12] The angular momentum designations will be discussed more carefully in Chapters 8 and 9, but the notation is straightforward and you may already have met it in your general modern physics background.[13] The orbital angular momentum is indicated by its usual letter, S for $l = 0$, P for $l = 1$. The quarks have spin $\frac{1}{2}$ each and can combine to give a total spin of unity or zero. If the total spin is unity, it can, like any unit angular momentum, have three orientations, corresponding to $m = -1, 0$, or $+1$, and the state is called a *triplet*. If the total spin is zero, then only $m = 0$ is possible, and the state is called a *singlet*. The small index on the upper left of the orbital angular momentum letter is three for triplets and one for singlets, so that ^3S, "triplet S," means $l = 0$ and total spin unity. Only triplet states are shown in Figure 7.4-1 although preliminary identifications of singlet charmonium states have been made. The total angular momentum j is the vector sum of orbital and spin angular momentum, and is shown as the subscript to the right of the orbital angular momentum letter. For S states, the resultant must equal the spin, and so the four states at the left of Figure 7.4-1 are all labeled $^3S_1$. For P states, the spin can add to the orbital angular momentum to give a total $j = 2$, or it can be at such an angle as to give a resultant $j = 1$, or it can cancel the orbital angular momentum to give $j = 0$. The P states are therefore $^3P_2$, $^3P_1$, and $^3P_0$, and their relatively small differences in energy show that the interaction depends slightly on the angle between spin and orbital angular momentum.

The physically accessible quantities are the masses M_n, or the rest energies M_nc^2, of the mesons, where $n = 1, 2, 3$, and 4 for the four S states shown in Figure 7.4-1. Each of the M_nc^2 is given by the rest energy of the charmed quark, Mc^2, plus the rest energy of its antiparticle, also Mc^2, plus the E_n given by (5),

$$M_nc^2 = 2Mc^2 + |\xi_n| \left(\frac{g^2\hbar^2}{M}\right)^{1/3} - V_0. \tag{7.4-6}$$

[11] The references in footnote 8 describe much of the data. Details and recent results are given in reviews of particle properties that are prepared periodically; see, for example, *Rev. Mod. Phys.* **48**, S1 (1976); *Phys. Lett.* **68B**, 1 (1977).

[12] For a discussion of these P states see M. S. Chanowitz and F. J. Gilman, *Phys. Lett.* **63B**, 178 (1976). Some familiarity with the field is required to follow their argument in detail, but it is not hard to understand the kinds of considerations used to identify these levels.

[13] K. W. Ford, *Classical and Modern Physics* (Lexington, Mass.: Xerox College Publishing, 1972), Chap. 24.

The reduced mass μ is $M/2$, as it is for any system that consists of two particles that have the same mass M. The choice

$$M = \frac{1.84 \text{ GeV}}{c^2}, \qquad g = \frac{1.11 \text{ GeV}}{f}, \qquad V_0 = 1.27 \text{ GeV}$$

fits the ^3S states well, as shown in Figure 7.4-1. Since there are really only two parameters, $2Mc^2 - V_0$ and g^2/M, that can be adjusted, it is encouraging that (6) can fit four experimental numbers.

PROBLEM 7.11

a. Predict the mass of the meson that is the $^3S_1$ charmonium state next above the one at 4.414 GeV.

b. Estimate the average separation of the two quarks in the J/ψ by computing half the distance from $r = 0$ to the classical turning point.

c. Make a crude guess of the energy of the lowest ^3P state of charmonium as follows. The difference between it and the J/ψ is the presence of the "centrifugal potential" in (2). Evaluate the centrifugal potential for the average separation estimated in (**b**) and add it to the energy of the J/ψ.

The treatment given here is incomplete. The energies are such that relativistic corrections should be included. States other than those with ^3S angular momentum have to be examined. There are theoretical reasons for expecting deviations from the linear potential near $r = 0$. Furthermore, the field is so young and is developing so rapidly that much of our discussion may be obsolete very soon. However, the central idea seems to be sound: Elementary particle physics has found its tractable two-body system.

Isolated free particles teach us very little, and three-body systems are too complicated. Classical mechanics became essentially what it is today when Newton learned to use the earth-moon system as the test of his thoughts. The basic ideas of atomic physics grew out of studies of the hydrogen atom. Nuclear physics became quantitative, and the nature of the nuclear force began to be understood, when the deuteron could be analyzed. It appears that charmonium may be the fourth of the great seminal two-body systems.

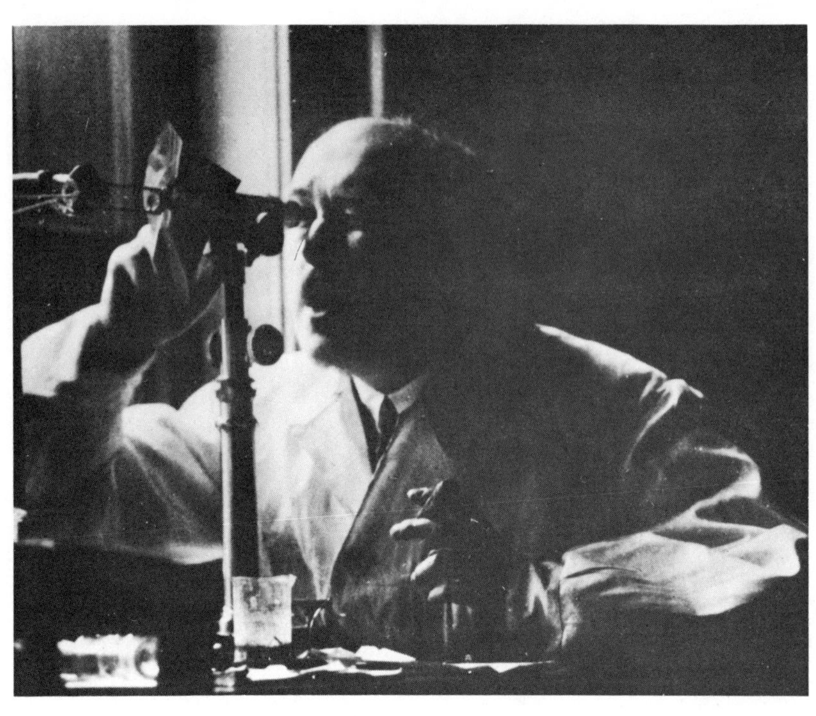

Otto Stern in his laboratory in Hamburg around 1930. After shorter appointments at several institutions, he worked at the University of Hamburg, 1923–33, and Carnegie–Mellon University, 1933–45. He and his collaborators used atomic and molecular beams to do several crucial experiments, including the demonstration of the wave nature of atoms and molecules, the first measurements of the magnetic moments of the proton and of the deuteron, and the Stern–Gerlach experiment discussed in Section 8.1. For the history of these experiments as described by one of his principal co-workers, see I. Estermann (edited by S. N. Foner), Am. J. Phys. **43**, *661(1975).* (*Courtesy of the American Institute of Physics Niels Bohr Library, Segrè Collection*)

8

Spin

The possibility $j = \frac{1}{2}$ appears in Section 6.1, but Section 6.2 points out that $j = \frac{1}{2}$ cannot describe *orbital* angular momentum. Nature has, however, utilized the $j = \frac{1}{2}$ possibility to provide electrons and many other particles with *spin*. The matter of spin is mentioned repeatedly in the preceding chapters, and we now examine it more carefully.

8.1 *Qualitative Background*

If the state of an electron were completely specified by a $\Psi(\vec{r}, t)$, then Section 7.3 would provide a nearly complete description of the hydrogen atom. To achieve great accuracy, one would still have to calculate relativistic and quantum electrodynamic corrections. However, even if all such refinements were taken into account, something would obviously be missing from the theory. In the spectrum of hydrogen many of the levels are split; that is, many levels that one might expect to be described completely by the quantum numbers n and l actually consist of two components, separated in energy by about 10^{-5} eV. The needed corrections suggest an additional degree of freedom which is quantized so that it can take on only two values.

In 1925 the essence of the explanation was developed by Goudsmit and Uhlenbeck and by Pauli: The electron has an intrinsic angular momentum, or *spin*, and an associated magnetic moment.

The magnitude of the spin is determined by the fact that the additional degree of freedom can have two and only two values. Look again at Section 6.1. There it turned out that angular momenta could be described by a quantum number j, which is either integral or half-integral, and a z-component quantum number m_j which can take on the $2j + 1$ values $-j$ to $+j$ in integral steps. If the two components of energy levels are identified with the possible orientations labeled by m_j, then $2j + 1 = 2$ and $j = \frac{1}{2}$. We will henceforth use s, rather than the general j, as the dimensionless spin quantum number. With $s = \frac{1}{2}$, $m_s = +\frac{1}{2}$ or $-\frac{1}{2}$.

This discussion has focused on hydrogen, but the concept of spin is the key to the understanding of a great variety of phenomena. The spectra of all elements would be incomprehensible without it. For example, the line in the sodium spectrum that accounts for the characteristic yellow of sodium flames is a doublet associated with the two orientations of electron spin. As mentioned in Section 4.2, the periodic table is understood on the basis of the Pauli exclusion principle plus the additional degree of freedom of spin. The count of states in Section 5.1 used the same ideas. Our pictures of the deuteron, Section 7.1, and of "charmonium," Section 7.4, require that neutrons, protons, and quarks also have $s = \frac{1}{2}$.

Associated with spin, there is the magnetic moment of the electron. A rotating little blob of negative charge appears not only as a charge, but also as current loops that produce a magnetic moment in the direction opposite that of the spin. The magnitude of the electron's magnetic moment cannot be derived by an integration over a rotating charge distribution because there is no suitable model of the electron, but the negative moment does exist. For the present we take the magnetic dipole moment as one of the basic empirically obtained properties of a particle, like its mass, charge, and spin. The magnetic moment $\vec{M}_s$ and spin angular momentum $\vec{S}$ are related by

$$\vec{M}_s = \gamma \vec{S}, \tag{8.1-1}$$

where γ is called the *gyromagnetic ratio*, and the energy of interaction with a magnetic field $\vec{B}$ is the *spin Hamiltonian*

$$H_s = -\vec{M}_s \cdot \vec{B} = -\gamma \vec{S} \cdot \vec{B}. \tag{8.1-2}$$

Suppose that we take the direction of $\vec{B}$ as the z direction and determine the sign and magnitude of the difference in the energies of the two states $m_s = +\frac{1}{2}$ and $m_s = -\frac{1}{2}$, or $S_z = +\hbar/2$ and $S_z = -\hbar/2$. The sign is such that if $\vec{S}$ and $\vec{B}$ point in the same direction, then the energy is raised, while if $\vec{S}$ and $\vec{B}$ point in opposite directions, the energy is lowered. The energy of a magnetic moment is minimized if it points along $\vec{B}$, and so the constant γ must be negative. If the spin points up, the magnetic moment points down, as is reasonable for a spinning negative charge distribution. The magnitude of the difference in energy between the $S_z = +\hbar/2$ and $-\hbar/2$ states is $-2\gamma(\hbar/2)|\vec{B}|$, and the experimental result is described by

$$\begin{aligned}\gamma\hbar/2 &= -0.928 \times 10^{-23} \text{ amp m}^2 \\ &= -0.579 \times 10^{-8} \frac{\text{eV}}{\text{gauss}} = -0.579 \times 10^{-4} \frac{\text{eV}}{(\text{weber/m}^2)}\end{aligned} \tag{8.1-3}$$

for all magnetic field strengths that can be explored. An electron in a state with $S_z = +\hbar/2$ has a magnetic moment with z component given by (3). Iron saturates around 2 webers/m^2, so that in easily obtainable laboratory fields, the energy of orientation of an electron is about 10^{-4} eV. The largest available man-made fields produce about a hundred times as much; even these values are less than room temperature thermal energies, and very small compared to optical transition energies.

The splitting of the energy levels of hydrogen is present even if there is no external field, and is caused by the internal magnetic field created by the motion of charges inside the atom. To estimate such fields, picture a hydrogen atom in an $n = 2$, $l = 1$, $m_l = 1$ state semiclassically as having an electron in a circular orbit at a distance r from the z axis. To have $L_z = \hbar$, the electron speed must be $\hbar/\mu r$. An observer traveling with, but not spinning with, the electron would say that the nucleus is moving in a circle of radius r with speed $v = \hbar/\mu r$ around the electron. This motion looks like a current loop of radius r with a current $I = ev/2\pi r$, or $\hbar e/2\pi\mu r^2$. Such a loop produces at its center a magnetic field

$$B = 4\pi \times 10^{-7} \frac{\text{weber}}{\text{amp meter}} \frac{I}{2r} = 4\pi \times 10^{-7} \frac{\text{weber}}{\text{amp meter}} \frac{\hbar e}{4\pi\mu} \frac{1}{r^3}.$$

In an $n = 2$, $l = 1$ state,

$$\psi(r, \theta, \varphi) = Y_{11}(\theta, \varphi)R_{21}(r) = Y_{11}(\theta, \varphi)\frac{re^{-r/2a}}{a\sqrt{24a^3}}.$$

Although it mixes classical and quantum ideas, a way to guess the size of the $1/r^3$ term is to find its expectation value for this state,

$$\left\langle \frac{1}{r^3} \right\rangle = \iiint \psi^* \frac{1}{r^3} \psi \, d\tau = \int_0^\infty \frac{1}{r^3} \frac{r^2 e^{-r/a}}{24a^5} r^2 \, dr = \frac{1}{24a^3}.$$

With this choice, $B = 0.5$ webers/m^2 $= 5000$ gauss. The magnetic moment in this field then has an energy of orientation of a few times 10^{-5} eV, in rough agreement with the observed level splittings.

What has been estimated is the size of the *spin-orbit interaction* for the $2p$ state of hydrogen. There is much that is incomplete here because the argument is semiclassical. Also, the spin precesses in such a way ("Thomas precession") as to halve the interaction. However, the estimate of magnitudes is adequate.

The gyromagnetic ratio γ was introduced as a physical constant that had to be determined empirically, but there is more to be said. Consider first the magnetic moment $\vec{M}_l$ associated with orbital angular momentum. Use again the semiclassical picture of an electron in a circular orbit with orbital angular momentum described by l and $m_l = l$. The magnetic moment of a current loop is the area of the loop times the current running around its edge. Since the speed is $l\hbar/\mu r$,

$$|\vec{M}_l| = \pi r^2\left(\frac{e}{2\pi r}\right)\frac{l\hbar}{\mu r} = \frac{e}{2\mu}\,l\hbar$$

so that here, the gyromagnetic ratio is $-e/2\mu$. The γ determined by (3) to be -0.928×10^{-23} amp m$^2/(\hbar/2)$ is, to good accuracy, $-e/\mu$. Precision measurements yield,[1] with M_e as the electron mass,

$$|\vec{M}_s| = 1.00116\left(\frac{e}{M_e}\right)\frac{\hbar}{2}.$$

In other words, one gets twice as much magnetic moment per unit of angular momentum out of spin as out of orbital angular momentum. The spin gyromagnetic ratio of charge/mass has a theoretical foundation. Both in relativistic classical mechanics and in the relativistic quantum mechanics of spin $\frac{1}{2}$ particles, the basic equations take their most simple form if spinning particles are assigned a gyromagnetic ratio of charge/mass. In both theories, however, it is possible to add further terms that allow one to adjust the gyromagnetic ratio to any value one wants. Quantum electrodynamic calculations produce to fantastic accuracy the observed $|\vec{M}_s|$ by treating the difference between $|\vec{M}_s|$ and $e\hbar/2M_e$ as the result of radiative corrections.

A hydrogen atom in its ground state has a magnetic moment very close to $\vec{M}_s$, the moment of the electron. There is no contribution from orbital motion because $l = 0$, and the proton contributes only about 10^{-3} times as much as the electron.[2] The force on the moment in a field $\vec{B}$ is $(\vec{M}_s \cdot \vec{\nabla})\vec{B}$. In a nonuniform magnetic field, hydrogen atoms will therefore be accelerated, as in the Stern–Gerlach experiment.[3] With the arrangement shown in Figure 8.1-1 this experiment measures the component of the spin in the z direction. It therefore forces the state to be an eigenstate of S_z. The incident beam is unpolarized,

[1] Actually they yield $(1.0011596567 \pm 35 \times 10^{-10})(e\hbar/2M_e)$. It is possible to measure the difference between $|\vec{M}_s|$ and $e\hbar/2M_e$ to great accuracy.

[2] For the proton the magnetic moment is 2.79 $e\hbar/2M_p$, where M_p is the proton mass. All spin $\frac{1}{2}$ particles with nonzero rest mass M—that is, all known spin $\frac{1}{2}$ particles except neutrinos—have magnetic moments of the order of $e\hbar/2M$.

[3] K. W. Ford, *Classical and Modern Physics* (Lexington, Mass.: Xerox College Publishing, 1972), Sec. 24.1. The original Stern–Gerlach experiment used silver rather than hydrogen atoms. Silver atoms stick wherever they strike a glass plate and are easy to detect.

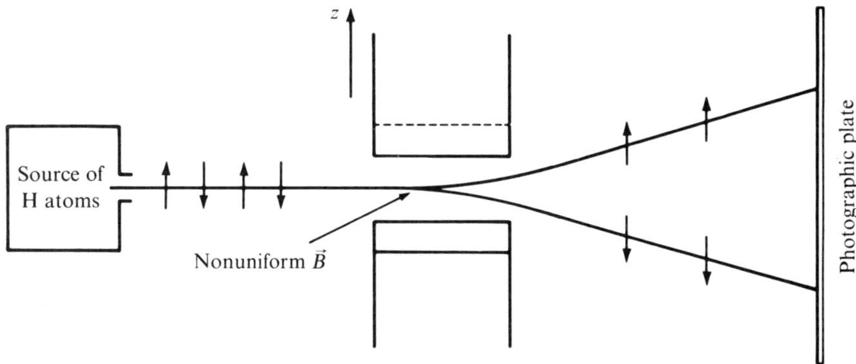

Fig. 8.1-1. The Stern–Gerlach experiment. An unpolarized beam is separated into two polarized beams. Each atom in the polarized beams is in an eigenstate of S_z. The arrows represent the spin directions.

and two polarized beams emerge. Classical physics predicts a continuous smear on the photographic plate, while quantum physics predicts correctly that there will be only two polarization directions and the two corresponding beams. The quantum number m_s can be $+\frac{1}{2}$ or it can be $-\frac{1}{2}$; it can never assume any other value.

PROBLEM 8.1

In Figure 8.1-1, which is the north pole and which is the south pole of the Stern–Gerlach magnet?

The effects of the electron spin and associated magnetic moment have been discussed in terms of the energy levels and motions of complete atoms, not in terms of free electrons. The reason is that the uncertainty principle foils many attempts to see the magnetic properties of free electrons. Various thought experiments illustrate this situation; here are two examples.[4]

Suppose that we try to measure the magnetic field of a free electron. If the field measuring device is a distance r from the electron, it will see something of the order of

$$\frac{|\vec{M}_s|}{r^3} \times \text{function of angles.}$$

[4] The measurability of a free electron's magnetic moment is discussed by J. Kalckar, *Nuovo Cimento* **8A**, 759, (1972).

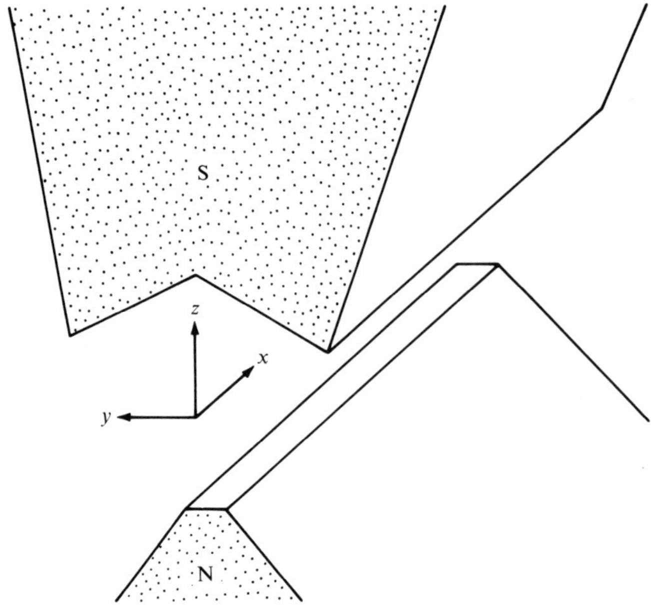

Fig. 8.1-2. The center of a Stern–Gerlach magnet.

If the position of the electron is uncertain by an amount Δx, it is necessary to measure the field outside of Δx, for otherwise even the sign of the field will fluctuate. Therefore the experiment must be designed so that $r \gg \Delta x$. If the electron has an uncertainty in speed Δv, it will produce by virtue of this motion an uncertain additional magnetic field $e\Delta v/r^2$. Therefore we want

$$\frac{|\vec{M}_s|}{r^3} \gg \frac{e\,\Delta v}{r^2},$$

or

$$|\vec{M}_s| \gg er\,\Delta v \gg e\,\Delta x\,\Delta v = e\,\Delta x\,\frac{\Delta p}{M_e} \gtrsim \frac{e\hbar}{2M_e}.$$

But $e\hbar/2M_e$ is $|\vec{M}_s|$, so that there is no way to satisfy this condition.

Can one do a Stern–Gerlach experiment with a beam of free electrons? Figure 8.1-2 shows the center of a Stern–Gerlach magnet with the inhomogeneous field pointing predominately in the z-direction and the electrons traveling in the x direction. The separating force is then

$$\pm|\vec{M}_s|\frac{\partial B_z}{\partial z} = \pm\frac{e\hbar}{2M_e}\frac{\partial B_z}{\partial z}.$$

Those electrons that do not travel exactly in the (x, z) plane experience an additional up or down force because they experience a nonzero B_y. This additional force is evB_y. Since B_y is zero at $y = 0$,

$$B_y \simeq \frac{\partial B_y}{\partial y}\,\Delta y.$$

Since $\vec{\nabla} \cdot \vec{B} = 0$ and $B_x \simeq 0$,

$$\frac{\partial B_y}{\partial y} + \frac{\partial B_z}{\partial z} = 0.$$

Therefore the additional unwanted force is

$$-\frac{\partial B_z}{\partial z}\,\Delta y e v.$$

Therefore

$$\frac{|\text{Wanted separating force}|}{|\text{Additional unwanted force}|} = \frac{\hbar}{2M_e v\,\Delta y} = \frac{\lambda}{4\pi\,\Delta y}.$$

But if the beam is to have any definition, the wavelength λ must be much less than the collimation Δy. The wanted force is then much less than the unwanted force, and the experiment is impossible. By the same argument it is impossible for free protons. It is, however, possible for neutral systems, including hydrogen atoms and neutrons.

The next section contains a development of spin operators and state functions based on the general angular momentum formalism of Section 6.1. First, however, there is a question that you might ask if you have the right kind of sophisticated simplicity.

This thing called spin has the twoness of $j = \frac{1}{2}$, it can be oriented in space, and it is associated with a magnetic moment that seems plausible for a spinning charged particle. However, how do we really know that it is an angular momentum?

In elementary physics you first learned about energy in mechanics. Later, you studied heat and decided that it also was a form of energy. Why? Because there were processes in which the sum of mechanical energy and heat was conserved, but neither was conserved separately. If we want to retain the principle of energy conservation, we *must* call heat a form of energy.

In an isolated hydrogen atom the total angular momentum should be conserved. In many states of such an atom, neither orbital nor spin angular momentum is conserved separately, but the vector sum of the two is. For example, suppose an electron state has total angular momentum $j = \frac{1}{2}$, with $m_j = +\frac{1}{2}$, and with $l = 1$. Such a state turns out to contain a contribution in which $m_l = 1$ and another contribution in which $m_l = 0$. If spin were not an angular momentum, then angular momentum would not be conserved. If, however, spin is an angular momentum, then there is no mystery: Both ($m_l = +1, m_s = -\frac{1}{2}$) and ($m_l = 0, m_s = +\frac{1}{2}$) produce $m_j = +\frac{1}{2}$, and it is the total given by m_j that must be conserved. If we want to retain the principle of angular momentum conservation, we *must* call spin a form of angular momentum.

PROBLEM 8.2

Use the discussion of the deuteron in Section 7.1 to argue that the neutron and proton spins are angular momenta.

8.2 *Spin Eigenfunctions and Operators*

The procedures of Section 6.1 can be applied directly to the special case $j = s = \frac{1}{2}$. This spin quantum number is fixed; if $s \neq \frac{1}{2}$, the particle is not an electron. The orbital angular momentum quantum number l can change as an electron goes from one state to another, but spin cannot change its magnitude; it can change only its direction.

The spin vector is

$$\vec{S} = \hat{e}_x S_x + \hat{e}_y S_y + \hat{e}_z S_z.$$

The square of the angular momentum is the operator S^2, and, for a general spin function, or *spinor*, χ,

$$S^2\chi = \hbar^2 s(s+1)\chi = \frac{3\hbar^2}{4}\chi. \tag{8.2-1}$$

Call the two S_z eigenstates χ_+ and χ_-. Since $m_s = \pm\frac{1}{2}$,

$$S_z\chi_+ = +\frac{\hbar}{2}\chi_+, \qquad S_z\chi_- = -\frac{\hbar}{2}\chi_-. \tag{8.2-2}$$

The three commutation relations (6.1-5) were first obtained from the properties of orbital angular momentum $\vec{L} = \vec{r} \times \vec{p}$. For spin there is no mechanical model from which these commutation relations follow. Rather, as anticipated in Section 6.1, the commutation relations are accepted as the essential properties of any angular momentum, and the components of spin satisfy

$$[S_x, S_y] = i\hbar S_z, \qquad [S_z, S_x] = i\hbar S_y, \qquad [S_y, S_z] = i\hbar S_x. \tag{8.2-3}$$

The raising and lowering operators developed in Section 6.1 are here

$$S_+ = S_x + iS_y, \qquad S_- = S_x - iS_y.$$

The normalization constants $N^{\pm}_{jm}$ are particularly simple. Since the raising operator produces a nonzero result only if it operates on χ_-, (6.1-26) needs to be evaluated only for $j = s = \frac{1}{2}$ and $m = m_s = -\frac{1}{2}$, and gives $N^{+}_{1/2\,-1/2} = \hbar$. Similarly, (6.1-28) also gives $\hbar$ for the only case of interest, that is, for $j = s = \frac{1}{2}$ and $m = m_s = +\frac{1}{2}$, $N^{-}_{1/2\,+1/2} = \hbar$. There are then the four equations

$$S_+\chi_+ = 0, \qquad S_+\chi_- = \hbar\chi_+,$$

$$S_-\chi_+ = \hbar\chi_-, \qquad S_-\chi_- = 0.$$

Since

$$S_x = \frac{S_+ + S_-}{2}, \qquad S_y = \frac{S_+ - S_-}{2i},$$

it follows that

$$S_x\chi_+ = \frac{\hbar}{2}\chi_-, \qquad S_x\chi_- = \frac{\hbar}{2}\chi_+, \tag{8.2-4}$$

$$S_y\chi_+ = \frac{i\hbar}{2}\chi_-, \qquad S_y\chi_- = -\frac{i\hbar}{2}\chi_+. \tag{8.2-5}$$

The equations (2), (4), and (5) specify what each component of spin does to each eigenfunction, and therefore these six equations contain everything needed to do a calculation if we impose in the usual way the expansion postulate that any spinor χ can be written in the form

$$\chi = C_+\chi_+ + C_-\chi_-, \tag{8.2-6}$$

normalization,

$$\int \chi^\dagger\chi = 1, \tag{8.2-7}$$

orthogonality,

$$\int \chi_+^\dagger\chi_- = \int \chi_-^\dagger\chi_+ = 0, \tag{8.2-8}$$

and the prescription for expectation values,

$$\langle\Omega\rangle = \int \chi^\dagger\Omega\chi. \tag{8.2-9}$$

The statement of the expansion postulate contains nothing new. Since the two spinors χ_+ and χ_- form the complete set of eigenfunctions of the Hermitian operator S_z, any χ can be expanded in terms of them. The normalization and orthogonality statements contain two symbols that may seem strange. There is the dagger † to indicate that $\chi^\dagger$ is the conjugate of χ, and there is the integral sign $\int$, but there has not been specified any structure for χ that lets us explicitly conjugate or integrate anything. The useful approach is to let (7), (8), and (9) *define* these operations. Then things can be calculated and proved in a very familiar-looking way. For example, if the χ in (6) is normalized,

$$1 = \int \chi^\dagger \chi = \int (C_+^* \chi_+^\dagger + C_-^* \chi_-^\dagger)(C_+ \chi_+ + C_- \chi_-)$$

$$= |C_+|^2 \int \chi_+^\dagger \chi_+ + |C_-|^2 \int \chi_-^\dagger \chi_- + C_+^* C_- \int \chi_+^\dagger \chi_- + C_-^* C_+ \int \chi_-^\dagger \chi_+$$

$$= |C_+|^2 + |C_-|^2; \tag{8.2-10}$$

the sum of the squares of magnitudes of the expansion coefficients is unity. Here are two examples of expectation value calculations. If $\chi = \chi_-$,

$$\langle S_z \rangle = \int \chi_-^\dagger S_z \chi_- = \int \chi_-^\dagger \left(-\frac{\hbar}{2}\right)\chi_- = -\frac{\hbar}{2},$$

$$\langle S_x \rangle = \int \chi_-^\dagger S_x \chi_- = \int \chi_-^\dagger \left(\frac{\hbar}{2}\right)\chi_+ = 0.$$

The discussion of (6.2-20) shows that there are no ordinary functions of θ and φ that can describe angular momentum eigenstates with $j = \frac{1}{2}$, and formally, there is no need to introduce an explicit representation for χ and $\vec{S}$. However, a customary notation involving the Pauli spin matrices will be given here. It is more comfortable to have an explicit representation, and the notation is so widely used that one should know it.

Since S_z has only two eigenfunctions, the scheme described by (3.8-5) and (3.8-6) suggests that we choose

$$\chi_+ = \begin{pmatrix}1\\0\end{pmatrix}, \qquad \chi_- = \begin{pmatrix}0\\1\end{pmatrix}, \qquad \chi = C_+\begin{pmatrix}1\\0\end{pmatrix} + C_-\begin{pmatrix}0\\1\end{pmatrix} = \begin{pmatrix}C_+\\C_-\end{pmatrix}, \tag{8.2-11}$$

and require that the rules of matrix multiplication apply. The notation and results of Section 3.8 are then applicable; review them. Once the choice (11) has been made, the representation is determined. The dagger † is the symbol for Hermitian conjugation; that is, it indicates interchange of rows and columns and complex conjugation as defined by (3.8-14) and (3.8-16),

$$\chi^\dagger = (C_+^*, C_-^*).$$

This choice leads to the correct handling of normalization. With the symbol $\int$ no longer necessary if matrix multiplication of adjacent terms is implied, the explicit form of (10) becomes the two-component form of (3.8-4),

$$1 = \chi^\dagger \chi = (C_+^*, C_-^*)\begin{pmatrix} C_+ \\ C_- \end{pmatrix} = |C_+|^2 + |C_-|^2.$$

The statement of orthogonality becomes

$$\chi_+^\dagger \chi_- = (1, 0)\begin{pmatrix} 0 \\ 1 \end{pmatrix} = 0, \qquad \chi_-^\dagger \chi_+ = (0, 1)\begin{pmatrix} 1 \\ 0 \end{pmatrix} = 0.$$

To get explicit forms for the spin operators, introduce the *Pauli spin matrices* $\sigma_x, \sigma_y, \sigma_z$ through

$$\vec{S} = \frac{\hbar}{2}\vec{\sigma}, \tag{8.2-12}$$

that is,

$$S_x = \frac{\hbar}{2}\sigma_x, \qquad S_y = \frac{\hbar}{2}\sigma_y, \qquad S_z = \frac{\hbar}{2}\sigma_z.$$

To operate on a χ represented as in (11), the S_i and the σ_i must be 2×2 matrices. From (4),

$$S_x\chi_+ = \frac{\hbar}{2}\chi_-, \qquad \sigma_x\chi_+ = \chi_-,$$

$$\begin{pmatrix} \sigma_{x11} & \sigma_{x12} \\ \sigma_{x21} & \sigma_{x22} \end{pmatrix}\begin{pmatrix} 1 \\ 0 \end{pmatrix} = \begin{pmatrix} 0 \\ 1 \end{pmatrix}, \qquad \begin{pmatrix} \sigma_{x11} \\ \sigma_{x21} \end{pmatrix} = \begin{pmatrix} 0 \\ 1 \end{pmatrix}.$$

$$S_x\chi_- = \frac{\hbar}{2}\chi_+, \qquad \sigma_x\chi_- = \chi_+,$$

$$\begin{pmatrix} \sigma_{x11} & \sigma_{x12} \\ \sigma_{x21} & \sigma_{x22} \end{pmatrix}\begin{pmatrix} 0 \\ 1 \end{pmatrix} = \begin{pmatrix} 1 \\ 0 \end{pmatrix}, \qquad \begin{pmatrix} \sigma_{x12} \\ \sigma_{x22} \end{pmatrix} = \begin{pmatrix} 1 \\ 0 \end{pmatrix}.$$

Therefore

$$\sigma_x = \begin{pmatrix} 0 & 1 \\ 1 & 0 \end{pmatrix}. \tag{8.2-13}$$

PROBLEM 8.3

Use a similar approach to get

$$\sigma_y = \begin{pmatrix} 0 & -i \\ +i & 0 \end{pmatrix} \tag{8.2-14}$$

and

$$\sigma_z = \begin{pmatrix} 1 & 0 \\ 0 & -1 \end{pmatrix}. \tag{8.2-15}$$

The result for σ_z could have been written down without any calculations; why? Are σ_x, σ_y, and σ_z Hermitian?

The Pauli spin matrices (13), (14), and (15) were introduced in Problem 3.24 to provide practice in matrix algebra. The results of that problem are directly applicable here.

PROBLEM 8.4

a. Check that

$$\sigma_x^2 = \sigma_y^2 = \sigma_z^2 = I \equiv \begin{pmatrix} 1 & 0 \\ 0 & 1 \end{pmatrix} \tag{8.2-16}$$

b. Write $\vec{S}$ as a matrix, and check that

$$S^2\chi = \hbar^2\tfrac{1}{2}(1 + \tfrac{1}{2})\chi \text{ for any } \chi.$$

c. Check explicitly that $[S_x, S_y] = i\hbar S_z$.

The eigenvalue problem for matrices was treated in Section 3.8, and can also be discussed as follows. Look at the example $S_z\chi = \omega\chi$.

$$\begin{pmatrix} \hbar/2 & 0 \\ 0 & -\hbar/2 \end{pmatrix}\begin{pmatrix} C_+ \\ C_- \end{pmatrix} = \omega\begin{pmatrix} C_+ \\ C_- \end{pmatrix} = \begin{pmatrix} \omega & 0 \\ 0 & \omega \end{pmatrix}\begin{pmatrix} C_+ \\ C_- \end{pmatrix},$$

$$\begin{pmatrix} \hbar/2 - \omega & 0 \\ 0 & -\hbar/2 - \omega \end{pmatrix}\begin{pmatrix} C_+ \\ C_- \end{pmatrix} = 0. \tag{8.2-17}$$

This equation represents a pair of linear homogeneous algebraic equations. You may have first met matrices in this context. If there is to be a nontrivial solution, the determinant of the matrix of coefficients must be zero,

$$\begin{vmatrix} \hbar/2 - \omega & 0 \\ 0 & -\hbar/2 - \omega \end{vmatrix} = \left(\frac{\hbar}{2} - \omega\right)\left(-\frac{\hbar}{2} - \omega\right) = 0, \; \omega = +\frac{\hbar}{2} \text{ or } -\frac{\hbar}{2}.$$

This determinant is called the *secular determinant*. If $\omega = \hbar/2$, (17) becomes

$$\begin{pmatrix} 0 & 0 \\ 0 & -\hbar \end{pmatrix}\begin{pmatrix} C_+ \\ C_- \end{pmatrix} = 0, \qquad 0C_+ = 0, \qquad \hbar C_- = 0,$$

so that C_+ can be anything and C_- must be zero. Of course, this approach is completely equivalent to that of Problem 3.24 and it is really no easier for this example. It is, however, convenient for dealing with more complicated matrices.

PROBLEM 8.5

Obtain the eigenvalues of S_x and S_y by the secular determinant approach. Discuss the physical content of the results.

Suppose that the spin function is χ_+. Then a measurement of S_z must give $+\hbar/2$, so that $\langle S_z \rangle = +\hbar/2$ and $\Delta S_z = 0$. What can be said about measurement of S_x?

$$\langle S_x \rangle = \chi_+^\dagger S_x \chi_+ = \chi_+^\dagger \frac{\hbar}{2} \chi_- = 0.$$

Because σ_x^2 is the identity, $\langle S_x^2 \rangle = \hbar^2/4$, and

$$\Delta S_x = \sqrt{\langle S_x^2 \rangle - \langle S_x \rangle^2} = \frac{\hbar}{2}.$$

One might guess these results. An eigenstate of S_z describes an angular momentum vector that lies somewhere on a cone which surrounds the z axis symmetrically, as illustrated in Figure 6.1-1. Positive and negative values of S_x are equally likely so that $\langle S_x \rangle$ must be zero. If S_x is measured, the electron is forced into one of the two eigenstates of S_x, and the result of the measurement is either $+\hbar/2$ or $-\hbar/2$. The root mean squared deviation is therefore $\hbar/2$. These questions are examined in slightly different language in the example at the end of Section 6.1.

If the spin function is

$$\chi = \begin{pmatrix} 1/\sqrt{2} \\ 1/\sqrt{2} \end{pmatrix},$$

then

$$\langle S_z \rangle = \left(\frac{1}{\sqrt{2}}, \frac{1}{\sqrt{2}}\right) \frac{\hbar}{2} \begin{pmatrix} 1 & 0 \\ 0 & -1 \end{pmatrix} \begin{pmatrix} 1/\sqrt{2} \\ 1/\sqrt{2} \end{pmatrix} = 0,$$

$$\langle S_z^2 \rangle = \frac{\hbar^2}{4}, \qquad \Delta S_z = \frac{\hbar}{2}.$$

Is the electron therefore unpolarized? Not at all:

$$\langle S_x \rangle = \left(\frac{1}{\sqrt{2}}, \frac{1}{\sqrt{2}}\right) \frac{\hbar}{2} \begin{pmatrix} 0 & 1 \\ 1 & 0 \end{pmatrix} \begin{pmatrix} 1/\sqrt{2} \\ 1/\sqrt{2} \end{pmatrix} = \frac{\hbar}{2},$$

$$\Delta S_x = 0.$$

The spinor is an eigenstate of S_x with eigenvalue $\hbar/2$;

$$\frac{\hbar}{2}\begin{pmatrix}0 & 1\\ 1 & 0\end{pmatrix}\begin{pmatrix}1/\sqrt{2}\\ 1/\sqrt{2}\end{pmatrix} = \frac{\hbar}{2}\begin{pmatrix}1/\sqrt{2}\\ 1/\sqrt{2}\end{pmatrix}.$$

If there is a definite phase difference between the upper and lower components of χ, the electron is polarized in some direction. To describe an unpolarized particle, there must be between the two components a *random* phase of the kind discussed in Section 2.8.

PROBLEM 8.6

Calculate $\langle S_x\rangle$, ΔS_x, $\langle S_y\rangle$, ΔS_y, $\langle S_z\rangle$, and ΔS_z for the spinors

a. $\begin{pmatrix}i/\sqrt{2}\\ 1/\sqrt{2}\end{pmatrix}$ **b.** $\begin{pmatrix}e^{i\delta}/\sqrt{2}\\ 1/\sqrt{2}\end{pmatrix}$.

For (**b**), write the results directly as they are obtained, and then average $\langle\vec{S}\rangle$ over δ, $0 \le \delta \le 2\pi$.

EXAMPLE

In the Stern–Gerlach experiment shown in Figure 8.1-1, the upper beam is described by χ_+. What happens if this upper beam is sent through a second Stern–Gerlach magnet that separates spin orientations along the y axis, perpendicular to z and to the direction of motion?

SOLUTION

The second separation produces two beams, each described by an eigenstate of S_y. The eigenstates are given by

$$S_y = -\frac{\hbar}{2}:\quad a_-\begin{pmatrix}i/\sqrt{2}\\ 1/\sqrt{2}\end{pmatrix}, \qquad S_y = +\frac{\hbar}{2}:\quad a_+\begin{pmatrix}i/\sqrt{2}\\ -1/\sqrt{2}\end{pmatrix},$$

where a_- and a_+ are constants. Since the experiment does nothing to the original χ_+ except to pull it apart into two contributions,

$$a_-\begin{pmatrix}i/\sqrt{2}\\ 1/\sqrt{2}\end{pmatrix} + a_+\begin{pmatrix}i/\sqrt{2}\\ -1/\sqrt{2}\end{pmatrix} = \begin{pmatrix}1\\ 0\end{pmatrix},$$

$$\frac{i}{\sqrt{2}}(a_- + a_+) = 1, \qquad \frac{1}{\sqrt{2}}(a_- - a_+) = 0, \qquad a_- = a_+ = -\frac{i}{\sqrt{2}}.$$

Therefore the two beams are described by

$$-\frac{i}{\sqrt{2}}\begin{pmatrix} i/\sqrt{2} \\ 1/\sqrt{2} \end{pmatrix} = \begin{pmatrix} \frac{1}{2} \\ -i/2 \end{pmatrix}, \qquad -\frac{i}{\sqrt{2}}\begin{pmatrix} i/\sqrt{2} \\ -1/\sqrt{2} \end{pmatrix} = \begin{pmatrix} \frac{1}{2} \\ +i/2 \end{pmatrix}.$$

Notice that the square of the magnitude of each of these spinors is $\frac{1}{2}$:

$$\left(\frac{1}{2}, \ \pm\frac{i}{2}\right)\begin{pmatrix} \frac{1}{2} \\ \mp i/2 \end{pmatrix} = \tfrac{1}{4} + \tfrac{1}{4} = \tfrac{1}{2}.$$

The flux described by the normalized χ_+ is divided equally between the two eigenfunctions of S_y.

PROBLEM 8.7

Suppose that in the example above, the beams with $S_y = -\hbar/2$ and $S_y = +\hbar/2$ are each sent through a Stern–Gerlach magnet that again separates spin orientations along the z axis. Write the four spinors that result from this third separation, and examine their relation to the original χ_+.

It is important to remember how the spin formalism fits into the complete picture, and to recall that the state function depends on position, time, and spin. The quantities C_+ and C_- are therefore, in general, functions of $\vec{r}$ and t and are called $\Psi_+(\vec{r}, t)$ and $\Psi_-(\vec{r}, t)$ when this aspect is to be emphasized. Thus

$$\Psi(\vec{r}, t, \vec{S}) = \begin{pmatrix} \Psi_+(\vec{r}, t) \\ \Psi_-(\vec{r}, t) \end{pmatrix}. \tag{8.2-18}$$

For example, in the Stern–Gerlach experiment, Ψ_+ would be large in the region of the upper beam, while Ψ_- would be large in the lower beam (Figure 8.2-1).

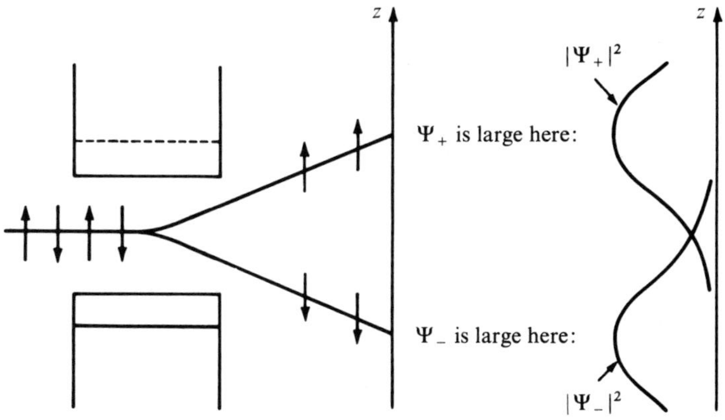

Fig. 8.2-1. The behavior of Ψ_+ and Ψ_- in a Stern–Gerlach experiment.

Suppose that the Hamiltonian for an electron is

$$H = H_{\vec{r}} + H_s,$$

where $H_{\vec{r}}$ is shorthand for the kinetic and potential energy terms that do not involve the spin, and H_s is the spin-dependent contribution to the energy given by (8.1-2);

$$H_s = -\gamma\vec{S}\cdot\vec{B} = -\gamma\frac{\hbar}{2}\vec{\sigma}\cdot\vec{B}.$$

Then the Schrödinger equation is

$$\left(H_{\vec{r}} - \frac{\gamma\hbar}{2}\vec{\sigma}\cdot\vec{B}\right)\begin{pmatrix}\Psi_+\\ \Psi_-\end{pmatrix} = i\hbar\frac{\partial}{\partial t}\begin{pmatrix}\Psi_+\\ \Psi_-\end{pmatrix}. \tag{8.2-19}$$

With (13), (14), and (15) for σ_x, σ_y, and σ_z,

$$\begin{aligned}\vec{\sigma}\cdot\vec{B} &= \begin{pmatrix}0 & 1\\ 1 & 0\end{pmatrix}B_x + \begin{pmatrix}0 & -i\\ i & 0\end{pmatrix}B_y + \begin{pmatrix}1 & 0\\ 0 & -1\end{pmatrix}B_z\\ &= \begin{pmatrix}B_z & B_x - iB_y\\ B_x + iB_y & -B_z\end{pmatrix}.\end{aligned} \tag{8.2-20}$$

Since it operates on the $\Psi(\vec{r}, t, \vec{S})$ given in (18), $H_{\vec{r}}$ has to be taken as

$$H_{\vec{r}}I = \begin{pmatrix}H_{\vec{r}} & 0\\ 0 & H_{\vec{r}}\end{pmatrix},$$

where I is the 2×2 identity matrix. Then (19) becomes

$$\begin{pmatrix}H_{\vec{r}} - (\gamma\hbar/2)B_z & -(\gamma\hbar/2)(B_x - iB_y)\\ -(\gamma\hbar/2)(B_x + iB_y) & H_{\vec{r}} + (\gamma\hbar/2)B_z\end{pmatrix}\begin{pmatrix}\Psi_+\\ \Psi_-\end{pmatrix} = i\hbar\frac{\partial}{\partial t}\begin{pmatrix}\Psi_+\\ \Psi_-\end{pmatrix}. \tag{8.2-21}$$

If B_x and B_y are both zero, the equations for Ψ_+ and Ψ_- are uncoupled, but in general, they are not.

Suppose that $\vec{B}$ does not depend on position, so that the spin contribution to the energy does not depend on position. With (19) written as

$$(H_{\vec{r}} + H_s)\Psi(\vec{r}, t, \vec{S}) = i\hbar\frac{\partial}{\partial t}\Psi(\vec{r}, t, \vec{S}), \tag{8.2-22}$$

one can separate space and spin variables. Substitute $\Psi(\vec{r}, t, \vec{S}) = \Psi(\vec{r}, t)\chi(t)$.

$$\chi\left[H_{\vec{r}}\Psi(\vec{r}, t) - i\hbar\frac{\partial}{\partial t}\Psi(\vec{r}, t)\right] + \Psi(\vec{r}, t)\left[H_s\chi - i\hbar\frac{\partial\chi}{\partial t}\right] = 0.$$

Multiply by $\chi^\dagger/\Psi$. Since χ is normalized so that $\chi^\dagger\chi = 1$,

$$\frac{1}{\Psi(\vec{r}, t)}\left[H_{\vec{r}}\Psi(\vec{r}, t) - i\hbar\frac{\partial}{\partial t}\Psi(\vec{r}, t)\right] + \chi^\dagger\left[H_s\chi - i\hbar\frac{\partial\chi}{\partial t}\right] = 0.$$

The first term depends on position and time, while the second term depends on spin and time. Each term can therefore depend at most on the time. If the first term is set equal to some $F(t) \neq 0$, then the effect is to shift the zero of energy for the spatial part in one direction and for the spin part in the opposite direction by the same amount. With both terms set equal to zero, the spin alone obeys a Schrödinger equation of the form

$$H_s\chi = i\hbar\dot{\chi}. \tag{8.2-23}$$

Since χ does not depend on $\vec{r}$, it is correct and unambiguous to write the ordinary time derivative $d\chi/dt$ or $\dot{\chi}$ rather than $\partial\chi/\partial t$. Section 8.4 describes applications of (23).

8.3 *Spinor Geometry*

If $\langle\vec{S}\rangle$ points in the direction given by the polar angles θ, φ, that is, if

$$\begin{aligned}\langle\vec{S}\rangle &= \frac{\hbar}{2}\chi^\dagger\vec{\sigma}\chi \\ &= \frac{\hbar}{2}(\hat{e}_x \sin\theta\cos\varphi + \hat{e}_y \sin\theta\sin\varphi + \hat{e}_z\cos\theta),\end{aligned} \tag{8.3-1}$$

then what can be said about χ? If χ has the components C_+, C_-, as in (8.2-11),

$$\begin{aligned}\chi^\dagger\sigma\chi &= (C_+^*, C_-^*)\left[\hat{e}_x\begin{pmatrix}0 & 1\\ 1 & 0\end{pmatrix} + \hat{e}_y\begin{pmatrix}0 & -i\\ i & 0\end{pmatrix} + \hat{e}_z\begin{pmatrix}1 & 0\\ 0 & -1\end{pmatrix}\right]\begin{pmatrix}C_+\\ C_-\end{pmatrix} \\ &= \hat{e}_x(C_+^* C_- + C_-^* C_+) + \hat{e}_y(iC_-^* C_+ - iC_+^* C_-) + \hat{e}_z(C_+^* C_+ - C_-^* C_-).\end{aligned} \tag{8.3-2}$$

The coefficients of the unit vectors in (1) and (2) must be equal.

$$\begin{aligned}C_+^* C_- + C_-^* C_+ &= \sin\theta\cos\varphi \\ iC_-^* C_+ - iC_+^* C_- &= \sin\theta\sin\varphi \\ C_+^* C_+ - C_-^* C_- &= \cos\theta\end{aligned}$$

The general solution to these equations is

$$C_+ = e^{i\delta}\cos\frac{\theta}{2}e^{-i\varphi/2}, \qquad C_- = e^{i\delta}\sin\frac{\theta}{2}e^{+i\varphi/2}. \tag{8.3-3}$$

The quantity δ can be any real number; as always, the overall phase of the state function cannot be fixed by any measurement. Normalization is assured by the requirement that the expectation value must have the right size, and $|C_+|^2 + |C_-|^2 = 1$ is satisfied by (3).

PROBLEM 8.8

Check that the expressions given for C_+ and C_- in (3) are the most general solutions. A method is to substitute

$$C_+ = A_+ e^{i\delta} \cos\frac{\theta}{2} e^{-i\varphi/2}, \qquad C_- = A_- e^{i\delta} \sin\frac{\theta}{2} e^{+i\varphi/2},$$

and find what A_+ and A_- must be to satisfy the three conditions.

The occurrence of half angles is an important feature of (3). Suppose that $\langle\vec{S}\rangle$ points along the positive z axis. Then $\theta = 0$, $|C_+|^2 = 1$, and $|C_-|^2 = 0$. Leave the particle undisturbed but rotate the coordinate system around the x axis through an angle π, so that the new z axis points down, opposite to $\langle\vec{S}\rangle$. In the new coordinate system, $\theta' = \pi$, $|C'_+|^2 = \cos^2 \pi/2 = 0$, and $|C'_-|^2 = \sin^2 \pi/2 = 1$. This result is just what it should be. An observer in the original coordinate system says that the particle is described by χ_+, while an observer in the new coordinate system is standing on his head and says that the particle is described by χ_-.

Now start with some arbitrary $\langle\vec{S}\rangle$, and consider various possible rotations through 2π. If θ is replaced by $\theta + 2\pi$, since

$$\cos\left(\frac{\theta + 2\pi}{2}\right) = -\cos\frac{\theta}{2}, \qquad \sin\left(\frac{\theta + 2\pi}{2}\right) = -\sin\frac{\theta}{2},$$

both C_+ and C_- change sign. Also, if φ is replaced by $\varphi + 2\pi$, since

$$e^{i(\varphi + 2\pi)/2} = -e^{i\varphi/2},$$

C_+ and C_- change sign. Under such rotations any spinor changes sign, and so spinors are not single valued. Intimations of this conclusion are visible in Section 6.2, where the occurrence of $e^{im\varphi}$ shows that wave functions are single valued only for *integral m*. Double-valued state functions are something new, but no disaster, because the measurable quantities, the expectation values $\chi^\dagger\Omega\chi$, are all single valued.

The same conclusions are right for all half-integral ($\frac{1}{2}$, $\frac{3}{2}$, ...) angular momentum systems; all show the same double valuedness under rotations through 2π. The special word *spinors* was introduced because this behavior under rotations is entirely different from that of the scalars, vectors, and tensors of classical physics.

The change in sign caused by 2π rotations has been observed. For example, a beam of slow, long-wavelength neutrons can be split into two beams which show interference in the usual way when they are brought back together. As described in the next section, the spin of the neutrons can be rotated a controlled amount by a magnetic field. If the neutrons that are in one of the beams have their spins turned around through 2π, while nothing is done to the other beam, constructive interference occurs where previously there was destructive interference, and *vice versa.*[5] In general, because spinors are functions of $\theta/2$ and $\varphi/2$, the change in interference patterns occurs for rotations through 2π, 6π, 10π, ..., while no change is seen for 4π, 8π, 12π, ...

Functions that describe states with half-integral angular momentum change sign under 2π rotations, while those for states with integral angular momentum are single valued. A consequence is that the various states of a given system must either all have integral angular momentum, or all have half-integral angular momentum. Suppose that there were a nucleus that had one state Ψ_0 with zero angular momentum and another state $\Psi_{1/2}$ with angular momentum $\hbar/2$. The nucleus could be put into a state which is a superposition of Ψ_0 and $\Psi_{1/2}$,

$$\Psi = a_0 \Psi_0 + a_{1/2} \Psi_{1/2}.$$

If the entire nucleus is rotated through 2π, Ψ would become

$$a_0 \Psi_0 - a_{1/2} \Psi_{1/2}.$$

Since one piece of the wave function changes sign and the other does not, observables such as $|\Psi|^2$ and expectation values $\int \Psi^\dagger \Omega \Psi$ would change, contrary to the requirement that rotations of an entire system through 2π must be undetectable. It is indeed true that the states of a given atom or of a given nucleus either all have integral angular momentum or all have half-integral angular momentum. Integral angular momentum is always found if the system contains an even number of spin $\frac{1}{2}$ particles, and half-integral angular momentum is always found if the system contains an odd number of spin $\frac{1}{2}$ particles.

PROBLEM 8.9

Two hundred footballs are rolling around on the floor. How many of them have double-valued state functions?

8.4 *Magnetic Moments in Magnetic Fields*

For any $\vec{B}$ that does not depend on position, the space and spin variables can be separated to give the spin Schrödinger equation (8.2-23). With $\vec{\sigma} \cdot \vec{B}$ written in the form (8.2-20), this equation is

$$-\frac{\gamma\hbar}{2}\begin{pmatrix} B_z & B_x - iB_y \\ B_x + iB_y & -B_z \end{pmatrix}\begin{pmatrix} C_+ \\ C_- \end{pmatrix} = i\hbar\begin{pmatrix} \dot{C}_+ \\ \dot{C}_- \end{pmatrix}. \tag{8.4-1}$$

[5] S. A. Werner, R. Colella, A. W. Overhauser, and C. F. Eagen, *Phys, Rev. Lett.* **35**, 1053 (1975); H. Rauch, A. Zeilinger, G. Badurek, A. Wilfing, W. Bauspiess, and U. Bouse, *Phys. Lett.* **54A**, 425 (1975); A. G. Klein and G. I. Opat, *Phys. Rev. Lett.* **37**, 238 (1976).

If the system is in an energy eigenstate with eigenvalue E,

$$i\hbar\dot{C}_\pm = EC_\pm,$$

and (1) becomes

$$\begin{pmatrix} -(\gamma\hbar/2)B_z - E & -(\gamma\hbar/2)(B_x - iB_y) \\ -(\gamma\hbar/2)(B_x + iB_y) & +(\gamma\hbar/2)B_z - E \end{pmatrix}\begin{pmatrix} C_+ \\ C_- \end{pmatrix} = 0. \tag{8.4-2}$$

This expression is equivalent to two homogeneous equations for the two unknowns C_+ and C_-, and a nontrivial solution exists if the determinant of the coefficients is zero,

$$E^2 - \left(\frac{\gamma\hbar B_z}{2}\right)^2 - \left(\frac{\gamma\hbar}{2}\right)^2(B_x^2 + B_y^2) = 0$$

or

$$E = \pm\frac{\gamma\hbar B}{2},$$

where B is the magnitude of the field. For $E = -\gamma\hbar B/2$, either of the two equations contained in (2) determines the ratio

$$\frac{C_+}{C_-} = \frac{B_x - iB_y}{B - B_z}.$$

If the direction of $\vec{B}$ is given by the polar angles θ, φ, then

$$\frac{C_+}{C_-} = \frac{\sin\theta\cos\varphi - i\sin\theta\sin\varphi}{1 - \cos\theta}$$

$$= \frac{\sin\frac{\theta}{2}\cos\frac{\theta}{2}e^{-i\varphi}}{\frac{(1-\cos\theta)}{2}} = \frac{\cos\frac{\theta}{2}e^{-i\varphi/2}}{\sin\frac{\theta}{2}e^{+i\varphi/2}}. \tag{8.4-3}$$

The "eigen spinor" for $E = -\gamma\hbar B/2$ is therefore the same as the spinor found in (8.3-3), as one should expect.

PROBLEM 8.10

a. Explain why one should expect the spinor determined by (3) to be the same as that in (8.3-3).

b. Find C_+/C_- for the eigen spinor with eigenvalue $E = +\gamma\hbar B/2$. Can you understand the result by applying a rotation to the spinor given by (3)?

For the special case $\vec{B} = B\hat{e}_z$, the two eigenstates are χ_+ and χ_-, that is, with time dependence included,

$$\begin{pmatrix}1\\0\end{pmatrix}e^{-iE_+t/\hbar} = \begin{pmatrix}e^{i\gamma Bt/2}\\0\end{pmatrix} \tag{8.4-4}$$

and

$$\begin{pmatrix}0\\1\end{pmatrix}e^{-iE_-t/\hbar} = \begin{pmatrix}0\\e^{-i\gamma Bt/2}\end{pmatrix}. \tag{8.4-5}$$

Be careful with the signs. Note that E_+, the energy for $m_s = +\frac{1}{2}$, is $-\gamma\hbar B/2$. Any solution to (1) for $\vec{B} = B\hat{e}_z$ can be written as a linear combination, C_+ times (4) plus C_- times (5), or

$$\chi(t) = \begin{pmatrix}C_+e^{i\gamma Bt/2}\\C_-e^{-i\gamma Bt/2}\end{pmatrix}. \tag{8.4-6}$$

PROBLEM 8.11

Obtain (4) and (5) by substituting $C_\pm = A_\pm e^{i\omega t}$ into (1) with $B = B_z$, $B_x = B_y = 0$.

If the system is not in an eigenstate of energy because $\langle\vec{S}\rangle$ is at some angle with respect to $\vec{B}$, then $\langle\vec{S}\rangle$ precesses around $\vec{B}$. The behavior is much like that of a classical gyroscope that is subjected to a torque. Suppose that, at $t = 0$, $\langle\vec{S}\rangle$ points in the x direction. Then $\chi(0)$ is an eigen spinor of S_x with eigenvalue $+\hbar/2$; that is, as is shown in Section 8.2, $\chi(t)$ is subject to the initial condition

$$\chi(0) = \frac{1}{\sqrt{2}}\begin{pmatrix}1\\1\end{pmatrix}.$$

Then

$$\chi(t) = \frac{1}{\sqrt{2}}\begin{pmatrix}e^{i\gamma Bt/2}\\e^{-i\gamma Bt/2}\end{pmatrix},$$

and the expectation value of spin at any time is

$$\begin{aligned}\langle\vec{S}\rangle = \chi^\dagger(t)\vec{S}\chi(t) &= \frac{\hbar}{4}(e^{-i\gamma Bt/2}, e^{+i\gamma Bt/2})\begin{pmatrix}\hat{e}_z & \hat{e}_x - i\hat{e}_y\\ \hat{e}_x + i\hat{e}_y & -\hat{e}_z\end{pmatrix}\begin{pmatrix}e^{i\gamma Bt/2}\\e^{-i\gamma Bt/2}\end{pmatrix}\\ &= \frac{\hbar}{2}(\hat{e}_x \cos\gamma Bt - \hat{e}_y \sin\gamma Bt).\end{aligned} \tag{8.4-7}$$

As in Chapter 3 examples such as Problems 3.18 and 3.48, a superposition of two energy eigenstates with E_1 and E_2 gives expectation values that oscillate with angular frequency $(E_2 - E_1)/\hbar$.

PROBLEM 8.12

Verify that, for the $\langle\vec{S}\rangle$ in (7),

$$\frac{d}{dt}\langle\vec{S}\rangle = \gamma\langle\vec{S}\rangle \times \vec{B}, \tag{8.4-8}$$

and show that classical mechanics also gives this result.

The precession of spins in magnetic fields is an important phenomenon in a great variety of contexts. For example, the behavior of muons provides an interesting application of elementary particle physics techniques to solid state problems.[6] It is possible to bring polarized positive muons, μ^+, to rest in samples of solid materials. The muons decay with a mean life of 2.20×10^{-6} seconds into a positron and two neutrinos,

$$\mu^+ \rightarrow e^+ + \bar{\nu}_\mu + \nu_e.$$

The decay interaction is such that, while the e^+ has some chance of heading in any direction, it comes out preferentially in the direction of polarization, in the direction of the μ^+ spin. Suppose that the muons are all polarized in the x direction, and that an e^+ counter views the sample along the y axis. The counter records the number of e^+ detected as a function of the time between the arrival of the muon and its decay. If there is no magnetic field, the μ^+ retains its spin direction until it decays, and the e^+ counter gives a simple exponential with a $1/e$ time of 2.20 microseconds. If, however, there is a magnetic field $\vec{B} = B\hat{e}_z$ inside the sample, each muon begins to precess according to (7) as soon as it arrives. The $\langle\vec{S}\rangle$ of the μ^+ points in the y direction toward the e^+ counter whenever γBt is $3\pi/2$, $7\pi/2$, $11\pi/2, \ldots$. A decay e^+ emitted at these times has a relatively good chance of hitting the counter, while a decay e^+ emitted when γBt is $\pi/2$, $5\pi/2$, $9\pi/2, \ldots$ has a relatively poor chance of hitting the counter. With the field turned on, the decay record is no longer a simple exponential; there is superposed an oscillation with period equal to $2\pi/\gamma B$. The known value of γ, e/M_μ, gives a period of 7.39×10^{-9} (webers/m^2) sec/B, so that fields of around 0.01 webers/m^2 or 100 gauss give periods of around a microsecond. The result of the measurement is the value of B at the place where the muons stop. Since the muons generally come to rest between the atoms, this result is of interest in solid state physics and is hard to determine by less exotic methods.

A different class of applications of (1) is provided by *magnetic resonance* experiments. Suppose that there is a strong and constant field $B_0\hat{e}_z$, and a small

[6] W. J. Kossler, "μ^+ Precession Studies of Magnetic Materials," in *Magnetism and Magnetic Materials–1974*, AIP Conference Proceedings No. 24, C. D. Graham, Jr., G. H. Lander, and J. J. Rhyne, eds. (New York: American Institute of Physics, Inc., 1975), pp. 788–792.

rotating field $\vec{B}_r$ perpendicular to $\hat{e}_z$, so that the resultant is

$$\vec{B}(t) = B_r(\hat{e}_x \cos \omega t - \hat{e}_y \sin \omega t) + B_0 \hat{e}_z. \tag{8.4-9}$$

With this field the Hamiltonian has explicit time dependence. Equation (3.10-4) shows that energy conservation is not applicable. There are no energy eigenstates, and (1) has to be solved from the beginning as a pair of coupled first-order differential equations. With the field (9), (1) becomes

$$\frac{i\gamma}{2}(B_0 C_+ + B_r e^{+i\omega t} C_-) = \dot{C}_+, \tag{8.4-10}$$

$$\frac{i\gamma}{2}(B_r e^{-i\omega t} C_+ - B_0 C_-) = \dot{C}_-. \tag{8.4-11}$$

Look for a solution by trying

$$C_\pm = A_\pm e^{i\omega_\pm t}.$$

With $\omega_0 \equiv \gamma B_0$, $\omega_r \equiv \gamma B_r$, (10) and (11) become

$$(\omega_0 - 2\omega_+)A_+ + \omega_r A_- e^{i(\omega_- - \omega_+ + \omega)t} = 0,$$

$$\omega_r A_+ e^{-i(\omega_- - \omega_+ + \omega)t} - (\omega_0 + 2\omega_-)A_- = 0.$$

These equations must be valid for all time if the trial solution works. Since a constant term cannot equal a term that varies in time,

$$\omega_+ = \omega + \omega_-$$

and the two equations become

$$(\omega_0 - 2\omega - 2\omega_-)A_+ + \omega_r A_- = 0,$$

$$\omega_r A_+ - (\omega_0 + 2\omega_-)A_- = 0.$$

A nontrivial solution exists if the determinant of the coefficients is zero.

$$\therefore \quad \omega_- = -\frac{\omega}{2} \pm \eta, \qquad \eta \equiv \tfrac{1}{2}\sqrt{(\omega - \omega_0)^2 + \omega_r^2}, \tag{8.4-12}$$

and the general solution for C_- is

$$C_-(t) = e^{-i\omega t/2}(A_{1-} e^{+i\eta t} + A_{2-} e^{-i\eta t}). \tag{8.4-13}$$

Take as initial condition that the spin is definitely "up" at $t = 0$; that is, choose $C_-(0) = 0$. Then $A_{1-} = -A_{2-}$, and with $2iA_{1-} \equiv A$,

$$C_-(t) = Ae^{-i\omega t/2} \sin \eta t.$$

This value substituted into (11) gives

$$C_+(t) = \frac{Ae^{+i\omega t/2}}{\omega_r}[(\omega_0 - \omega)\sin \eta t - 2i\eta \cos \eta t].$$

If $C_+(0) = 1$, then $A = i\omega_r/2\eta$, so that

$$|C_-(t)|^2 = \frac{\omega_r^2}{4\eta^2} \sin^2 \eta t.$$

This result means that a fraction

$$\frac{\omega_r^2}{4\eta^2} = \frac{\omega_r^2}{(\omega - \omega_0)^2 + \omega_r^2} \tag{8.4-14}$$

of the spins oscillates up and down with angular frequency 2η.

PROBLEM 8.13

Repeat the development above with the initial condition that the spin is definitely "down" at $t = 0$.

The expression (14) has the Lorentz or Breit–Wigner form discussed in Section 3.11 and graphed in Figure 3.11-3. The maximum value, unity, occurs when the resonance condition $\omega = \omega_0$ is satisfied. The full width at half maximum is $2\omega_r$.

It is easy to understand the location and width of the peak. With $B_0 \gg B_r$ the difference in energy between states with magnetic moment opposite and along $\hat{e}_z$ is $\hbar\gamma B_0/2 - (-\hbar\gamma B_0/2) = \hbar\omega_0$. The rotating field is equivalent to photons with angular frequency ω and energy $\hbar\omega$. For perfect resonance, $\hbar\omega = \hbar\omega_0$, and the photons have the right energy to induce transitions between the spin up and spin down states. The spins oscillate with angular frequency 2η, which is about ω_r near resonance. The mean life in either state is roughly $1/\omega_r$, and so the width in energy, Γ in the language of Section 3.11, should be roughly $\hbar\omega_r$. A width in frequency of $2\omega_r$ is therefore reasonable.

Nuclear magnetic resonance experiments[7] are a particularly significant application. If a sample is placed in a field described by (9), the resonance curve (14) can be traced out by varying either the angular frequency ω or the field B_0. The signal that is actually observed may be the power absorbed in the oscillating field, or it may be the voltage induced in a pick-up coil by the changing magnetization of the sample (Figure 8.4-1). Resonance occurs when $\omega = \gamma B_0$, and the experiment can be used to measure γ. Since only a frequency and a magnetic field need to be measured, great precision is possible. Samples that

[7] An introductory survey and good lists of basic references are given by R. Schumacher, *Magnetic Resonance* (New York: W. A. Benjamin, Inc., 1970). See also C. Kittel, *Introduction to Solid State Physics*, 5th ed. (New York: John Wiley & Sons, Inc., 1976), Chap. 16, and the discussion of magnetic resonance for nuclei and for electrons in A. C. Melissinos, *Experiments in Modern Physics* (New York: Academic Press, Inc., 1966), Chap. 8.

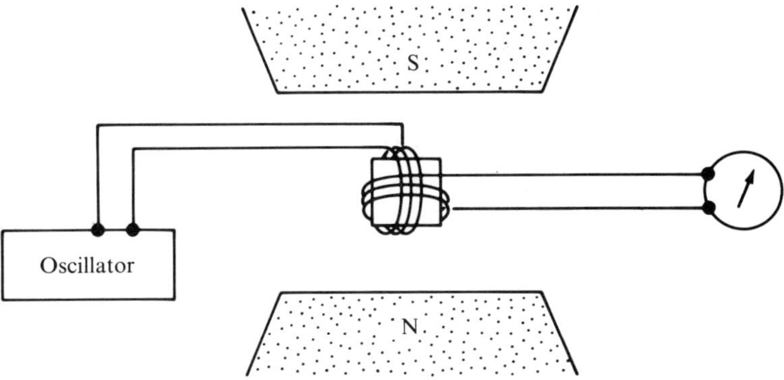

Fig. 8.4-1. A magnetic resonance experiment. The field $B_0\hat{e}_z$ is supplied by the poles of a large magnet. The oscillating field is usually a linear oscillation rather than the rotating field assumed in the calculation. One can trace out the resonance either by changing the frequency of the oscillator or by changing the vertical field B_0.

contain hydrogen, such as water or paraffin, give for the magnetic moment of the proton

$$\gamma_p \frac{\hbar}{2} = (2.7928456 \pm 0.0000011)\frac{e\hbar}{2M_p}. \tag{8.4-15}$$

To measure the magnetic moment of the neutron, a somewhat different technique is needed because it is impossible to have a high-density sample. One can send a beam of polarized neutrons through a field of the kind described by (9) and measure the frequency at which the neutrons are depolarized. The result for the magnetic moment of the neutron is

$$\gamma_n \frac{\hbar}{2} = -(1.913148 \pm 0.000066)\frac{e\hbar}{2M_p}. \tag{8.4-16}$$

The results (15) and (16) help support our meson picture of nuclear forces. If the proton and neutron were uncomplicated particles like the electron, one might expect the proton moment to be $e\hbar/2M_p$, and the moment of the uncharged neutron to be zero. However, as is mentioned in Section 1.3, the proton undergoes virtual disintegration with a high probability and spends some of its time being a neutron and a positive pion,

$$p \rightleftarrows n + \pi^+.$$

The π^+ is a relatively light particle, with mass about $\frac{1}{7}$ of the proton mass. It therefore forms a relatively effective current loop, and one can understand qualitatively the proton moment as containing $e\hbar/2M_p$ contributed by the proton itself, and about $+1.8\ e\hbar/2M_p$ contributed by the π^+. The neutron spends some of its time being a proton and a negative pion,

$$n \rightleftarrows p + \pi^-.$$

Its moment can be understood qualitatively as containing zero contributed by the neutron itself, and about $-1.9e\hbar/2M_p$ contributed by the negative pion. The fact that the coefficients of the pion contribution, $+1.8$ and -1.9, have similar magnitudes and opposite signs supports our pictures of the pions, π^+ and π^-, being emitted into orbital angular momentum $l = 1$ states with about the same probabilities.

Measurements of the kinds described for the nucleons have also been used to determine the magnetic moments of more complicated nuclei, and the results are valuable nuclear structure information. The calculations of this section are directly applicable to nuclei with total angular momentum $\hbar/2$, such as ^{3}H, ^{3}He, ^{15}N, ^{117}Sn, and ^{205}Tl. Nuclei with total angular momentum zero, such as ^{4}He, ^{16}O, and ^{238}U, have no vector to which a magnetic moment can be tied, they cannot know which way is up, and their moments are all zero. Nuclei with total angular momentum $1\hbar$, $3\hbar/2$, $2\hbar$, $5\hbar/2$, and higher require calculations that differ from the $\hbar/2$ case only in detail. For example, $1\hbar$ nuclei, such as the deuteron ^{2}H and ^{14}N, can have three orientations, with $m = -1, 0,$ or $+1$, and a field of the form (9) can induce transitions between these three states. For the deuteron the magnetic moment is found to be 0.85739 $e\hbar/2M_p$. The sum of (15) and (16) is 0.87970 $e\hbar/2M_p$. The closeness of the two numbers supports the picture that the deuteron is primarily a state of a neutron and a proton with parallel spins and no other angular momenta, with room for a few percent of more complicated behavior, as described in Section 7.1 and considered more carefully in Section 9.3.

Once the magnetic moment of a nucleus is known accurately, a measurement of its resonant frequency $\omega_0 = \gamma B_0$ becomes a measurement of the field. In solids and liquids nuclear magnetic resonance techniques are therefore used widely to determine internal fields. Once the internal fields in a compound are known, the resonance techniques become an analytic tool for identifying the compound. Nuclear magnetic resonance is today used in this way in chemistry and biology, and has become one of the many developments at the frontiers of physics that turn into standard tools in other fields.

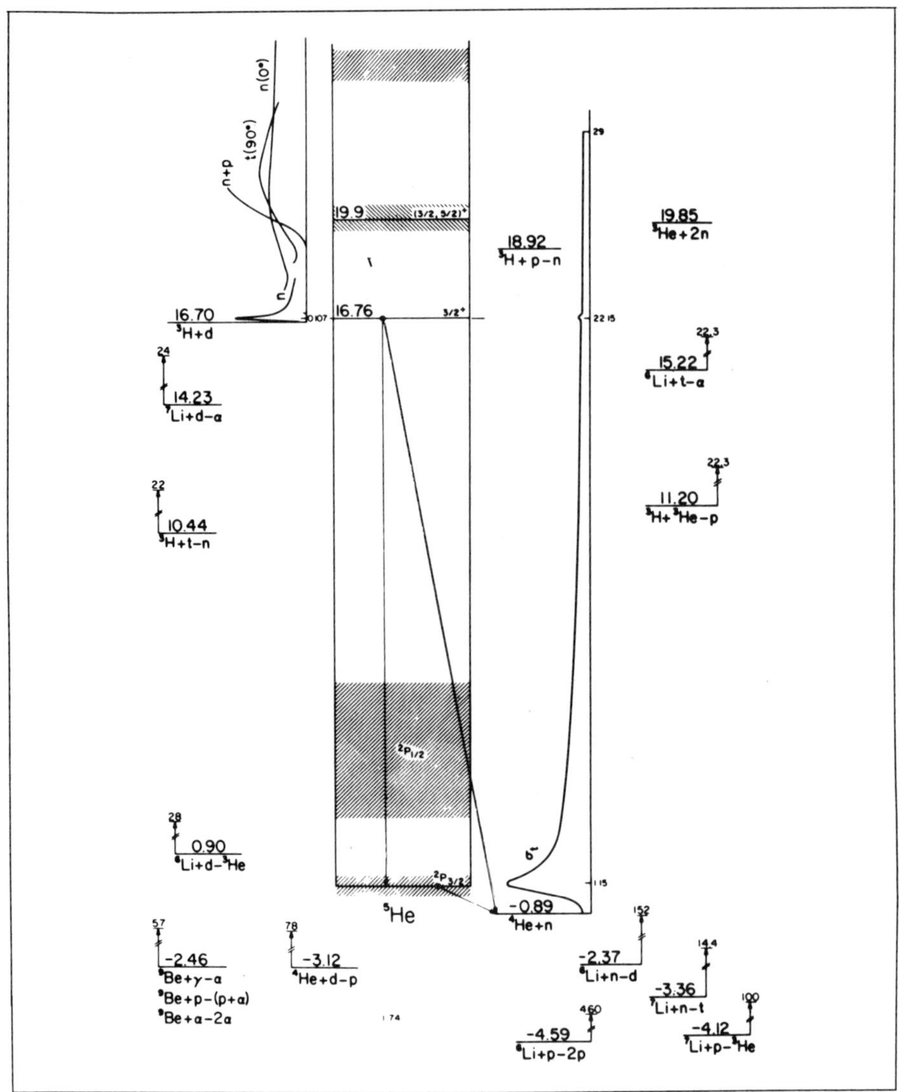

Energy levels of ^{5}He, that is, of the nuclear system formed by two protons and three neutrons. This system is not bound because the energy of its lowest state lies 0.89 MeV above the energy of the nucleus ^{4}He and a separate neutron. The shading indicates levels that have large natural widths because of their short life. To good approximation, the two lowest levels can be described as consisting of two neutrons and two protons in $l = 0$, or s, states, plus a third neutron that is forced by the exclusion principle to be in an $l = 1$, or p, state. This third neutron then accounts for all the angular momentum of the system, and its spin $\frac{1}{2}$ can combine with its orbital angular momentum to give either $j = \frac{3}{2}$, described as $P_{3/2}$, or $j = \frac{1}{2}$, described as $P_{1/2}$. Reactions used to study the system are also indicated. All energies are in MeV. Further details are given by F. Ajzenberg-Selove and T. Lauritsen, Energy Levels of Light Nuclei, A = 5 − 10, *in Nuclear Physics* **A227** *(1974) 1. Updated data of this kind for all nuclei are regularly published in the journal* Nuclear Data Sheets *(New York: Academic Press). (With permission of the North-Holland Publishing Company and of F. Ajzenberg-Selove)*

9

The Addition of Angular Momenta

A neutron and a proton are interacting because they are bound together to make a deuteron, or because the neutron is being scattered by the proton. Neutrons and protons have spin $\frac{1}{2}$. What can be said about the *total* spin of the two-particle system? An electron is in a hydrogen atom and has orbital angular momentum quantum number $l = 1$. What can be said about the sum of its orbital and spin angular momentum? To answer questions of this kind, we need to learn something about the addition of angular momenta.

Quick guesses can be poor guides in this area. Quantal angular momenta are not fixed vectors; all that can be known is that they lie in cones with opening angle $\cos^{-1}(m/\sqrt{j^2 + j})$, and their addition has to be seen as an assembly of such cones. Many "modern physics" texts present geometric treatments of such additions.[1] Here the development will be based on the angular momentum formalism of Chapters 6 and 8.

[1] K. W. Ford, *Classical and Modern Physics* (Lexington, Mass.: Xerox College Publishing, 1972), Sec. 24.4.

9.1 Spin $\frac{1}{2}$ Plus Spin $\frac{1}{2}$

As described in Chapter 8, the spin state of a neutron can be written in the form

$$\chi(n) = C_{n+}\chi_+(n) + C_{n-}\chi_-(n),$$

where $\chi_+(n)$ and $\chi_-(n)$ are the eigenstates of S_{nz} with eigenvalues $+\hbar/2$ and $-\hbar/2$. Similarly, the spin state of a proton is

$$\chi(p) = C_{p+}\chi_+(p) + C_{p-}\chi_-(p).$$

How should the spin state for the two particles together be written? First, suppose we know that S_{nz} has eigenvalue $+\hbar/2$ and that S_{pz} has eigenvalue $-\hbar/2$. Then the state is described correctly by the product

$$\chi_+(n)\chi_-(p).$$

To check this claim, note that components of $\vec{S}_n$ do not operate on $\chi(p)$, that is, they treat $\chi(p)$ as if it were a constant; and components of $\vec{S}_p$ do not operate on $\chi(n)$. Therefore

$$S_{nz}[\chi_+(n)\chi_-(p)] = [S_{nz}\chi_+(n)]\chi_-(p) = +\frac{\hbar}{2}\chi_+(n)\chi_-(p),$$

$$S_{pz}[\chi_+(n)\chi_-(p)] = \chi_+(n)[S_{pz}\chi_-(p)] = -\frac{\hbar}{2}\chi_+(n)\chi_-(p);$$

the product is an eigenstate of S_{nz} with eigenvalue $+\hbar/2$, and an eigenstate of S_{pz} with eigenvalue $-\hbar/2$. Also, the product is normalized:

$$[\chi_+(n)\chi_-(p)]^\dagger[\chi_+(n)\chi_-(p)] = [\chi_+^\dagger(n)\chi_+(n)][\chi_-^\dagger(p)\chi_-(p)] = 1 \times 1 = 1.$$

Next, suppose that we want the most general spin state for the two particles. This state is the product of the $\chi(n)$ and the $\chi(p)$ given above:

$$\begin{aligned}\zeta(n, p) &\equiv \chi(n)\chi(p) \\ &= C_{++}\chi_+(n)\chi_+(p) + C_{+-}\chi_+(n)\chi_-(p) \\ &\quad + C_{-+}\chi_-(n)\chi_+(p) + C_{--}\chi_-(n)\chi_-(p),\end{aligned} \tag{9.1-1}$$

where the products of constants have been renamed, $C_{n\pm}C_{p\pm} \equiv C_{\pm\pm}$. The spin of the neutron can be up or down, the spin of the proton can be up or down, there are four different possible results of a measurement of both spins, and each term in (1) corresponds to one of these results. The four terms

$$\chi_+(n)\chi_+(p), \qquad \chi_+(n)\chi_-(p), \qquad \chi_-(n)\chi_+(p), \qquad \chi_-(n)\chi_-(p)$$

are *complete*, and (1) is indeed the most general possibility. These ideas are implicit in Chapter 4 where several-particle wave functions are written as sums of products of single-particle wave functions.

The total spin is the sum of the neutron and proton spins,

$$\vec{S}_T = \vec{S}_n + \vec{S}_p, \qquad S_{Tz} = S_{nz} + S_{pz}. \tag{9.1-2}$$

Let us now find the two-particle spin states that are eigenfunctions of S_{Tz} and S_T^2. For S_{Tz} the eigenvalue equation is

$$S_{Tz}\zeta(n, p) = \hbar m\zeta(n, p). \tag{9.1-3}$$

With

$$(S_{nz} + S_{pz})\chi_+(n)\chi_\pm(p) = \frac{\hbar}{2}\chi_+(n)\chi_\pm(p) \pm \frac{\hbar}{2}\chi_+(n)\chi_\pm(p),$$

$$(S_{nz} + S_{pz})\chi_\pm(n)\chi_+(p) = \pm\frac{\hbar}{2}\chi_\pm(n)\chi_+(p) + \frac{\hbar}{2}\chi_\pm(n)\chi_+(p),$$

(1) substituted into (3) gives

$$\begin{aligned}(1 - m)C_{++}\chi_+(n)\chi_+(p) &- mC_{+-}\chi_+(n)\chi_-(p)\\ - mC_{-+}\chi_-(n)\chi_+(p) &- (1 + m)C_{--}\chi_-(n)\chi_-(p) = 0.\end{aligned} \tag{9.1-4}$$

The four spin functions in this equation are orthogonal, and therefore their four coefficients must all be zero. If any of the $C_{\pm\pm}$ are different from zero, then m must be $+1$, -1, or 0. If $m = +1$, then C_{+-}, C_{-+}, and C_{--} must be zero, and, with the choice $C_{++} = 1$, the spin state is

$$\zeta_{++}(n, p) = \chi_+(n)\chi_+(p). \tag{9.1-5}$$

This result contains the reasonable conclusion that if both particles have z component of spin angular momentum equal to $+\hbar/2$, then the combination has a z component $m\hbar = +1\hbar$. If $m = -1$, C_{++}, C_{+-}, and C_{-+} must be zero, and, with $C_{--} = 1$,

$$\zeta_{--}(n, p) = \chi_-(n)\chi_-(p) \tag{9.1-6}$$

has a z component of total spin $-\hbar$. If $m = 0$, any spin function

$$\zeta_{+-}(n, p) = C_{+-}\chi_+(n)\chi_-(p) + C_{-+}\chi_-(n)\chi_+(p) \tag{9.1-7}$$

satisfies (4). This result says that any superposition of spin states with one spin up and the other spin down has a z component of total spin equal to zero.

PROBLEM 9.1

a. Multiply (4) by $\chi_\pm(n)^\dagger\chi_\pm(p)^\dagger$ and show formally that $(1 - m)C_{++} = 0$, $mC_{+-} = 0$, $mC_{-+} = 0$, and $(1 + m)C_{--} = 0$.

b. Are (5), (6), and (7) normalized?

To examine the eigenvalue equation for the square of the total spin,

$$S_T^2\zeta(n, p) = \hbar^2\lambda\zeta(n, p), \tag{9.1-8}$$

it is useful to express S_T^2 as follows.

$$\begin{aligned} S_T^2 &= (\vec{S}_n + \vec{S}_p)\cdot(\vec{S}_n + \vec{S}_p) = S_n^2 + S_p^2 + 2\vec{S}_n\cdot\vec{S}_p \\ &= S_n^2 + S_p^2 + 2S_{nx}S_{px} + 2S_{ny}S_{py} + 2S_{nz}S_{pz} \\ &= S_n^2 + S_p^2 + (S_{nx} + iS_{ny})(S_{px} - iS_{py}) + (S_{nx} - iS_{ny})(S_{px} + iS_{py}) + 2S_{nz}S_{pz}. \end{aligned}$$

For any state, S_n^2 and S_p^2 each give $3\hbar^2/4$. Also,

$$S_{nx} \pm iS_{ny} = S_{n\pm}$$

$$S_{px} \pm iS_{py} = S_{p\pm}$$

are the raising and lowering operators for the neutron and proton z components of spin, as defined in general by (6.1-25) and (6.1-27), and used for spin in Section 8.2. The operator S_T^2 can then be written as a sum of terms, each of which does something simple to the spin functions.

$$S_T^2 = \frac{3\hbar^2}{2} + S_{n+}S_{p-} + S_{n-}S_{p+} + 2S_{nz}S_{pz}. \tag{9.1-9}$$

The algebra is the same as that discussed for a single spin in Section 8.2;

$$S_{nz}\chi_\pm(n) = \pm\frac{\hbar}{2}\chi_\pm(n), \qquad S_{n\pm}\chi_\pm(n) = 0, \qquad S_{n\pm}\chi_\mp(n) = \hbar\chi_\pm(n),$$

and similar relations hold for the proton spinors and operators. When (9) is applied to the $\zeta_{\pm\pm}$ given by (5) and (6), the results are

$$S_T^2\zeta_{++} = \frac{3\hbar^2}{2}\zeta_{++} + 0 + 0 + 2\left(\frac{\hbar}{2}\right)\left(\frac{\hbar}{2}\right)\zeta_{++} = 2\hbar^2\zeta_{++},$$

$$S_T^2\zeta_{--} = \frac{3\hbar^2}{2}\zeta_{--} + 0 + 0 + 2\left(-\frac{\hbar}{2}\right)\left(-\frac{\hbar}{2}\right)\zeta_{--} = 2\hbar^2\zeta_{--}.$$

These two spin functions satisfy the eigenvalue equation (8) with eigenvalue $\hbar^2 \times 1(1 + 1)$, that is, with a total spin quantum number $S_T = 1$. Because ζ_{++} is a state with $S_T = 1$ and $m = +1$, and ζ_{--} is a state with $S_T = 1$ and $m = -1$, it makes sense to rename them,

$$\zeta_{11} \equiv \chi_+(n)\chi_+(p), \tag{9.1-10}$$

$$\zeta_{1-1} \equiv \chi_-(n)\chi_-(p). \tag{9.1-11}$$

The operator (9) applied to the spin-up–spin-down superposition (7) gives

$$S_T^2\zeta_{+-} = \hbar^2(C_{+-} + C_{-+})[\chi_+(n)\chi_-(p) + \chi_-(n)\chi_+(p)].$$

If ζ_{+-} is to satisfy the eigenvalue equation (8), then

$$(C_{+-} + C_{-+})[\chi_+(n)\chi_-(p) + \chi_-(n)\chi_+(p)] \\ - \lambda[C_{+-}\chi_+(n)\chi_-(p) + C_{-+}\chi_-(n)\chi_+(p)] = 0.$$

Because $\chi_+(n)\chi_-(p)$ and $\chi_-(n)\chi_+(p)$ are orthogonal functions, their coefficients must separately be zero.

$$(1 - \lambda)C_{+-} + C_{-+} = 0,$$

$$C_{+-} + (1 - \lambda)C_{-+} = 0.$$

The determinant of the coefficients, $\lambda^2 - 2\lambda$, must be zero, and λ must be either 0 or 2. If $\lambda = 2$, $C_{+-} = C_{-+}$, and ζ_{+-} becomes

$$\zeta_{10} \equiv \frac{1}{\sqrt{2}}[\chi_+(n)\chi_-(p) + \chi_-(n)\chi_+(p)], \qquad (9.1\text{-}12)$$

where C_{+-} has been set equal to $1/\sqrt{2}$ to normalize ζ_{10}. The designation ζ_{10} makes sense because $\lambda = 2$ means that the total spin quantum number S_T is unity, and the m value has already been shown to be zero. If $\lambda = 0$, $C_{+-} = -C_{-+}$, and ζ_{+-} becomes

$$\zeta_{00} \equiv \frac{1}{\sqrt{2}}[\chi_+(n)\chi_-(p) - \chi_-(n)\chi_+(p)], \qquad (9.1\text{-}13)$$

where the indices on ζ_{00} mean that S_T and m are zero.

This rather ponderous development needs a summary. The most general spin description for two spin ½ particles is given by (1). Requiring the system to be in an eigenstate of the z component of total spin reduces the possibilities to states in which both spins point up, states in which both spins point down, and states that are arbitrary superpositions of one spin pointing up and one spin pointing down. Requiring further that the system must be in an eigenstate of S_T^2 determines the states completely.

The four linearly independent and experimentally distinguishable possibilities in (1),

$$\chi_+(n)\chi_+(p), \qquad \chi_+(n)\chi_-(p), \qquad \chi_-(n)\chi_+(p), \qquad \chi_-(n)\chi_-(p), \qquad (9.1\text{-}14)$$

are not all eigenstates of S_T^2. These terms can be arranged to give the four linearly independent and experimentally distinguishable possibilities (10), (11), (12), and (13),

$$\zeta_{1+1}, \qquad \zeta_{10}, \qquad \zeta_{1-1}, \qquad \zeta_{00}, \qquad (9.1\text{-}15)$$

which are eigenstates of S_T^2. The replacement of (14) by (15) is both a gain and a loss of information. In the state $\chi_+(n)\chi_-(p)$, the spin of the neutron is definitely up and the spin of the proton is definitely down, but the total spin is not definite. In the state ζ_{10}, the total spin is definitely unity, but the neutron and the proton are both in a superposition of a spin up state and a spin down state.

The three states ζ_{1+1}, ζ_{10}, and ζ_{1-1} are the states with total spin unity and the three orientations that are possible for any $j = 1$ state, and are called triplet states. The state ζ_{00} has total spin zero, the only possibility is $m = 0$ as for any $j = 0$ state, and it is called a singlet state. The words *triplet* for $S_T = 1$ and *singlet* for $S_T = 0$ are used in discussions of all systems of two spin $\frac{1}{2}$ particles, such as the two electrons in helium, a neutron and a proton scattering each other, and "charmonium."

Figure 9.1-1 may help to clarify the nature of the four eigenfunctions. The states ζ_{1+1} and ζ_{1-1} simply show that if the two spins both point up, or both point down, then the resultant is an angular momentum $j = 1$ that points up or down. For the $j = 0$ state ζ_{00}, the spins cancel to give zero. The state ζ_{10} causes the greatest discomfort at first glance. How can a spin that points up plus a spin that points down give a resultant with $j = 1$? As Figure 9.1-1 shows, the answer is that the spins are not fixed vectors, but lie in cones, and can therefore be given a relative phase that produces a horizontal resultant.

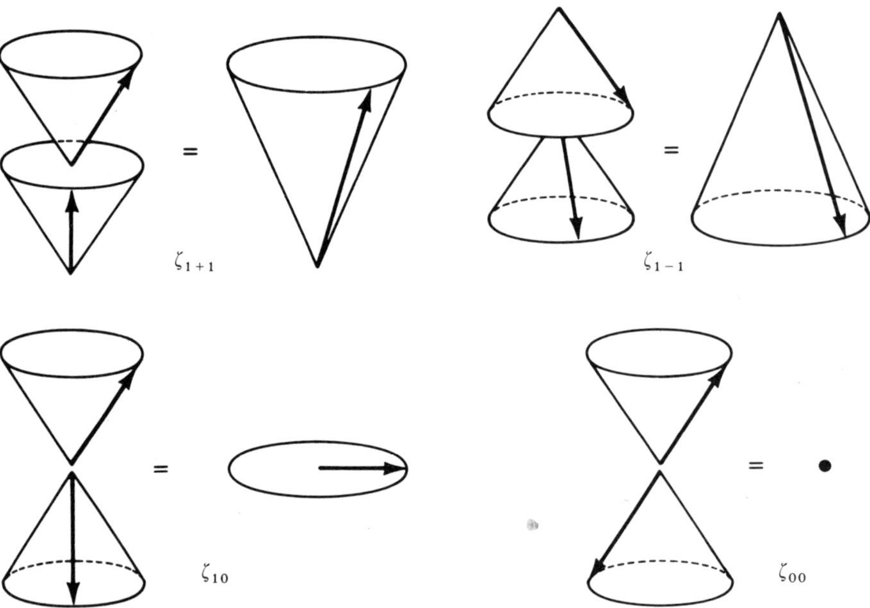

Fig. 9.1-1. The geometric interpretation of the four angular momentum eigenfunctions for two spin $\frac{1}{2}$ particles. Note particularly how a χ_+ and a χ_-, depending on their relative phase, can combine to give either ζ_{10} or ζ_{00}.

PROBLEM 9.2

Obtain the four quantities

$$\langle S_{nz} \rangle, \qquad \Delta S_{nz}, \qquad \langle S_T^2 \rangle, \qquad \Delta(S_T^2)$$

for the states

a. ζ_{11},
b. $\chi_+(n)\chi_-(p)$,
c. ζ_{10}.

Look at the behavior of the four eigenfunctions under particle exchange. The three triplet functions $\zeta_{1+1}, \zeta_{10}, \zeta_{1-1}$ are unchanged if n is put in the place of p, and p is put in the place of n. The singlet function changes sign in such an exchange. Suppose that the two particles are identical fermions rather than a neutron and a proton, so that the considerations of Section 4.2 apply. The complete two-particle state function depends on the positions of the two fermions and their spins, and must change sign under exchange of all coordinates. If the spatial dependence of the state function is even under particle exchange, then the spin dependence must be described by an odd function, and *vice versa.* The two electrons in the ground state of helium are both described by the same $\psi(r, \theta, \varphi)$, so that nothing changes if their spatial coordinates are exchanged. An elementary statement of the Pauli exclusion principle is, "Because the two electrons are in the same spatial state, one electron must have spin up and the other must have spin down." Talk about the one electron and the other electron is not very clear, and it is better to say, "Because the two electrons are in the same, and therefore even, spatial state, they must be in the odd spin state ζ_{00} with total spin quantum number zero."

Much of what is known about the interactions between particles has been learned from scattering experiments. For example, many studies of the neutron-proton force involve scattering a beam of neutrons, polarized or unpolarized, off the protons, polarized or unpolarized, in a hydrogen target. To analyze such experiments, one needs to know what spin states are formed when two spin $\frac{1}{2}$ particles are thrown together. Suppose first that the incident neutrons are unpolarized. It is wrong to use

$$\frac{\chi_+(n) + \chi_-(n)}{\sqrt{2}}$$

to describe the unpolarized beam. This spin function is normalized and gives equal likelihood of finding spin up and spin down, but, as observed in Section 8.2, it describes a neutron polarized in the x direction; see Problem 8.6. It is necessary to write

$$\frac{e^{i\delta_+}\chi_+(n) + e^{i\delta_-}\chi_-(n)}{\sqrt{2}} \tag{9.1-16}$$

and to average all measurable quantities over the random phases $\delta_\pm$. The discussion of random phases in Section 2.8 is relevant here. If the protons in the target are also unpolarized, they are described by

$$\frac{e^{i\eta_+}\chi_+(p) + e^{i\eta_-}\chi_-(p)}{\sqrt{2}}, \tag{9.1-17}$$

also subject to the requirement that one must average over the $\eta_\pm$. The spin function for the two particles together is the product of these two terms,

$$\begin{aligned}\zeta = \tfrac{1}{2}[&e^{i(\delta_+ + \eta_+)}\chi_+(n)\chi_+(p) + e^{i(\delta_- + \eta_-)}\chi_-(n)\chi_-(p)\\ &+ e^{i(\delta_+ + \eta_-)}\chi_+(n)\chi_-(p) + e^{i(\delta_- + \eta_+)}\chi_-(n)\chi_+(p)]\end{aligned}$$

The first two terms can be interpreted as they stand because they contain ζ_{11} and ζ_{1-1}. The second two terms inside the square bracket can be written in the form

$$\begin{aligned}&\frac{e^{i(\delta_+ + \eta_-)} + e^{i(\delta_- + \eta_+)}}{\sqrt{2}} \cdot \frac{\chi_+(n)\chi_-(p) + \chi_-(n)\chi_+(p)}{\sqrt{2}}\\ &\quad + \frac{e^{i(\delta_+ + \eta_-)} - e^{i(\delta_- + \eta_+)}}{\sqrt{2}} \cdot \frac{\chi_+(n)\chi_-(p) - \chi_-(n)\chi_+(p)}{\sqrt{2}}.\end{aligned}$$

The combined spin function therefore is the linear combination of the four eigenfunctions of total spin,

$$\zeta = C_{11}\zeta_{1+1} + C_{10}\zeta_{10} + C_{1-1}\zeta_{1-1} + C_{00}\zeta_{00}, \tag{9.1-18}$$

with

$$C_{11} = \frac{e^{i(\delta_+ + \eta_+)}}{2}, \qquad C_{10} = \frac{e^{i(\delta_+ + \eta_-)} + e^{i(\delta_- + \eta_+)}}{2\sqrt{2}}$$

$$C_{1-1} = \frac{e^{i(\delta_- + \eta_-)}}{2}, \qquad C_{00} = \frac{e^{i(\delta_+ + \eta_-)} - e^{i(\delta_- + \eta_+)}}{2\sqrt{2}}.$$

The probability of finding an eigenfunction is the square of the magnitude of its coefficient. Multiplying C_{10} by its complex conjugate gives

$$|C_{10}|^2 = \frac{1 + \cos(\delta_- + \eta_+ - \delta_+ - \eta_-)}{4}.$$

The observed result is $|C_{10}|^2$ *averaged over the random phases*, or $\frac{1}{4}$. In the same way $|C_{00}|^2$ averaged over the random phases gives $\frac{1}{4}$. The other two coefficients give $|C_{11}|^2 = \frac{1}{4}$, $|C_{1-1}|^2 = \frac{1}{4}$, without any need for averaging. The result is that, in collisions between unpolarized spin $\frac{1}{2}$ particles, the four possible eigenfunctions of total spin occur with equal probability so that the chance of the particles interacting in the triplet state is $\frac{3}{4}$, and the chance of their interacting in the singlet state is $\frac{1}{4}$.

Suppose that the incident neutrons are polarized and described by $e^{i\delta_+}\chi_+(n)$, and that the protons are unpolarized and described by (17). The product of the two terms can be written

$$\frac{e^{i(\delta_+ + \eta_+)}}{\sqrt{2}}\chi_+(n)\chi_+(p) + \frac{e^{i(\delta_+ + \eta_-)}}{\sqrt{2}}\chi_+(n)\chi_-(p) = \frac{e^{i(\delta_+ + \eta_+)}}{\sqrt{2}}\zeta_{1+1} + \frac{e^{i(\delta_+ + \eta_-)}}{2}\zeta_{10} + \frac{e^{i(\delta_+ + \eta_-)}}{2}\zeta_{00}.$$

The result has the form (18) with $|C_{11}|^2 = \frac{1}{2}$, $|C_{10}|^2 = \frac{1}{4}$, and $|C_{00}|^2 = \frac{1}{4}$. The coefficient C_{1-1} is zero because there is no way to reach a z component of $-\hbar$. The neutron contributes $+\hbar/2$, and the proton cannot take away more than $\hbar/2$. The probability of finding the triplet state is still $\frac{3}{4}$, but now this total is made up out of $\frac{1}{2}$ for $m = +1$ and $\frac{1}{4}$ for $m = 0$.

PROBLEM 9.3

Suppose that the incident neutrons are polarized with spin up, and that the target contains polarized protons. What are the probabilities of the four eigenstates of total spin if the protons are polarized

a. up,
b. down?

9.2 *The General Problem of Angular Momentum Addition*

The addition of any two different angular momenta, $\vec{J}_1$ and $\vec{J}_2$, is treated in this section. These angular momenta might be the contributions of two different particles, as in Section 9.1, or they might be the spin and orbital angular momentum of a single particle, or they might be more complicated constructs such as the combined spins of several particles and their combined orbital angular momenta.

It is necessary to assume that $\vec{J}_1$ and $\vec{J}_2$ commute,

$$[\vec{J}_1, \vec{J}_2] = 0. \tag{9.2-1}$$

This condition is implied by the statement that $\vec{J}_1$ and $\vec{J}_2$ are "different." Section 9.1 deals with a neutron spin $\vec{S}_n$ that operates only on $\chi(n)$ and a proton spin $\vec{S}_p$ that operates only on $\chi(p)$. Since these two spins operate on different functions, they satisfy $[\vec{S}_n, \vec{S}_p] = 0$. If $\vec{J}_1$ and $\vec{J}_2$ are the orbital and spin angular momenta of the same particle, they commute because $\vec{J}_1$ contains spatial derivatives while $\vec{J}_2$ contains spin matrices. If $\vec{J}_1$ and $\vec{J}_2$ are the orbital angular momenta of two different particles, then $\vec{J}_1$ contains derivatives with respect to the coordinates of the first particle, $\vec{J}_2$ contains derivatives with respect to the coordinates of the second particle, and again (1) is right.

As in Section 9.1, the problem is to add $\vec{J}_1$ and $\vec{J}_2$ so that the resultant is an eigenstate of

$$J_z = J_{1z} + J_{2z} \tag{9.2-2}$$

and of

$$\begin{aligned} J^2 &= (\vec{J}_1 + \vec{J}_2)\cdot(\vec{J}_1 + \vec{J}_2) \\ &= J_1^2 + J_2^2 + 2\vec{J}_1 \cdot \vec{J}_2. \end{aligned} \tag{9.2-3}$$

That is, the problem is to construct angular momentum states that are as well defined as quantum physics permits. Condition (1) has already been used; without it, (3) would contain the terms $\vec{J}_1 \cdot \vec{J}_2 + \vec{J}_2 \cdot \vec{J}_1$ rather than $2\vec{J}_1 \cdot \vec{J}_2$. With the raising and lowering operators

$$J_{n\pm} = J_{nx} \pm iJ_{ny}, \qquad n = 1, 2,$$

the same steps as those used to obtain (9.1-9) turn (3) into

$$J^2 = J_1^2 + J_2^2 + 2J_{1z}J_{2z} + J_{1+}J_{2-} + J_{1-}J_{2+}. \tag{9.2-4}$$

The problem and its notation are defined as follows. There are angular momentum states $|j_1 m_1\rangle$ and $|j_2 m_2\rangle$ that satisfy

$$J_{nz}|j_n m_n\rangle = \hbar m_n |j_n m_n\rangle,$$

$$J_n^2 |j_n m_n\rangle = \hbar^2 j_n(j_n + 1)|j_n m_n\rangle.$$

An eigenstate $|jm\rangle$ of J_z and J^2,

$$J_z|jm\rangle = \hbar m|jm\rangle, \tag{9.2-5}$$

$$J^2|jm\rangle = \hbar^2 j(j + 1)|jm\rangle, \tag{9.2-6}$$

is to be constructed out of products $|j_1 m_1\rangle|j_2 m_2\rangle$. The quantum numbers j_1 and j_2 describe the total angular momenta of the two contributions and are fixed; in Section 9.1 both are always $\frac{1}{2}$. These contributions can, however, have various orientations, that is, various values of m_1 and m_2; in Section 9.1 both could be $+\frac{1}{2}$ or $-\frac{1}{2}$. The task is then to evaluate the coefficients $\langle j_1 j_2 m_1 m_2|jm\rangle$ in

$$|jm\rangle = \sum_{m_1, m_2} \langle j_1 j_2 m_1 m_2|jm\rangle |j_1 m_1\rangle |j_2 m_2\rangle. \tag{9.2-7}$$

Only those values of j that satisfy

$$|j_1 - j_2| \le j \le j_1 + j_2$$

are of interest; the smallest possible resultant of the two component angular momenta has $j = |j_1 - j_2|$, and the largest has $j = j_1 + j_2$. The constants $\langle j_1 j_2 m_1 m_2 | jm \rangle$ are called *Clebsch–Gordan* coefficients, and the rather imposing notation[2] is designed to show that they are overlap integrals, as in (3.8-29), that measure how much of the product state $|j_1 m_1\rangle |j_2 m_2\rangle$, or $|j_1 j_2 m_1 m_2\rangle$, is contained in $|jm\rangle$.

The first step is to arrange for J_z to have the eigenvalue $\hbar m$. Since

$$\begin{aligned} J_z |j_1 m_1\rangle |j_2 m_2\rangle &= (J_{1z} + J_{2z}) |j_1 m_1\rangle |j_2 m_2\rangle \\ &= \hbar(m_1 + m_2) |j_1 m_1\rangle |j_2 m_2\rangle, \end{aligned}$$

the sum of terms (7) satisfies (5) if in every term

$$m_1 + m_2 = m, \tag{9.2-8}$$

that is, if the double sum over m_1 and m_2 is really a single sum because m_2 is restricted to be $m - m_1$. This result is nothing but the requirement that the z component of the sum is the sum of the z components. Condition (8), combined with the conditions $m_1 \leq j_1$ and $m_2 \leq j_2$, implies that there are only a few nonzero Clebsch–Gordan coefficients for a given choice of j_1, j_2, j, and m.

The next step is to impose the requirement that J^2 must have the eigenvalue $\hbar^2 j(j + 1)$. Before looking at other examples, be sure to recognize that Section 9.1 does precisely what is needed here for the special case $j_1 = \frac{1}{2}, j_2 = \frac{1}{2}$. In the notation of (7), (9.1-12) is

$$\begin{aligned} |1\ 0\rangle &= \sum_{m_1, m_2} \langle \tfrac{1}{2}\ \tfrac{1}{2}\ m_1\ m_2 | 1\ 0\rangle |\tfrac{1}{2}\ m_1\rangle |\tfrac{1}{2}\ m_2\rangle \\ &= \langle \tfrac{1}{2}\ \tfrac{1}{2}\ +\tfrac{1}{2}\ -\tfrac{1}{2} | 1\ 0\rangle |\tfrac{1}{2}\ +\tfrac{1}{2}\rangle |\tfrac{1}{2}\ -\tfrac{1}{2}\rangle \\ &\quad + \langle \tfrac{1}{2}\ \tfrac{1}{2}\ -\tfrac{1}{2}\ +\tfrac{1}{2} | 1\ 0\rangle |\tfrac{1}{2}\ -\tfrac{1}{2}\rangle |\tfrac{1}{2}\ +\tfrac{1}{2}\rangle, \end{aligned}$$

with

$$\langle \tfrac{1}{2}\ \tfrac{1}{2}\ +\tfrac{1}{2}\ -\tfrac{1}{2} | 1\ 0\rangle = \frac{1}{\sqrt{2}}, \qquad \langle \tfrac{1}{2}\ \tfrac{1}{2}\ -\tfrac{1}{2}\ +\tfrac{1}{2} | 1\ 0\rangle = \frac{1}{\sqrt{2}}.$$

All other coefficients in this expression are zero because ($m_1 = +\frac{1}{2}, m_2 = -\frac{1}{2}$) and ($m_1 = -\frac{1}{2}, m_2 = +\frac{1}{2}$) are the only ways to reach $m = 0$.

[2] An even more explicit and widely used notation is $\langle j_1 j_2 m_1 m_2 | j_1 j_2 jm\rangle$ for the coefficients and $|j_1 j_2 jm\rangle$ for the state that we call $|jm\rangle$. Although it is cumbersome, it makes good sense because the quantum numbers j_1 and j_2 are part of the description of $|jm\rangle$. More detailed discussions and mention of other nomenclatures are in E. Merzbacher, *Quantum Mechanics*, 2d ed. (New York: John Wiley & Sons, Inc., 1970), Chap. 16, Secs. 6 and 7; L. I. Schiff, *Quantum Mechanics*, 3d ed. (New York: McGraw-Hill Book Company, 1968), Sec. 28. An extensive treatment of these matters and tabulations of Clebsch–Gordan coefficients are in A. R. Edmonds, *Angular Momentum in Quantum Mechanics*, 2d ed., revised printing (Princeton: Princeton University Press, 1968); and M. Rotenberg, R. Bivins, N. Metropolis, and J. K. Wooten, Jr., *The 3j and 6j Symbols* (Cambridge: The Technology Press. M.I.T., 1959).

PROBLEM 9.4
List all nonzero Clebsch–Gordan coefficients that are implicitly evaluated in Section 9.1.

One way to proceed is to substitute the general sum (7) into the requirement (6), with J^2 written in the form (4). The behavior of the raising and lowering operators is described by (6.1-25) and (6.1-27),

$$J_{\pm}|j\ m\rangle = N^{\pm}_{jm}|j\ m\pm 1\rangle,$$

where the normalization constants

$$N^{\pm}_{jm} = \hbar\sqrt{j(j+1) - m(m \pm 1)}$$

are given by (6.1-26) and (6.1-28). Equation (6) then determines the coefficients in the same way as the spin functions (9.1-15) are obtained in Section 9.1.

EXAMPLE
Construct a state with $j = 1$ and $m = 1$ out of $j_1 = 1$ and $j_2 = 2$.

SOLUTION
There are only three ways to reach $m = 1$, $(m_1 = -1, m_2 = +2)$, $(m_1 = 0, m_2 = +1)$, and $(m_1 = +1, m_2 = 0)$. With C_-, C_0, and C_+ as shorthand for the three nonzero Clebsch–Gordan coefficients, (7) is

$$|1\ +1\rangle = C_-|1\ -1\rangle|2\ +2\rangle + C_0|1\ 0\rangle|2\ +1\rangle + C_+|1\ +1\rangle|2\ 0\rangle.$$

With (4) the eigenvalue equation (6) becomes

$$\begin{aligned}
C_-\Big\{&[1(1+1) + 2(2+1) + 2(-1)(+2)]|1\ -1\rangle|2\ +2\rangle \\
&+ \frac{N^+_{1-1}N^-_{2+2}}{\hbar^2}|1\ 0\rangle|2\ +1\rangle + 0\Big\} \\
+ C_0\Big\{&[1(1+1) + 2(2+1) + 2(0)(+1)]|1\ 0\rangle|2\ +1\rangle \\
&+ \frac{N^+_{10}N^-_{21}}{\hbar^2}|1\ +1\rangle|2\ 0\rangle + \frac{N^-_{10}N^+_{21}}{\hbar^2}|1\ -1\rangle|2\ +2\rangle\Big\} \\
+ C_+\Big\{&[(1+1) + 2(2+1) + 2(+1)(0)]|1\ +1\rangle|2\ 0\rangle + 0 \\
&+ \frac{N^-_{11}N^+_{20}}{\hbar^2}|1\ 0\rangle|2\ 1\rangle\Big\} \\
= 1(1+1)[&C_-|1\ -1\rangle|2\ +2\rangle + C_0|1\ 0\rangle|2\ +1\rangle + C_+|1\ +1\rangle|2\ 0\rangle].
\end{aligned}$$

The terms are organized so that their origins are visible; check them carefully because this example illustrates the major possibilities that can arise in calculations of this kind. Evaluate normalization constants and collect the coefficients of each $|j_1 m_1\rangle|j_2 m_2\rangle$.

$$(\sqrt{8}C_0 + 2C_-)|1\ -1\rangle|2\ +2\rangle + (\sqrt{8}C_- + 6C_0 + \sqrt{12}C_+)|1\ 0\rangle|2\ +1\rangle + (\sqrt{12}C_0 + 6C_+)|1\ +1\rangle|2\ 0\rangle = 0.$$

The three states $|j_1 m_1\rangle|j_2 m_2\rangle$ are orthogonal, and the coefficient of each must be zero. The first and third give

$$C_- = -\sqrt{2}C_0, \qquad C_+ = -\sqrt{\frac{1}{3}}C_0.$$

These values gove zero for the coefficient of $|1\ 0\rangle|2\ +1\rangle$, so that the three equations are consistent. To normalize $|jm\rangle$, the sum of the squares of its Clebsch–Gordan coefficients must be unity,

$$|C_-|^2 + |C_0|^2 + |C_+|^2 = 1.$$

These conditions are satisfied by the set of constants

$$\begin{aligned}\langle 1\ 2\ -1\ +2|1\ +1\rangle &= C_- = +\sqrt{\frac{3}{5}}\\ \langle 1\ 2\ 0\ +1|1\ +1\rangle &= C_0 = -\sqrt{\frac{3}{10}}\\ \langle 1\ 2\ +1\ 0|1\ +1\rangle &= C_+ = +\sqrt{\frac{1}{10}}\end{aligned} \qquad (9.2\text{-}9)$$

PROBLEM 9.5

An important example is the addition of the orbital and spin angular momenta of an electron in an atom, or of a nucleon in a nucleus. Derive the Clebsch–Gordan coefficients needed to construct, for $m = +\frac{1}{2}$, a $P_{1/2}$ state, that is, a state with $l = 1, j = \frac{1}{2}$. Show that the resulting $|jm\rangle$ can be written

$$|\tfrac{1}{2}\ +\tfrac{1}{2}\rangle = -\sqrt{\frac{1}{4\pi}}\begin{pmatrix}\cos\theta\\ \sin\theta e^{+i\varphi}\end{pmatrix}.$$

Look at the special case $j = j_1 + j_2$ and $m = j$. There is only one way to satisfy (8). Both of the angular momenta that are being added must point as close to the z axis as they can, with $m_1 = j_1$, and $m_2 = j_2$, because there is no other way to reach m. The general expression (7) can have only one term,

$$|j\ +j\rangle = \langle j_1\ j_2\ +j_1\ +j_2|j\ +j\rangle|j_1\ +j_1\rangle|j_2\ +j_2\rangle.$$

Normalization requires that the coefficient have the square of its magnitude be unity, and it is conventional to choose

$$\langle j_1 \; j_2 \; +j_1 \; +j_2 | j \; +j \rangle = 1, \tag{9.2-10}$$

$$|j \; +j\rangle = |j_1 \; +j_1\rangle |j_2 \; +j_2\rangle. \tag{9.2-11}$$

The spin state ζ_{1+1} given in (9.1-10) is an example of (11).

PROBLEM 9.6

For $j = j_1 + j_2$, evaluate $\langle j_1 \; j_2 \; -j_1 \; -j_2 | j \; -j \rangle$.

Some Clebsch–Gordan coefficients can be generated conveniently by applying lowering operators to (11). For example, $P_{3/2}$ states for a spin $\frac{1}{2}$ particle have orbital angular momentum quantum number $l = 1$, and orbital and spin contributions combining to give $j = \frac{3}{2}$. If $m = \frac{3}{2}$, there is only one possibility, $m_l = +1$ and $m_s = +\frac{1}{2}$. The state is then

$$|\tfrac{3}{2} \; +\tfrac{3}{2}\rangle = \chi_+ Y_{11}(\theta, \varphi) = \begin{pmatrix} 1 \\ 0 \end{pmatrix}\left(-\sqrt{\frac{3}{8\pi}} \sin\theta e^{+i\varphi}\right) = -\sqrt{\frac{3}{8\pi}} \begin{pmatrix} \sin\theta e^{i\varphi} \\ 0 \end{pmatrix}.$$

The lowering operator is given by

$$\begin{aligned} J_- &= J_x - iJ_y = (L_x + S_x) - i(L_y + S_y) \\ &= L_- + S_- \end{aligned}$$

When applied to $|\frac{3}{2} \; +\frac{3}{2}\rangle$, J_- gives $|\frac{3}{2} \; +\frac{1}{2}\rangle$ times the normalization constant $N^-_{3/2\,3/2} = \hbar\sqrt{3}$. Therefore

$$\begin{aligned} |\tfrac{3}{2} \; +\tfrac{1}{2}\rangle &= \frac{1}{\hbar\sqrt{3}} (L_- + S_-)\chi_+ Y_{11} \\ &= \frac{1}{\hbar\sqrt{3}} (\chi_+ \hbar\sqrt{2}\, Y_{10} + \hbar\chi_- Y_{11}) \end{aligned}$$

where $\hbar\sqrt{2}$ and $\hbar$ are the normalization constants N^-_{11} and $N^-_{1/2\,1/2}$. The two coefficients are

$$\langle \tfrac{1}{2} \; 1 \; +\tfrac{1}{2} \; 0 | \tfrac{3}{2} \; +\tfrac{1}{2}\rangle = \sqrt{\frac{2}{3}}, \qquad \langle \tfrac{1}{2} \; 1 \; -\tfrac{1}{2} \; +1 | \tfrac{3}{2} \; +\tfrac{1}{2}\rangle = \sqrt{\frac{1}{3}}.$$

The state can be written in the various forms

$$|\tfrac{3}{2}\ +\tfrac{1}{2}\rangle = \sqrt{\frac{2}{3}}\chi_+ Y_{10} + \sqrt{\frac{1}{3}}\chi_- Y_{11}$$

$$= \sqrt{\frac{2}{3}}\begin{pmatrix}1\\0\end{pmatrix}\left(\sqrt{\frac{3}{4\pi}}\cos\theta\right) + \sqrt{\frac{1}{3}}\begin{pmatrix}0\\1\end{pmatrix}\left(-\sqrt{\frac{3}{8\pi}}\sin\theta e^{+i\varphi}\right)$$

$$= \frac{1}{\sqrt{8\pi}}\begin{pmatrix}2\cos\theta\\ \sin\theta e^{+i\varphi}\end{pmatrix} \tag{9.2-12}$$

PROBLEM 9.7

By applying lowering operators to (12), obtain the $P_{3/2}$ states with $m = -\frac{1}{2}$ and $m = -\frac{3}{2}$. Could you have guessed the result for $m = -\frac{3}{2}$?

Although there are more elegant techniques than those of this section, computing Clebsch–Gordan coefficients is a dreary business. One should work out a few to understand them, but in extensive calculations, one should reach for tabulations of the kind mentioned in footnote 2 with celerity and without apology.

9.3 *The Magnetic Moment of the Deuteron*

Section 7.1 mentions the somewhat complicated angular momentum mixture of the deuteron.[3] The neutron and proton are in a triplet spin state because the nuclear force is spin dependent; in the singlet state the attraction is too weak to hold the particles together. The dominant component is a $^3S_1$ state, an $l = 0$ state in which the neutron and proton spins give the total angular momentum $j = 1$. With the choice $m = 1$, such a state is

$$|s_T = 1, l = 0, j = 1, m = 1\rangle = \chi_+(n)\chi_+(p)Y_{00} = \zeta_{1+1}Y_{00}. \tag{9.3-1}$$

[3] A qualitative discussion of the topics in this section is given by H. Frauenfelder and E. M. Henley, *Subatomic Physics* (Englewood Cliffs, N.J.: Prentice-Hall, Inc., 1974), Sec. 12.5. A clear and readable quantitative discussion of the deuteron is given by S. DeBenedetti, *Nuclear Interactions* (New York: John Wiley & Sons, Inc., 1964), Sec. 1.2. Section 1.1 contains a good summary of the essential angular momentum ideas and tables of some Clebsch–Gordan coefficients. See also R. R. Roy and B. P. Nigam, *Nuclear Physics* (New York: John Wiley & Sons, Inc., 1967), Sec. 3.11.

If the interaction between the neutron and proton could be described completely by a potential energy $V(r)$, that is, if the interaction were strictly *central*, then the angular momentum properties would be given completely by (1).

The force is, however, not central. The interaction is stronger if the line joining the nucleons is parallel to the spin directions than if it is perpendicular to them. There are, as a consequence, torques that mix in orbital angular momenta other than $l = 0$. The two nucleons prefer to be close to the spin axis, one above the other, and to form a cigar-shaped charge distribution. This charge distribution is described by saying that the deuteron has an electric *quadrupole moment*, and the energy of orientation of this quadrupole moment can be measured. If (1) were a complete description, then, as for any $l = 0$ state, the charge distribution would be spherical; Y_{00} is a constant. The nonzero quadrupole moment is proof that there must be some $l \neq 0$ contribution.

Although *orbital* angular momentum need not be conserved if there are internal torques, the *total* angular momentum must be. The deuteron as a nucleus of a deuterium atom feels only forces that are negligible on the nuclear scale. It is therefore an isolated system, and its total angular momentum must be constant. The dominant term (1) has $j = 1$, and all other contributions must also have $j = 1$. In addition to $l = 0$, it is possible to consider $l = 1$ and $l = 2$, because the total spin of unity can be added to give a resultant $j = 1$. However, $l > 2$ is impossible. For example, if $l = 3$, the smallest possible j is $3 - 1 = 2$.

The nuclear interaction, to excellent approximation, is invariant under reflection; it *conserves parity*. In other words, if a system starts out in a state that is even under the replacement $\vec{r} \to -\vec{r}$, it will remain in an even state. To very great accuracy the parity of every nuclear state is definitely even or definitely odd, and is not a mixture of the two possibilities. As is pointed out in (6.2-44), the parity of any orbital angular momentum state is $(-1)^l$. The dominant $l = 0$ state has $(-1)^0 = +1$, or even parity, and the only states that can be mixed in are other even parity states, with $l = 2, 4, 6, \ldots$. The $l > 2$ possibilities have already been eliminated by angular momentum conservation, and so an $l = 2$, or D state, is all that can be mixed with the $l = 0$, or S state (1).

The deuteron can then be described by

$$\begin{aligned} \psi = {} & \sqrt{1 - |C_D|^2}\, R_0(r) |\, s_T = 1, l = 0, j = 1, m = 1\rangle \\ & + C_D R_2(r) |\, s_T = 1, l = 2, j = 1, m = 1\rangle, \end{aligned} \tag{9.3-2}$$

where the normalized functions R_0 and R_2 describe the radial dependence of the two terms. The admixture of D state is described by C_D, that is, $|C_D|^2$ is the probability of finding $l = 2$, and $1 - |C_D|^2$ is the probability of finding $l = 0$. The problem of constructing the D state is solved in Section 9.2, where the necessary Clebsch–Gordan coefficients are given in (9.2-9). In the notation of

Sections 9.1, 9.2, and 6.2,

$$|s_T = 1, l = 2, j = 1, m = 1\rangle = \sqrt{\frac{3}{5}}\zeta_{1-1}Y_{22}(\theta, \varphi)$$

$$- \sqrt{\frac{3}{10}}\zeta_{10}Y_{21}(\theta, \varphi) + \sqrt{\frac{1}{10}}\zeta_{1+1}Y_{20}(\theta, \varphi). \tag{9.3-3}$$

Our task is to calculate the expectation value of the z component of the magnetic moment $\langle M_z \rangle$ for the state (2). The contribution of the proton spin is

$$\gamma_p S_{pz}, \qquad \gamma_p = 2.7928\frac{e}{M_p};$$

that is, from (8.4-15), $2.7928e\hbar/2M_p$ if $S_{pz} = +\hbar/2$. The contribution of the neutron spin is, from (8.4-16),

$$\gamma_n S_{nz}, \qquad \gamma_n = -1.9131\frac{e}{M_p}.$$

As described in Section 8.1, the gyromagnetic ratio associated with orbital angular momentum is, for a proton, $e/2M_p$. The deuteron is a neutron and a proton moving around their center of mass. Their masses are nearly equal, and half the orbital angular momentum comes from each. The orbital angular momentum of the uncharged neutron does not produce any magnetic moment, but the proton motion does create a current loop. The orbital contribution is therefore

$$\gamma_l L_z, \qquad \gamma_l = \frac{1}{2}\frac{e}{2M_p}.$$

The operator for the z component of the magnetic moment is then

$$M_z = \gamma_l L_z + \gamma_p S_{pz} + \gamma_n S_{nz}$$

$$= \gamma_l L_z + \frac{\gamma_p + \gamma_n}{2}(S_{pz} + S_{nz}) + \frac{\gamma_p - \gamma_n}{2}(S_{pz} - S_{nz}). \tag{9.3-4}$$

The second way of writing M_z is useful because all states of the deuteron are triplet spin states, for which $S_{pz} - S_{nz}$ has zero expectation value. Also, ζ_{1+1}, ζ_{10}, and ζ_{1-1} are eigenstates of $S_{Tz} = S_{pz} + S_{nz}$, with eigenvalues $+\hbar$, 0, and $-\hbar$.

PROBLEM 9.8

Show explicitly that, in the triplet states $\zeta_{1\pm 1,0}$, the expectation value of $S_{pz} - S_{nz}$ is zero.

For the S state, with state function (1), since $L_z Y_{00} = 0$,

$$\langle M_z, l = 0\rangle = \int \zeta_{1+1}^{\dagger} Y_{00}^{*}\left[\frac{\gamma_p + \gamma_n}{2}(S_{pz} + S_{nz})\right]\zeta_{1+1} Y_{00}$$

$$= \left(\frac{\gamma_p + \gamma_n}{2}\right)\hbar = (2.7928 - 1.9131)\frac{e\hbar}{2M_p} \qquad (9.3\text{-}5)$$

$$= 0.8797\frac{e\hbar}{2M_p}.$$

The algebraic sum of the proton and neutron moments gives the resultant, as one should expect. For the D state, with state function (3), three terms must be calculated.

$$\int \zeta_{1-1}^{\dagger} Y_{22}^{*}\left[\gamma_l L_z + \frac{\gamma_p + \gamma_n}{2}(S_{pz} + S_{nz})\right]\zeta_{1-1} Y_{22} = 2\gamma_l\hbar - \left(\frac{\gamma_p + \gamma_n}{2}\right)\hbar$$

$$\int \zeta_{10}^{\dagger} Y_{21}^{*}\left[\gamma_l L_z + \frac{\gamma_p + \gamma_n}{2}(S_{pz} + S_{nz})\right]\zeta_{10} Y_{21} = \gamma_l\hbar + 0$$

$$\int \zeta_{1+1}^{\dagger} Y_{20}^{*}\left[\gamma_l L_z + \frac{\gamma_p + \gamma_n}{2}(S_{pz} + S_{nz})\right]\zeta_{1+1} Y_{20} = 0 + \left(\frac{\gamma_p + \gamma_n}{2}\right)\hbar.$$

The total is $\frac{3}{5}$ times the first of these terms, plus $\frac{3}{10}$ times the second, plus $\frac{1}{10}$ times the third.

$$\langle M_z, l = 2\rangle = \frac{3\gamma_l\hbar}{2} - \frac{1}{2}\left(\frac{\gamma_p + \gamma_n}{2}\right)\hbar.$$

A magnetic moment measurement on the combined state function (2) then has the expectation value

$$\langle M_z\rangle = (1 - |C_D|^2)\langle M_z, l = 0\rangle + |C_D|^2\langle M_z, l = 2\rangle$$

$$= \left(\frac{\gamma_p + \gamma_n}{2}\right)\hbar + \left[\frac{3\gamma_l\hbar}{2} - \frac{3}{2}\left(\frac{\gamma_p + \gamma_n}{2}\right)\hbar\right]|C_D|^2$$

$$= (0.8797 - 0.5696|C_D|^2)\frac{e\hbar}{2M_p}. \qquad (9.3\text{-}6)$$

PROBLEM 9.9

Cross terms between the three terms in $|s_T = 1,\ l = 2,\ j = 1,\ m = 1\rangle$, and cross terms between this state function and $|s_T = 1, l = 0, j = 1, m = 1\rangle$, were not considered. Why?

Nuclear magnetic resonance experiments with samples that contain deuterium give $0.8574e\hbar/2M_p$ for the magnetic moment of the deuteron. This value, when set equal to (6), gives

$$|C_D|^2 = 0.04;$$

that is, this calculation indicates that the deuteron is 96% in the S state and 4% in the D state.

Once again we must conclude with a description of important neglected effects. Some will despair because few of our calculations are accurate and definitive, while others will be delighted because so many nice things remain to be done. The fraction of D state, $|C_D|^2$, is determined by the small difference between the quantities 0.8797 and 0.8574. As is mentioned in Section 8.4, the magnetic moments of the neutron and the proton are determined by the behavior of their meson clouds. Their interaction is caused by the same meson clouds, and it is plausible that their magnetic moments are modified when they are close to each other. Also, relativistic effects cannot be neglected entirely because the kinetic energies of the nucleons are around 2% of their rest energies. The theory of these effects is far from complete, but it is believed that $|C_D|^2 \simeq 0.04$–0.06 is a reasonable guess.

PROBLEM 9.10

Calculate the magnetic moment for a neutron and proton in a $^3P_2$ state; that is, calculate $\langle M_z \rangle$ for $s_T = 1, l = 1, j = 2, m = 2$.

PROBLEM 9.11

Calculate the magnetic moment for a neutron and proton in a $^3P_1$ state; that is, calculate $\langle M_z \rangle$ for $s_T = 1$, $l = 1$, $j = 1$, $m = 1$. You will need to calculate or look up some Clebsch–Gordan coefficients. Try looking them up.

Radio telescopes at Jodrell Bank, England. Much of our information about remote regions of the universe comes from studies of microwaves and radio waves. In Section 10.2, it is shown that 21 cm wavelength radiation is emitted in transitions between the lowest triplet and singlet states of the hydrogen atom. There is very little absorption of such long wavelength photons by interstellar dust. The 21 cm radiation has therefore made it possible to map the distribution of hydrogen in our galaxy. See J. S. Hey, The Radio Universe, *2d ed. (Oxford: Pergamon Press, 1976). (Courtesy of the American Institute of Physics Niels Bohr Library)*

10

Approximation Methods for Trapped Particles

Regions of constant potential energy, harmonic oscillators, and Coulomb potentials are the only forms $V(\vec{r})$ that give sets of truly elementary eigenfunctions. Even a constant force (Sections 2.9 and 7.3) involves the introduction of a special function. It follows that most applications require approximation methods.

Quantum and Newtonian physics do not differ in this respect. Classical mechanics books contain very few problems that are solved exactly, and these few generally have the same forms $V(\vec{r})$ that are considered here in Chapters 2 and 7. The typical physicists' classical mechanics text treats approximations rather briefly because many applications have moved into other fields, but the need is the same.

This chapter presents some methods that are useful for calculating the approximate energy eigenvalues and eigenfunctions of particles trapped in potential wells. Chapter 11 also deals with approximations for obtaining energy eigenfunctions, but treats scattering processes in which the energy is at the experimenter's disposal, and it is the scattered wave that must be found. Chapter 12 deals with time-dependent interactions, and with the transitions that such interactions induce.

10.1 *Perturbation Theory for Nondegenerate States*

Suppose that one wants to do a Newtonian calculation of the orbit of Mercury, with the sun assumed to be at rest, but with the gravitational effects of Venus and Earth taken into account. It is impossible to solve the problem exactly, but there exists a good approximation scheme. Complete neglect of interactions between the planets yields immediately the familiar elliptic orbits for each planet and is the zeroth approximation. To get the first approximation, assume that Venus and Earth are moving as in the zeroth approximation and calculate the perturbation of Mercury's orbit by their fields. The details may be tedious, but the procedure is straightforward because the locations of Venus and Earth are treated as if they were known. To get the second approximation, do the first-order approximation not only for Mercury, but also for Venus and Earth. Then calculate the perturbations in the orbit of Mercury that are caused by the first-order orbits of Venus and Earth. The scheme can be taken to as many orders as the required accuracy dictates. In each step the orbits of Venus and Earth must be known to order p if the orbit of Mercury is to be calculated to order $p + 1$. The procedure converges rapidly, that is, the accuracy improves rapidly as the order or approximation increases, because the perturbing forces between the planets are small compared to the sun's force. Quantal perturbation procedures work much like this classical method, and it will help to compare the two in each major step of the following development.[1]

The task is to solve

$$H\psi_n = E_n\psi_n$$

for one of the eigenfunctions ψ_n. The Hamiltonian H contains a term, H_0, the *unperturbed Hamiltonian*, for which the exact eigenfunctions are known,

$$H_0\psi_{0n} = E_{0n}\psi_{0n}. \tag{10.1-1}$$

The exact Hamiltonian is $H = H_0 + H'$, where H' is the *perturbing Hamiltonian*, and H' complicates things so that it is necessary to resort to an approximation method.

It is a convenient device to write

$$H = H_0 + \alpha H', \tag{10.1-2}$$

so that $\alpha = 1$ describes the actual problem and $\alpha = 0$ describes the unperturbed problem. Assume that, as α goes continuously from one to zero, the eigenfunctions and energies ψ_n and E_n change smoothly into the unperturbed

[1] A qualitative comparison will suffice, but a quantitative comparison is more instructive. Treatments of perturbation theory in classical mechanics, arranged to look much like the quantal treatments, are given by K. R. Symon, *Mechanics* 3d ed. (Reading, Mass.: Addison-Wesley Publishing Co., Inc., 1971), Sec. 12.5; J. B. Marion, *Classical Dynamics of Particles and Systems*, 2d ed. (New York: Academic Press, Inc., 1970), Sec. 14.6.

eigenfunctions and energies ψ_{0n} and E_{0n}. Assume also that the ψ_{0n} are not degenerate; the presence of degeneracy introduces complications that are considered in the next section.

The unperturbed eigenfunctions are complete and can be used to expand ψ_n. Substitute

$$\psi_n = \sum_m C_{mn}\psi_{0m} \tag{10.1-3}$$

into

$$(H_0 + \alpha H')\psi_n = E_n\psi_n. \tag{10.1-4}$$

$$\sum_m C_{mn}E_{0m}\psi_{0m} + \alpha\sum_m C_{mn}H'\psi_{0m} = E_n\sum_m C_{mn}\psi_{0m}.$$

Multiply by $\psi_{0r}^\dagger$, integrate, and use $\int \psi_{0r}^\dagger\psi_{0m} = \delta_{rm}$.

$$\sum_m C_{mn}E_{0m}\delta_{rm} + \alpha\sum_m C_{mn}\int\psi_{0r}^\dagger H'\psi_{0m} = E_n\sum_m C_{mn}\delta_{rm},$$

$$C_{rn}(E_n - E_{0r}) = \alpha\sum_m H'_{rm}C_{mn}. \tag{10.1-5}$$

The *matrix element* of H',

$$H'_{rm} \equiv \int\psi_{0r}^\dagger H'\psi_{0m} \equiv \langle r|H'|m\rangle, \tag{10.1-6}$$

has the same structure as (3.8-26). The unperturbed eigenfunctions are the set of basis functions. The r, m element of the matrix that represents the operator H' is the integral of H' sandwiched between ψ_{0m} and the conjugate basis function $\psi_{0r}^\dagger$. The conjugate † is simply the complex conjugate if the basis functions are scalar functions, but is the Hermitian conjugate if the basis functions have two or more components, as when spin is being described. The integral sign means integration over functions of position and summation over components.

No approximations have been made, and no approximations are necessary if (5) can be solved exactly. However, the usual situation is that the range of the summations, the number of coefficients C_{mn}, and the number of equations (5) are infinite. To see what is involved, let us preview the example that is worked out at the end of this section, the change in the Coulomb bound state energies (7.3-9) caused by the finite size of the nucleus. The unperturbed states ψ_{0m} are the point-nucleus Coulomb wave functions that satisfy (7.3-1). The perturbing H' is the difference between the $1/r$ point charge potential energy and the potential energy caused by a charged sphere of finite radius. The perturbed states ψ_n will differ from the unperturbed ψ_{0m}, particularly near the origin, but can be expanded in terms of the ψ_{0m} as in (3). The expansion is possible because the ψ_{0m} form a complete set that consists of the infinitely many bound states $\psi_{nlm}(r, \theta, \varphi)$ considered in Section 7.3, and the continuum of the positive energy or scattering states in the Coulomb potential. The exact system of equations (5) is in this example even a little worse than it looks; there are discrete energies and a summation for $E < 0$, and continuously distributed energies that require $\sum$ to be treated as an integration for $E > 0$. Nevertheless, the perturbation scheme is useful because at each level of approximation, one or a few of the terms are much more important than the others. The important ψ_{0m} are those that in some sense are close to the state that is being expanded. If the lowest state for the finite nuclear size problem is expanded in terms of the point-nucleus states, then the lowest point-nucleus state should contribute the most if the perturbation is small.

In general, the limit as the perturbation is removed is described by

$$\alpha \to 0, \qquad \psi_n \to \psi_{0n}, \qquad C_{mn} \to \delta_{mn}, \qquad E_n \to E_{0n}. \tag{10.1-7}$$

If the perturbation is not too large, then, for $\alpha \neq 0$, C_{nn} should be close to unity and all the other C_{mn} should be small. The equations (5) divided by C_{nn} give

$$E_n = E_{0n} + \alpha H'_{nn} + \alpha \sum_{m \neq n} H'_{nm} \frac{C_{mn}}{C_{nn}}, \qquad r = n, \tag{10.1-8}$$

$$\frac{C_{rn}}{C_{nn}} = \frac{\alpha H'_{rn}}{E_n - E_{0r}} + \frac{\alpha}{E_n - E_{0r}} \sum_{m \neq n} H'_{rm} \frac{C_{mn}}{C_{nn}}, \qquad r \neq n. \tag{10.1-9}$$

The set of equations (5) is linear and homogeneous, and can only determine the ratios of the C_{rn}. Normalization adds the requirement that $\sum_m |C_{mn}|^2 = 1$, or

$$|C_{nn}|^2 = \left(1 + \sum_{m \neq n} \left|\frac{C_{mn}}{C_{nn}}\right|^2\right)^{-1}; \tag{10.1-10}$$

the perturbation reduces $|C_{nn}|^2$ below unity by introducing contributions from the $m \neq n$ states.

These equations can be used to calculate E_n and ψ_n to any order. The zeroth order is given by (7). To obtain the first order, substitute the zeroth order results into the right sides of (8), (9), and (10).

$$E_n = E_{0n} + \alpha H'_{nn},$$

$$\frac{C_{rn}}{C_{nn}} = \frac{\alpha H'_{rn}}{E_{0n} - E_{0r}}, \qquad r \neq n,$$

$$C_{nn} = 1.$$

To obtain the second-order energy, substitute the first-order results into the right side of (8),

$$E_n = E_{0n} + \alpha H'_{nn} + \alpha^2 \sum_{m \neq n} \frac{H'_{nm} H'_{mn}}{E_{0n} - E_{0m}}. \tag{10.1-11}$$

PROBLEM 10.1

a. Show that, to first order, the wave function is

$$\psi_n = \psi_{0n} + \alpha \sum_{m \neq n} \frac{H'_{mn}}{E_{0n} - E_{0m}} \psi_{0m}. \tag{10.1-12}$$

Is this wave function normalized?

b. Obtain C_{rn}/C_{nn} and C_{nn} to second order.

The expressions become cumbersome in higher orders, but there is no intrinsic difficulty in the construction of successive iterations. The parameter α is unity for the actual problem, but is worth retaining because in all iterations α^p multiplies terms that contain small quantities to the pth power. Note that the expansion coefficients, and therefore the wave function, must be known only to order p if the energy is being calculated to order $p + 1$. The energy denominators show why it is essential to exclude degeneracy. If one or more energy differences are of the order of H'_{mn} or smaller, the terms multiplied by α are large, and the scheme does not work.

EXAMPLE

Find the ground state energy for the one-dimensional potential energy function

$$\begin{aligned} V_1 &= gx, && 0 \leq x \leq a, \\ &= \infty && \text{otherwise.} \end{aligned} \tag{10.1-13}$$

SOLUTION

Take $H' = gx$ as the perturbing Hamiltonian. The unperturbed H_0 has potential energy zero in $0 \leq x \leq a$, and the unperturbed states and energies are

$$\psi_{0m} = \sqrt{\frac{2}{a}} \sin \frac{m\pi x}{a}, \qquad E_{0m} = \frac{\pi^2 \hbar^2 m^2}{2Ma^2}, \qquad m = 1, 2, 3, \ldots. \tag{10.1-14}$$

The first-order correction to each of the energies is

$$H'_{nn} = \int_0^a \psi_{0n}^* gx \psi_{0n}\, dx = \frac{ga}{2}. \tag{10.1-15}$$

Notice the units; the matrix element of an energy calculated with normalized state functions must be an energy. To understand the result, look at the slightly different problem,

$$V = g\left(x - \frac{a}{2}\right), \qquad 0 \le x \le a, \tag{10.1-16}$$
$$= \infty \qquad \text{otherwise,}$$

which differs from the first example by a downward shift in energy of $ga/2$. For this problem all first-order energy corrections are zero.

$$H'_{nn} = \int_0^a \psi^*_{0n} g\left(x - \frac{a}{2}\right)\psi_{0n}\,dx = 0$$

because the perturbation is odd with respect to $x = a/2$, the center of the well, while all $|\psi_{0n}|^2$ are even with respect to the center. The first problem (13) has to show the result (15); it has to give the shift in the origin of energy necessary to transform it into (16).

To obtain the second-order correction to the ground state energy with (16), it is necessary to calculate the matrix elements

$$H'_{1m} = \frac{2g}{a}\int_0^a \left(x - \frac{a}{2}\right)\sin\frac{\pi x}{a}\sin\frac{m\pi x}{a}\,dx$$
$$= \frac{2ga}{\pi^2}\int_{-\pi/2}^{+\pi/2} u\cos u \sin\left(mu + \frac{m\pi}{2}\right)du.$$

If m is odd, H'_{1m} is zero because the integrand is odd. If m is even,

$$H'_{1m} = \frac{2ga}{\pi^2}(-1)^{m/2}\int_{-\pi/2}^{+\pi/2} u\cos u \sin mu\,du$$
$$= \frac{2ga}{\pi^2}(-1)^{m/2}\left[\frac{\sin(m-1)\pi/2}{(m-1)^2} + \frac{\sin(m+1)\pi/2}{(m+1)^2}\right] \tag{10.1-17}$$
$$= -\frac{8ga}{\pi^2}\frac{m}{(m^2-1)^2}.$$

From (11), with $H'_{11} = 0$, the ground state energy is, to second order,

$$E_1 = E_{01} + 0 + \sum_{m\neq 1}\frac{H'_{1m}H'_{m1}}{E_{01} - E_{0m}}.$$

All second-order calculations of this kind can be simplified because H' is Hermitian, so that

$$H'_{nm} = H'^*_{mn},$$

and the numerator in the summation is $|H'_{1m}|^2$. With the E_{0m} given by (14) and the matrix elements (17),

$$E_1 = E_{01} + \sum_{\text{even } m>0} \frac{\left[\dfrac{8ga}{\pi^2}\dfrac{m}{(m^2-1)^2}\right]^2}{E_{01}(1-m^2)}$$
$$= E_{01}\left[1 - \left(\frac{8ga}{\pi^2 E_{01}}\right)^2 \sum_{\text{even } m>0} \frac{m^2}{(m^2-1)^5}\right]. \tag{10.1-18}$$

The summation converges very rapidly; just the $m = 2$ plus $m = 4$ terms give 0.016482, while the infinite sum gives 0.016483. The result is

$$E_1 = E_{01}\left[1 - 0.01083\left(\frac{ga}{E_{01}}\right)^2\right]. \tag{10.1-19}$$

Since $ga/2$ is the maximum size of the perturbation, the calculation should be accurate if ga/E_{01} is not too large. The accuracy for this example can be checked for special values of the parameters through an application of the function $\text{Ai}(\xi)$ discussed in Section 2.9 and graphed in Figure (2.9-1). A little arithmetic shows that with the change in variable (2.9-2), and with the special choice $ga/E_{01} = 0.54287$, the edges of the well are brought to the two zeros ξ_1 and ξ_2 of $\text{Ai}(\xi)$. The function $\text{Ai}(\xi)$ is then the exact solution to the problem defined by (16), and the exact energy turns out to be $E_{01}(1 - 0.00332)$. For the same ga/E_{01}, (19) gives $E_{01}(1 - 0.00319)$, so that the small correction is wrong by only 4%, even though the perturbation is not small.

PROBLEM 10.2

For the example (16) obtain the first-order ground state wave function from (12) with some reasonable choice for ga/E_{01}. Sketch the result and explain qualitatively why the perturbation *lowers* the energy, as shown by (19).

The examples (13) and (16) are artificial, but they do show features that occur in many perturbation calculations. Symmetries of the integrands in the H'_{nm} often make many of the matrix elements zero. Infinite sums arise in second- and higher-order calculations of the energy, and convergence must be studied in each case. However, the convergence is often rapid for two reasons. First, the high-energy, large-quantum-number states have short wavelengths and overlap poorly with smooth perturbations and low-lying states; the $\sim 1/m^3$ behavior of (17) shows this effect. Second, the energy denominators in (11) reduce the importance of states that are far in energy from the E_{0n} of the dominant state; the $1 - m^2$ in the denominator of (18) comes from this source.

PROBLEM 10.3

In a one-dimensional problem, $V = Kx^2/2 + f + gx$. Treat the harmonic oscillator potential as the large term, and $f + gx$ as the perturbation.

a. Obtain the energies of all states to first order.

b. Examine the behavior of the second-order correction to the energy. Calculate the most important term in this second-order correction.

c. Recognize that $V = (K/2)(x + g/K)^2 + (f - g^2/2K)$, obtain the exact values, and compare with (**a**) and (**b**).

PROBLEM 10.4

In a one-dimensional problem, $V = Kx^2/2 + K'x^2/2$. Treat $Kx^2/2$ as the large term, $K'x^2/2$ as the perturbation, and obtain the ground state energy to first order. Compare the result with the exact answer, which is that for a harmonic oscillator with spring constant $K + K'$.

PROBLEM 10.5

In a one-dimensional problem, $V = Kx^2/2 + Gx^4/4$. Treat $Kx^2/2$ as the large term and $Gx^4/4$ as the perturbation. Obtain the ground state and first excited state energies to first order. What happens to the accuracy of the first-order calculation for a given G as one goes to highly excited states?

The effect of finite nuclear size on Coulomb wave functions is previewed above. For a quantitative calculation it is necessary to use some form for the nuclear charge distribution. Assume that the nucleus is a uniformly charged sphere of radius R and total charge Ze, so that the electric charge density is given by

$$\begin{aligned}\rho_e &= \frac{Ze}{4\pi R^3/3}, \qquad r \le R,\\ &= 0, \qquad r > R.\end{aligned}$$

Gauss' law shows that the electric field is $Zer/4\pi\varepsilon_0 R^3$ for $r \le R$ and $Ze/4\pi\varepsilon_0 r^2$ for $r > R$. In this field the potential energy of a particle with charge $-e$ is

$$\begin{aligned}V(r) &= \zeta\left(-\frac{3}{2R} + \frac{r^2}{2R^3}\right), \qquad r \le R,\\ &= -\frac{\zeta}{r}, \qquad r > R, \qquad (10.1\text{-}20)\\ \zeta &\equiv \frac{Ze^2}{4\pi\varepsilon_0}.\end{aligned}$$

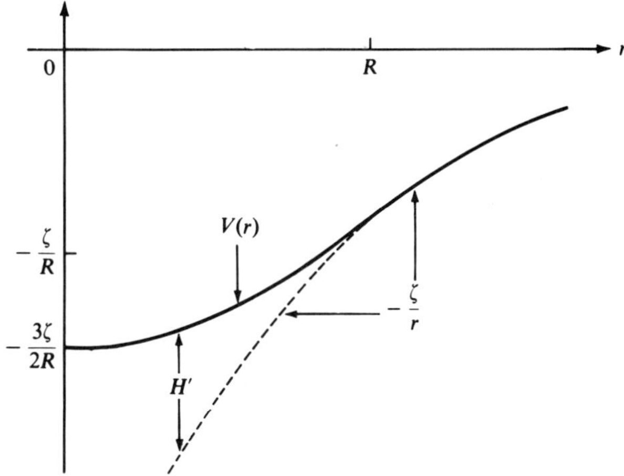

Fig. 10.1-1. The perturbation H' for the finite-nucleus problem. The solid curve is the $V(r)$ given by (20), the dashed curve is the point charge potential energy, and the difference is the perturbation.

The exact Hamiltonian H is $p^2/2\mu + V(r)$, the unperturbed Hamiltonian H_0 is the point-nucleus Hamiltonian $p^2/2\mu - \zeta/r$, and the perturbation is the difference shown in Figure 10.1-1:

$$\begin{aligned} H'(r) &= H - H_0 = V(r) - \left(-\frac{\zeta}{r}\right) \\ &= \zeta\left(-\frac{3}{2R} + \frac{r^2}{2R^3} + \frac{1}{r}\right), \qquad r \leq R, \\ &= 0, \qquad r > R. \end{aligned}$$

The unperturbed ground state wave function, as obtained in Section 7.3, is

$$\psi_{100} = \frac{e^{-r/a}}{a^{3/2}\sqrt{\pi}}, \qquad a = \frac{\hbar^2}{\zeta\mu}. \tag{10.1-21}$$

The first-order correction to the ground state energy is

$$\begin{aligned} H'_{00} &= \iiint |\psi_{100}|^2 H'(r) \sin\theta \, d\theta \, d\varphi \, r^2 \, dr \\ &= \frac{4\zeta}{a^3}\int_0^R e^{-2r/a}\left(-\frac{3r^2}{2R} + \frac{r^4}{2R^3} + r\right)dr. \end{aligned}$$

If the perturbation approach is valid, then only a small fraction of the probability distribution can lie inside R, $R/a \ll 1$, and $e^{-2r/a}$ can be replaced by unity.

Integration then gives

$$H'_{00} = \frac{2\zeta R^2}{5a^3} = \frac{4Z^2e^4\mu}{5(4\pi\varepsilon_0\hbar)^2}\frac{R^2}{a^2} = \frac{4}{5}(-E_{00})\frac{R^2}{a^2}, \tag{10.1-22}$$

where E_{00} is the unperturbed ground state energy, -13.6 eV for hydrogen. The answer is a positive number because the perturbation introduces a potential that is less attractive than the unperturbed $-\zeta/r$.

For the hydrogen atom, $a = 0.53$ Å and R, the radius of the proton's charge distribution, is about $0.9f$. The result is

$$H'_{00}(\text{hydrogen}) = \frac{4}{5}(13.6\text{ eV})\left(\frac{0.9 \times 10^{-13}\text{ cm}}{0.53 \times 10^{-8}\text{ cm}}\right)^2 = 3.1 \times 10^{-9}\text{ eV},$$

an undetectably small amount. For uranium, $Z = 92$, $-E_{00} \simeq (92)^2$ (13.6 eV) $= 115{,}000$ eV, and $a = 0.53$ Å$/92 = 5.8 \times 10^{-11}$ cm. A nucleus with A neutrons and protons has a radius of about $1.2A^{1/3}f$, so that for ^{238}U, $R = 7.4 \times 10^{-13}$ cm. Therefore

$$H'_{00}(\text{uranium}) = \frac{4}{5}(115{,}000\text{ eV})\left(\frac{7.4 \times 10^{-13}\text{ cm}}{5.8 \times 10^{-11}\text{ cm}}\right)^2 = 15\text{ eV},$$

a greater fraction of the unperturbed energy because more of the electron probability distribution lies inside the nuclear volume. The result is, however, only one of many effects that modify the energy of the K electrons in heavy elements. The binding energy is nearly a quarter of the rest energy, and relativistic effects are appreciable, though not hard to calculate. The major problem is that each electron is affected by all the other electrons. The resulting perturbations are much larger than that caused by the finite nuclear size, and it is very difficult to calculate them accurately.

It is much easier to untangle nuclear size effects in some "exotic" atoms.[2] Consider a muon in the Coulomb field of a carbon nucleus. The size a of the region in which the particle moves is inversely proportional to the mass. The muon mass is 206.77 times the electron mass, and the reduced mass of the $\mu^- - C^{12}$ system is 204.83 electron masses. The low-lying muon states are therefore so far inside the electron cloud that they are almost unaffected by it, and also R/a is appreciable. As the μ^- makes a transition from the $2P$ state ψ_{21m} to the $1S$ state ψ_{100}, an x-ray is emitted. The energy is proportional to Z^2 and to the reduced mass, so that the prediction of Section 7.3 for the x-ray energy is

$$13.606\text{ eV} \times 36 \times 204.83\left(\frac{1}{1^2} - \frac{1}{2^2}\right) = 75.25\text{ keV}$$

[2] Y. N. Kim, *Mesic Atoms and Nuclear Structure* (Amsterdam: North-Holland Publishing Company, 1971); R. Engfer et al., *Atomic Data and Nuclear Data Tables* **14**, 509 (1974).

Relativistic effects and vacuum polarization (see footnote 4 in Section 7.3) change the point-nucleus prediction to 75.67 keV. The experimental result is 75.25 ± 0.02 keV, and the difference, 0.42 ± 0.02 keV, is the nuclear size effect. The upper state, ψ_{21m}, is an $l = 1$ state, $|\psi_{21m}|^2 \sim r^2$ for small r, and the overlap with the nucleus is poor. The 0.42 ± 0.02 keV is therefore almost entirely H'_{00}, the perturbation of the ground state. With

$$-E_{00} = \tfrac{4}{3} \times 75.67 \text{ keV} = 100.89 \text{ keV},$$

$$a = \frac{0.529 \text{ Å}}{6 \times 204.83} = 4.31 \times 10^{-12} \text{ cm},$$

(22) gives

$$R = a\sqrt{\frac{5H'_{00}}{-4E_{00}}} = 3.1 \times 10^{-13} \text{ cm}.$$

This result is a rather good value for the size of the carbon nucleus. Various corrections can be made, and the uniformly charged sphere is not the best model, but refinements do not change the radius very much. Measurements of the x-rays from muonic atoms have provided considerable information regarding nuclear charge distributions.

10.2 *Perturbation Theory for Degenerate States*

There are no restrictions on the validity of the beginning of Section 10.1; the relation (10.1-5),

$$C_{rn}(E_n - E_{0r}) = \alpha \sum_m H'_{rm} C_{mn},$$

is exact. However, the development thereafter depends on C_{nn} being near unity, and on all other C_{rn} being small. Now suppose that ψ_{0n} and ψ_{0l} are either exactly degenerate or that the difference in their energies is small compared to the matrix elements of the perturbation H'. Then the perturbation can change ψ_{0n} drastically because it costs negligible energy to reach ψ_{0l}, and the constant C_{ln}, as well as the constant C_{nn}, might be large. Both C_{nn} and C_{ln} deserve special respect and attention, and it is useful to write, rather than (10.1-8) and (10.1-9), three equations that are special cases of (10.1-5).

$$C_{nn}(E_n - E_{0n}) = \alpha H'_{nn} C_{nn} + \alpha H'_{nl} C_{ln} + \alpha \sum_{m \neq n, l} H'_{nm} C_{mn}, \qquad r = n, \quad (10.2\text{-}1)$$

$$C_{ln}(E_n - E_{0l}) = \alpha H'_{ln} C_{nn} + \alpha H'_{ll} C_{ln} + \alpha \sum_{m \neq n, l} H'_{lm} C_{mn}, \qquad r = l, \quad (10.2\text{-}2)$$

$$C_{rn} = \frac{\alpha}{E_n - E_{0r}} \left[H'_{rn} C_{nn} + H'_{rl} C_{ln} + \sum_{m \neq n, l} H'_{rm} C_{mn} \right], \qquad r \neq n \text{ or } l. \quad (10.2\text{-}3)$$

Because $E_n - E_{0r}$ is not small for $r \neq n$ or l, (3) shows that the largest contributions to C_{rn} are first order in small quantities, and that therefore the summations in (1) and (2) contain only second-order terms. To first order, (1) and (2) become, with $H'_{ln} = H'^*_{nl}$, for the choice $\alpha = 1$,

$$\begin{aligned}(H'_{nn} - \Delta)C_{nn} + H'_{nl}C_{ln} &= 0,\\ H'^*_{nl}C_{nn} + (H'_{ll} - \Delta)C_{ln} &= 0,\end{aligned} \tag{10.2-4}$$

where

$$\Delta \equiv E_n - E_{0n} \simeq E_n - E_{0l}. \tag{10.2-5}$$

It is legitimate to use the same Δ in the two equations if $|E_{0n} - E_{0l}|$ is small compared to H'_{ll}. The equations (4) are linear homogeneous equations for the coefficients C_{nn} and C_{ln}. A nontrivial solution exists if

$$\begin{vmatrix} H'_{nn} - \Delta & H'_{nl} \\ H'^*_{nl} & H'_{ll} - \Delta \end{vmatrix} = 0. \tag{10.2-6}$$

There are then two solutions for Δ and two corresponding ratios C_{nn}/C_{ln}.

Suppose that the perturbation is an odd real function, such as a linear potential energy, and that ψ_{0n} and ψ_{0l} are functions with definite parity, such as two states $\psi_{nlm}(r, \theta, \varphi)$ of the hydrogen atom. Then

$$H'_{nn} = \int |\psi_{0n}|^2 H' d\tau, \qquad H'_{ll} = \int |\psi_{0l}|^2 H' d\tau$$

are both zero because the integrands are odd, and (6) becomes

$$\begin{vmatrix} -\Delta & H'_{nl} \\ H'^*_{nl} & -\Delta \end{vmatrix} = 0, \tag{10.2-7}$$

$$\Delta^2 - |H'_{nl}|^2 = 0, \qquad \Delta = \pm |H'_{nl}|.$$

In many important applications, H'_{nl} is real, so that $\Delta = \pm H'_{nl}$. Then, for $\Delta = +H'_{nl}$, $C_{nn} = C_{ln}$ from either of the equations (4). Normalization requires $|C_{nn}|^2 + |C_{ln}|^2 = 1$, and the wave function is

$$\psi_+ = \frac{\psi_{0n} + \psi_{0l}}{\sqrt{2}}. \tag{10.2-8}$$

For $\Delta = -H'_{nl}$, $C_{nn} = -C_{ln}$, and the wave function is

$$\psi_- = \frac{\psi_{0n} - \psi_{0l}}{\sqrt{2}}. \tag{10.2-9}$$

If the perturbation exceeds the energy difference $|E_{0n} - E_{0l}|$, the state is changed drastically from the original ψ_{0n} into either ψ_+ or ψ_-. If the states ψ_{0n} and ψ_{0l} are truly degenerate, then even a vanishingly small perturbation can produce this drastic change. There is no actual paradox here because no two states are completely degenerate; see the remarks at the end of Section 2.4.

PROBLEM 10.6

Suppose that there are three degenerate unperturbed states ψ_{0n}, ψ_{0l}, and ψ_{0k}.

a. Write out the development for this case that is the analog of the relations (1) through (6).

b. Suppose, as above, that the perturbation is an odd real function of position, and that the unperturbed states are real functions with definite parity. Unless you enjoy solving general cubic equations, suppose further that $H'_{lk} = 0$. Obtain the first-order energy shifts.

Suppose next that, in the two-state system described by (4), the off-diagonal matrix element H'_{nl} is zero. Then (6) becomes

$$\begin{aligned}(H'_{nn} - \Delta)(H'_{ll} - \Delta) &= 0,\\ \Delta &= H'_{nn} \text{ or } H'_{ll}.\end{aligned} \qquad (10.2\text{-}10)$$

For $\Delta = H'_{nn}$, $C_{ln} = 0$, and normalization requires C_{nn} to have magnitude unity. The system is still in state n, the first-order energy change is H'_{nn}, and everything looks just like the nondegenerate case studied in Section 10.1. Unless the off-diagonal elements are different from zero, the states are not mixed even if there is degeneracy. This observation permits the scheme that gives the condition (6) to be discussed in different language: The original set of basis states ψ_{0r} produces the matrix with elements H'_{rm}. This set of basis states is not "right" for the problem because H'_{rm} is not diagonal. To find the eigenvalues of the matrix, write (6), which is the *secular determinant* for the problem, as defined in Section 8.2. With Δ equal to one of the eigenvalues, the equations (4) determine the ratios of coefficients so that the state is an eigenfunction of the matrix. The perturbation drives the state into becoming one of these eigenfunctions, regardless of the size of the perturbation, if higher-order terms can be neglected.

PROBLEM 10.7

Review the conditions that give the states ψ_+ and ψ_- as defined by (8) and (9). Under these conditions show that if ψ_+ and ψ_-, rather than ψ_{0n} and ψ_{0l} are used as basis functions, then the corresponding 2×2 matrix of the perturbation is diagonal, with the possible values of Δ on the leading diagonal.

If the off-diagonal H'_{nl} are zero in the basis that one is led naturally to use, the simple solution (10) comes out without further effort. An interesting example is the interaction between the magnetic moments of the proton and the electron in the ground state of hydrogen. The spin-orbit interaction estimated in Section 8.1 splits states into components with slightly different energies and gives rise to complications that were named *fine structure* by spectroscopists. The interactions between nuclear and electron moments are still smaller because nuclear moments are small, the effect on spectra was named *hyperfine structure*, and the interaction is called the *hyperfine interaction.*

The state ψ_{100} of hydrogen found in Section 7.3 is separated by the hyperfine interaction into states that differ slightly in energy because their proton-electron spin states are different. To calculate the energy shifts, an expression is needed for the interaction between two magnetic moments. Pretend that the magnetic moment of the proton, $\vec{M}_p$, is caused by a current loop of radius $R \simeq 0.9$ f with current I,

$$|\vec{M}_p| = \pi R^2 I.$$

The field at the center of such a loop has magnitude

$$|\vec{B}(0)| = \frac{\mu_0 I}{2R} = \frac{\mu_0 |\vec{M}_p|}{2\pi R^3},$$

where the mks unit permeability constant $\mu_0 = 4\pi \times 10^{-7}$ weber/amp m. The direction of $\vec{B}(0)$ is that of the proton spin $\vec{S}_p$. With $\vec{M}_p = \gamma_p \vec{S}_p$, $\vec{M}_e = \gamma_e \vec{S}_e$, the perturbation at $r = 0$ is

$$H'(0) = -\vec{B}(0) \cdot \vec{M}_e = -\mu_0 \left(\frac{\gamma_p \vec{S}_p}{2\pi R^3} \right) \cdot (\gamma_e \vec{S}_e). \qquad (10.2\text{-}11)$$

There are four degenerate unperturbed states. All of them have the spatial dependence ψ_{100} as in (10.1-21), but they differ in their spin states. In the language of Section 9.1, three of them are the electron-proton triplet spin states ζ_{1+1}, ζ_{10}, and ζ_{1-1}, and one of them is the electron-proton singlet spin state ζ_{00}. It is physically inevitable, and the calculation will show, that the perturbation must shift the energy of the three triplet states by the same amount. Only the relative spin orientation of the two particles can matter; the absolute orientation of the resultant angular momentum cannot affect the energy. The matrix elements that must be calculated are then H'_{ss}, H'_{tt}, and $H'_{st} = H'^{\dagger}_{ts}$, where s and t stand for *singlet* and *triplet*. All of these quantities involve the spatial integration

$$\iiint |\psi_{100}|^2 \vec{B}(\vec{r}) \cdot \vec{M}_e \, d\tau.$$

Because the electron is in an $l = 0$ state, ψ_{100} is spherically symmetric. When averaged over angles, all contributions from the region $r > R$ cancel. If above and below the current loop the contribution is positive, then the contribution is negative near the plane of the loop. The integration need go only from $r = 0$ to $r = R$. Since the Bohr radius a is much greater than R, $\psi_{100} = e^{-r/a}/a^{3/2}\sqrt{\pi}$ is very close to constant inside R, and the integration becomes

$$|\psi_{100}(0)|^2 \iiint_{r<R} \vec{B}(\vec{r}) \cdot \vec{M}_e \, d\tau.$$

It can be shown that the mean value of $\vec{B}(r)$ in the spherical $r < R$ region is simply $\vec{B}(0)$, so that the integration gives

$$|\psi_{100}(0)|^2 \frac{4\pi R^3}{3} \vec{B}(0) \cdot \vec{M}_e.$$

The value (11) of $H'(0)$ is all that is needed. The artificially introduced radius R cancels, and the matrix elements become

$$\langle r|H'|m\rangle = -\tfrac{1}{3}\mu_0 \gamma_p \gamma_e |\psi_{100}(0)|^2 \langle r|2\vec{S}_p \cdot \vec{S}_e|m\rangle, \qquad (10.2\text{-}12)$$

with the indices r and m each equal to s or to t. This result is much better than the argument that produced it. It can be shown rigorously that the interaction of two point dipoles is

$$-\tfrac{1}{3}\mu_0 |\psi_{100}(0)|^2 \langle r|2\vec{M}_p \cdot \vec{M}_e|m\rangle \qquad (10.2\text{-}13)$$

if angle-dependent terms are averaged over the sphere, as they must be for all $l = 0$ states. The introduction of the current loop, the claim that there is no contribution for $r > R$, and the claim that the mean value for $r < R$ is $\vec{B}(0)$ are an attempt to make (12) plausible without invoking much electromagnetic theory.[3]

With $\vec{S}_T = \vec{S}_p + \vec{S}_e$, the sum of the proton and electron spins,

$$2\vec{S}_p \cdot \vec{S}_e = S_T^2 - S_p^2 - S_e^2.$$

It is conventional to use the quantum number F to denote the sum of the electron and nuclear spins, in the sense that the eigenvalues of S_T^2 are $\hbar^2 F(F + 1)$, with $F = 0$ in the singlet state and $F = 1$ in the triplet state. Because S_p^2 and S_e^2 are each $3\hbar^2/4$,

$$2\vec{S}_p \cdot \vec{S}_e|s\rangle = -\frac{3\hbar^2}{2}|s\rangle,$$

$$2\vec{S}_p \cdot \vec{S}_e|t\rangle = +\frac{\hbar^2}{2}|t\rangle.$$

[3] A rigorous derivation of (13) is given by J. D. Jackson, *Classical Electrodynamics*, 2d ed. (New York: John Wiley & Sons, Inc. 1975), Sec. 5.7. See also S. Gasiorowicz, *Quantum Physics* (John Wiley & Sons, Inc., 1974), Chap. 17.

Both $|s\rangle$ and $|t\rangle$ are eigenstates of H', and

$$\langle s|H'|t\rangle \sim \langle s|t\rangle = 0$$

because $|s\rangle$ and $|t\rangle$ are orthogonal. As promised, the off-diagonal elements are zero, the perturbation does not mix the states, and the solution has the form (10). With $\langle s|2\vec{S}_p \cdot \vec{S}_e|s\rangle = -3\hbar^2/2$, $\langle t|2\vec{S}_p \cdot \vec{S}_e|t\rangle = +\hbar^2/2$, and $|\psi_{100}(0)|^2 = 1/\pi a^3$, (12) gives the two nonzero matrix elements

$$H'_{ss} = \frac{2\mu_0}{\pi a^3}\left(\frac{\gamma_p \hbar}{2}\right)\left(\frac{\gamma_e \hbar}{2}\right) = -4.415 \times 10^{-6}\ \text{eV},$$
$$H'_{tt} = -\tfrac{1}{3}H'_{ss} = +1.472 \times 10^{-6}\ \text{eV} \qquad (10.2\text{-}14)$$

The difference in the energies of the triplet state and the singlet state is 5.887×10^{-6} eV, and the transition between the states gives rise to photons with wavelength $hc/5.887 \times 10^{-6}$ eV $= 21.06$ cm. The experimental result is, to four significant figures, 21.11 cm. The transition is particularly interesting for two reasons. First, it has been possible to measure its energy extremely accurately, to better than one part in 10^{10}. The quarter percent different between our 21.06 cm and the measured value is a testing ground for a great variety of subtle corrections. Second, the 21-cm radiation has provided much of our data on the distribution and motion of hydrogen in the universe.[4] The long wavelength permits the radiation to penetrate interstellar dust with negligible absorption.

Next, we shall look at an example with off-diagonal matrix elements that are not zero. Suppose that a hydrogen atom is in a uniform electric field $\mathscr{E}\hat{e}_z$. This field adds to the potential energy of the atom a perturbation $H' = e\mathscr{E}z$, where z is the z component of the position of the electron with respect to the proton, and e is the magnitude of the charge. This perturbation gives rise to a change in energy levels that is called the Stark effect, in honor of its discoverer J. Stark. Here also we consider only the first-order effect. The maximum value of H' within the volume of the atom is small for typical electric fields.

The unperturbed states are the $\psi_{nlm}(r, \theta, \varphi)$ of Section 7.3. Spin is irrelevant here; in the nonrelativistic domain, magnetic moments ignore electric fields. The perturbation is odd, all ψ_{nlm} have definite parity, and all elements on the leading diagonal of the H' matrix,

$$\iiint |\psi_{nlm}|^2 e\mathscr{E}z\, d\tau,$$

are zero because the integrand is odd. This example is of the kind described by (7). Since in its spatial behavior ψ_{100} is not degenerate, the $n = 1$ state cannot be shifted by $e\mathscr{E}z$ in first order. For the four $n = 2$ states ψ_{200}, ψ_{21-1}, ψ_{210}, and

[4] W. K. Hartmann, *Astronomy: The Cosmic Journey* (Belmont, Calif.: Wadsworth Publishing Co., Inc., 1978), pp. 312–313, 363–366; J. M. Pasachoff and M. L. Kutner, *University Astronomy* (Philadelphia: W. B. Saunders Co., 1978), Chaps. 24 and 25; C. Heiles, "The Structure of the Interstellar Medium," *Scientific American* **238** (1) 74 (1978).

ψ_{21+1}, there is degeneracy, and it is necessary to calculate the off-diagonal matrix elements that connect these states. Many of these terms are, however, zero. The z component of orbital angular momentum, $L_z = xp_y - yp_x$, commutes with z. If ψ_{nlm} is an eigenstate of L_z with eigenvalue $m\hbar$, then $z\psi_{nlm}$ is also an eigenstate of L_z with the same eigenvalue:

$$L_z(z\psi_{nlm}) = zL_z\psi_{nlm} = m\hbar z\psi_{nlm}.$$

The function $z\psi_{nlm}$ is therefore orthogonal to all eigenfunctions of L_z with different eigenvalues. It follows that a perturbation $\sim z$ gives nonzero matrix elements only between states that have the same m value:

$$\langle 2\ 0\ 0|z|2\ 1\ +1\rangle = \langle 2\ 1\ 0|z|2\ 1\ +1\rangle = \langle 2\ 1\ -1|z|2\ 1\ +1\rangle = 0,$$
$$\langle 2\ 0\ 0|z|2\ 1\ -1\rangle = \langle 2\ 1\ 0|z|2\ 1\ -1\rangle = \langle 2\ 1\ +1|z|2\ 1\ -1\rangle = 0,$$

neither $|2\ 1\ +1\rangle$ nor $|2\ 1\ -1\rangle$ can mix with any of the other $n = 2$ states, and they remain unshifted in first order. The problem becomes a two-state calculation, and only one matrix element remains to be calculated. The result is

$$\langle 2\ 0\ 0|e\mathscr{E}z|2\ 1\ 0\rangle = -3e\mathscr{E}a, \tag{10.2-15}$$

where a is the Bohr radius. As in (7), the energy shift Δ is plus or minus this matrix element. The states $|2\ 0\ 0\rangle$ and $|2\ 1\ 0\rangle$ are changed into the combinations

$$\frac{|2\ 0\ 0\rangle \pm |2\ 1\ 0\rangle}{\sqrt{2}},$$

one of which is shifted by $+3e\mathscr{E}a$ while the other is shifted by $-3e\mathscr{E}a$ (Figure 10.2-1).

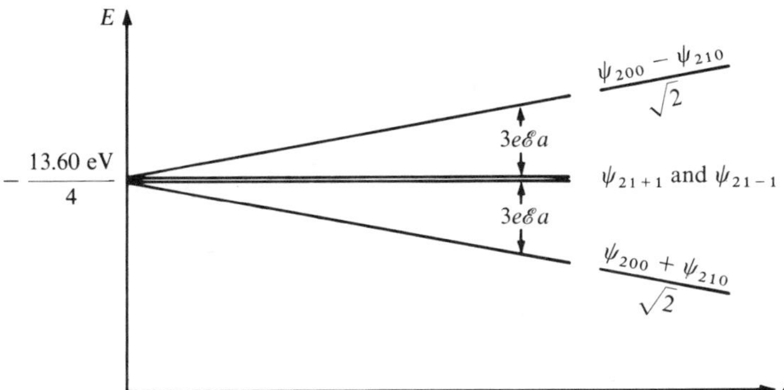

Fig. 10.2-1. The first-order Stark effect. The energies of the $n = 2$ states change as shown in an electric field $\mathscr{E}$. For $\mathscr{E} = 10^5$ volts/cm, $3e\mathscr{E}a$ is 1.6×10^{-5} eV.

PROBLEM 10.8

a. Obtain (15).

b. Write as functions of r, θ, φ the linear combinations of states produced by the perturbation. Examine how they behave at $\theta = 0$ and at $\theta = \pi$ around $r = 4a$ and try to extract a qualitative understanding of Figure 10.2-1.

10.3 *The WKB Approximation*

If the potential energy $V(\vec{r})$ changes so gradually that the wavelength is very well defined in any neighborhood, then classical mechanics is right. This limit is discussed and illustrated at the beginning of Section 1.4. If $V(\vec{r})$ changes enough to make the change in wavelength per wavelength noticeable but not large, then we are dealing with the transition region between classical and quantum physics. The WKB approximation method is appropriate for this region. G. Wentzel, H. A. Kramers, and L. Brillouin were associated with the application of the method to quantum physics in 1926, therefore the name. The method is, however, applicable to any wave propagating through a medium that is not homogeneous, and is much older than quantum theory.[5]

Begin with a one-dimensional definite energy Schrödinger equation

$$\frac{d^2\psi(x)}{dx^2} + \frac{p^2(x)}{\hbar^2}\psi(x) = 0,$$
$$p(x) = +\sqrt{2\mu[E - V(x)]}. \qquad (10.3\text{-}1)$$

The problem can involve the Cartesian coordinate x as indicated, but it can equally well treat the r dependence in a spherically symmetric potential as in Chapter 7, with $V(x)$ replaced by $V(r) + \hbar^2 l(l+1)/2\mu r^2$; the structure of the equation is the same and the mathematics cannot know the difference.

If V were constant, the description of a wave traveling in the $+x$ direction would be $Ae^{ipx/\hbar}$, with p constant. It is therefore sensible to try

$$\psi(x) = Ae^{iS(x)/\hbar} \qquad (10.3\text{-}2)$$

and let the deviation of $S(x)$ from a linear function reflect the changes in $V(x)$.

[5] References back to 1837 and a somewhat different treatment of the WKB approximation are given by L. I. Schiff, *Quantum Mechanics*, 3d ed. (New York: McGraw-Hill Book Company, 1968), Sec. 34. The way the approximation is introduced here is similar to that given by E. Merzbacher, *Quantum Mechanics*, 2d ed. (New York: John Wiley & Sons, Inc., 1970), Chap. 7.

Substituting (2) into (1) gives

$$i\hbar\frac{d^2S}{dx^2} - \left(\frac{dS}{dx}\right)^2 + p^2(x) = 0. \tag{10.3-3}$$

For a wave traveling in the $+x$ direction,

$$\frac{dS}{dx} = p(x)\left[1 + \frac{i\hbar}{p^2(x)}\frac{d^2S}{dx^2}\right]^{1/2}. \tag{10.3-4}$$

For slowly varying potentials, S differs only slightly from a linear function of x, and the magnitude of the second term in the square root is small. Neglect it entirely in the zeroth approximation;

$$\frac{dS_0}{dx} = p(x). \tag{10.3-5}$$

In the first approximation, use the zeroth approximation to evaluate d^2S/dx^2 in (4).

$$\frac{dS_1}{dx} = p(x)\left[1 + \frac{i\hbar}{p^2(x)}\frac{dp(x)}{dx}\right]^{1/2}.$$

If the second term is small, the square root can be expanded,

$$\frac{dS_1}{dx} = p(x)\left[1 + \frac{i\hbar}{2p^2(x)}\frac{dp(x)}{dx}\right]. \tag{10.3-6}$$

With a local wavelength $\lambda(x) \equiv h/p(x)$, as discussed in Section 1.4 and defined in (1.4-1),

$$\frac{i\hbar}{p^2(x)}\frac{dp(x)}{dx} = \frac{1}{2\pi i}\frac{1}{\lambda}\left(\frac{d\lambda}{dx}\lambda\right) = \frac{1}{2\pi i}\frac{\Delta\lambda}{\lambda}, \tag{10.3-7}$$

where $\Delta\lambda = \lambda(d\lambda/dx)$ is the change in wavelength produced by a change in x of one wavelength. This relation shows that

$$\left|\frac{1}{2\pi}\frac{\Delta\lambda}{\lambda}\right| \ll 1 \tag{10.3-8}$$

is a criterion for the validity of the method. With x_0 as some reference position, the integral of (6) gives

$$S_1 = \int_{x_0}^{x} p(x')dx' + \frac{i\hbar}{2}\ln\left(\frac{p(x)}{p(x_0)}\right).$$

Then (2) becomes

$$\begin{aligned}\psi_+(x) &= A\exp\left[\frac{i}{\hbar}\int_{x_0}^{x} p(x')dx' - \tfrac{1}{2}\ln\left(\frac{p(x)}{p(x_0)}\right)\right]\\ &= A\sqrt{\frac{p(x_0)}{p(x)}}\exp\left[i\int_{x_0}^{x} k(x')dx'\right],\end{aligned} \tag{10.3-9}$$

where $\hbar k(x) = p(x)$. The probability density $|\psi_+|^2$ is proportional to $1/v(x)$, where $v(x) = p(x)/M$ is the speed that a classical particle would have at x. The phase of the exponential is an integral that accounts for changes in $p(x)$ caused by changes in $V(x)$. If $V(x)$ is strictly constant, the phase becomes the familiar $ipx/\hbar$.

The result (9) describes a wave traveling toward $+x$ in an $E > V(x)$ region. For $E > V(x)$ and a wave traveling toward $-x$,

$$\psi_-(x) = A\sqrt{\frac{p(x_0)}{p(x)}}\exp\left[-i\int_{x_0}^{x} k(x')dx'\right]. \tag{10.3-10}$$

Bound states can be described by a superposition of (9) and (10), much as bound states in a region of constant potential can be described by superpositions of $e^{ipx/\hbar}$ and $e^{-ipx/\hbar}$ that give $\sin(px/\hbar)$ or $\cos(px/\hbar)$.

PROBLEM 10.9

Use the WKB approximation to obtain the energy levels for $V = gx$, $0 \leq x \leq a$; $V = \infty$ otherwise. The WKB approximation is not valid here unless $ga \ll E$, and it is both legitimate and helpful to make an expansion in powers of ga/E at the end of the calculation. This problem was done with first-order perturbation theory in Section 10.1; compare the results.

PROBLEM 10.10

Evaluate the probability current density for the $\psi_+(x)$ given by (9).

PROBLEM 10.11

Show that if $E < V(x)$, a procedure similar to that used to get (9) and (10) yields

$$\begin{aligned}\varphi_\pm(x) &= A\sqrt{\frac{\alpha(x_0)}{\alpha(x)}}\exp\left[\pm\int_{x_0}^{x} \alpha(x')dx'\right],\\ \hbar\alpha(x) &\equiv +\sqrt{2\mu[V(x) - E]}.\end{aligned} \tag{10.3-11}$$

Show that a criterion for validity is

$$\left|\frac{\Delta\alpha}{2\alpha}\right| \ll 1, \tag{10.3-12}$$

where $\Delta\alpha$ is the change in α in a distance $1/\alpha$.

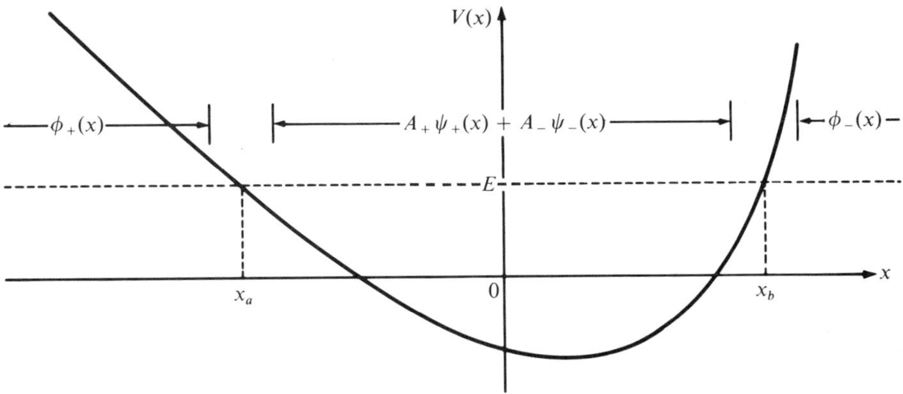

Fig. 10.3-1. The WKB treatment of a typical bound state problem. Near the center of the well a linear combination of ψ_+ and ψ_- may suffice, and the $\varphi_\pm$ may be good outside of the classical turning points x_a and x_b. Near x_a and x_b the WKB approximation must fail, and something else must be patched in.

Suppose that we want to find the energy levels for the $V(x)$ illustrated in Figure 10.3-1. If $V(x)$ varies sufficiently slowly to satisfy (8) and (12), the WKB approximation is applicable in the three regions that are separated by the classical turning points. Near these turning points the approximation cannot be valid. In (9) and (10), $p(x)$ goes to zero, and in (11), $\alpha(x)$ goes to zero, so that the $\psi_\pm$ and $\varphi_\pm$ diverge. The picture of nearly constant wavelengths must be false at points where $p(x)$ goes to zero, the wavelength goes to infinity, and the behavior of the wave function changes from oscillatory to exponential. The second term in (6) cannot be small wherever the exact wave function has an inflection point.

Figure 10.3-1 shows that near x_a and x_b, $V(x)$ can be approximated quite well by a straight line. The problem can be handled by patching in the solution for the $V(x) = gx - V_0$ potential, Section 2.9, and connecting the WKB solutions to its asymptotic forms. Consider the $x \simeq x_b$ neighborhood, where

$$-\frac{\hbar^2}{2\mu}\frac{d^2\psi}{dx^2} + (gx - V_0)\psi = E\psi.$$

Since x_b is a classical turning point, $gx_b - V_0 = E$,

$$-\frac{\hbar^2}{2\mu}\frac{d^2\psi}{dx^2} + g(x - x_b)\psi = 0.$$

As in (2.9-1) through (2.9-4), the solution that decreases for large $x - x_b$ is the Airy integral $\mathrm{Ai}(\xi_b)$, where

$$\xi_b = (x - x_b)\left(\frac{2\mu g}{\hbar^2}\right)^{1/3}.$$

This solution, extended into the $x < x_b$ region, rapidly approaches the asymptotic form (2.9-7),

$$\frac{1}{\sqrt{\pi}(-\xi_b)^{1/4}}\sin\left[\frac{2}{3}(-\xi_b)^{3/2}+\frac{\pi}{4}\right]. \tag{10.3-13}$$

In the $x < x_b$ region, near x_b,

$$\int_x^{x_b} k(x')dx' = \frac{1}{\hbar}\int_x^{x_b}\sqrt{2\mu g(x_b - x')}\,dx'$$

$$= \frac{2\sqrt{2\mu g}}{3\hbar}(x_b - x)^{3/2} = \tfrac{2}{3}(-\xi_b)^{3/2}.$$

Therefore the solution (13) extended into the $x < x_b$ region is that combination of the ψ_+ of (9) and the ψ_- of (10) that is proportional to

$$\sin\left[-\int_{x_b}^{x} k(x')dx' + \frac{\pi}{4}\right] = \sin\left[\int_{x_a}^{x_b} k(x')dx' - \int_{x_a}^{x} k(x')dx' + \frac{\pi}{4}\right]. \tag{10.3-14}$$

The amplitude $\sim(-\xi_b)^{-1/4} \sim (x_b - x)^{-1/4} \sim p(x)^{-1/2}$ comes out right, but is not needed to find the energy levels.

In the $x \simeq x_a$ neighborhood the problem can be approximated by

$$-\frac{\hbar^2}{2\mu}\frac{d^2\psi}{dx^2} + (V_0' - g'x)\psi = E\psi.$$

Since x_a is a classical turning point, $V_0' - g'x_a = E$.

$$-\frac{\hbar^2}{2\mu}\frac{d^2\psi}{dx^2} - g'(x - x_a)\psi = 0.$$

The solution that decreases for large $x_a - x$ is $\mathrm{Ai}(\xi_a)$, where

$$\xi_a = (x_a - x)\left(\frac{2\mu g'}{\hbar^2}\right)^{1/3}.$$

This solution, extended into the $x > x_a$ region, also approaches the asymptotic form (2.9-7); that is, it approaches expression (13) with the variable ξ_a rather than ξ_b. For $x > x_a$, near x_a,

$$\int_{x_a}^{x} k(x')dx' = \frac{1}{\hbar}\int_{x_a}^{x}\sqrt{2\mu g'(x' - x_a)}\,dx'$$

$$= \frac{2\sqrt{2\mu g'}}{3\hbar}(x - x_a)^{3/2} = \tfrac{2}{3}(-\xi_a)^{3/2}.$$

Therefore the solution extended into the $x > x_a$ region is that combination of ψ_+ and ψ_- that is proportional to

$$\sin\left[\int_{x_a}^{x} k(x')dx' + \frac{\pi}{4}\right]. \tag{10.3-15}$$

Between x_a and x_b, (14) and (15) must give the same answer. Since (14) and (15) can be multiplied by positive or negative amplitudes, the required match can be achieved if

$$\sin\left[\int_{x_a}^{x_b} k(x')dx' - \int_{x_a}^{x} k(x')dx' + \frac{\pi}{4}\right] = \pm\sin\left[\int_{x_a}^{x} k(x')dx' + \frac{\pi}{4}\right].$$

If $\sin A = \pm\sin B$, then either $A + B = m\pi$ or $A - B = m\pi$, $m = 0, \pm 1, \pm 2, \ldots$. The $A - B = m\pi$ possibility gives

$$\int_{x_a}^{x_b} k(x')dx' - 2\int_{x_a}^{x} k(x')dx' = m\pi$$

and must be rejected because it requires a function of x to be a constant. The $A + B = m\pi$ possibility gives

$$\int_{x_a}^{x_b} k(x')dx' = (m - \tfrac{1}{2})\pi = (n + \tfrac{1}{2})\pi, \qquad n = 0, 1, 2, \ldots. \tag{10.3-16}$$

The lowest value is $\pi/2$ because the integral cannot be negative.

Connecting the WKB solutions through the classical turning points[6] is a little painful, but the result is simple. If the regions $x < x_a$ and $x > x_b$ were totally impenetrable, the *phase integral* given by (16) would have to be $(n + 1)\pi$, $n = 0, 1, 2, \ldots$. Since ψ can spill into the classically forbidden regions, the phase does not have to change so much from x_a to x_b. It is as though the equivalent of $\frac{1}{8}$ wavelength spilled into the $V > E$ region at each side of the well.

Before modern quantum theory was developed, a rule called the Bohr-Sommerfeld-Wilson quantum condition[7] had some success in predicting energy levels. The rule is

$$\oint p(x')dx' = 2\pi(n + \tfrac{1}{2})\hbar, \tag{10.3-17}$$

[6] General formulas for making these connections are derived by L. I. Schiff, *Quantum Mechanics*, 3d ed. (New York: McGraw-Hill Book Company, 1968), Sec. 34; E. Merzbacher, *Quantum Mechanics*, 2d ed. (New York: John Wiley & Sons, Inc., 1970), Chap. 7; R. H. Dicke and J. P. Wittke, *Introduction to Quantum Mechanics* (Reading, Mass.: Addison-Wesley Publishing Co., Inc., 1970), Sec. 14–5.

[7] E. H. Wichmann, *Quantum Physics, Berkeley Physics Course*, Vol. 4 (New York: McGraw-Hill Book Company, 1971), Secs., 8.24–8.26. An elementary summary of the WKB approximation is also given there.

where the circle on the integral sign means that the integration is taken through one cycle of the classical motion. In this spirit,

$$\oint p(x')dx' = 2\int_{x_a}^{x_b} p(x')dx',$$

with the factor of two present because the particle has to go from x_a to x_b and back again to complete one oscillation. Since $p(x) = \hbar k(x)$, (17) is the same as (16). Both the successes and the failures of the old rule (17) can be understood now; although it is often a good approximation, it fails if $V(\vec{r})$ varies too rapidly.

PROBLEM 10.12

Use (16) to obtain the energy levels of the harmonic oscillator. (The energies are miraculously the same as the exact results, although the approximate wave functions are not the same as the exact wave functions.[8])

The WKB method is useful for calculating the transmission coefficient of barriers. Section 2.3 deals with rectangular and delta function barriers, but more complicated forms usually require approximations. The WKB approximation tends to be good for broad barriers, such as those encountered in field emission of electrons from metals and the penetration of the Coulomb field around a nucleus by positively charged particles. The general case[9] requires the solutions $\varphi_\pm(x)$ for the interior of the barrier, $\psi_+(x)$ and $\psi_-(x)$ to describe incident and reflected waves to the left of the barrier, and a $\psi_+(x)$ to describe the transmitted wave. It would be exactly right to use only $\varphi_-(x)$ inside the barrier if the barrier extended to infinity; if the transmission is small, it is a good approximation to use only $\varphi_-(x)$. The behavior of the transmission coefficient is then dominated by the roughly exponential decrease in φ_-, and numerical factors that come from detailed matching at the boundaries are unimportant. The order of magnitude of the transmission coefficient is given by

$$T \simeq \frac{|\varphi_-(\text{a little less than } x_t)|^2}{|\varphi_-(\text{a little more than } x_i)|^2},$$

where x_i is the classical turning point on the side of the incident beam, and x_t is the classical turning point on the side of the transmitted beam. It is necessary to evaluate φ_- at "a little more than x_i" and at "a little less than x_t" to avoid the infinities at the classical turning points. From (11), the dominant term is

$$T \simeq \exp\left[-2\int_{x_i}^{x_t} \alpha(x)dx\right]. \tag{10.3-18}$$

[8] J. L. Powell and B. Crasemann, *Quantum Mechanics* (Reading, Mass.: Addison-Wesley Publishing Co., Inc., 1961), Sec. 5.13, show in their Figure 5–23 a comparison between the exact, WKB, and Airy wave functions for this problem.

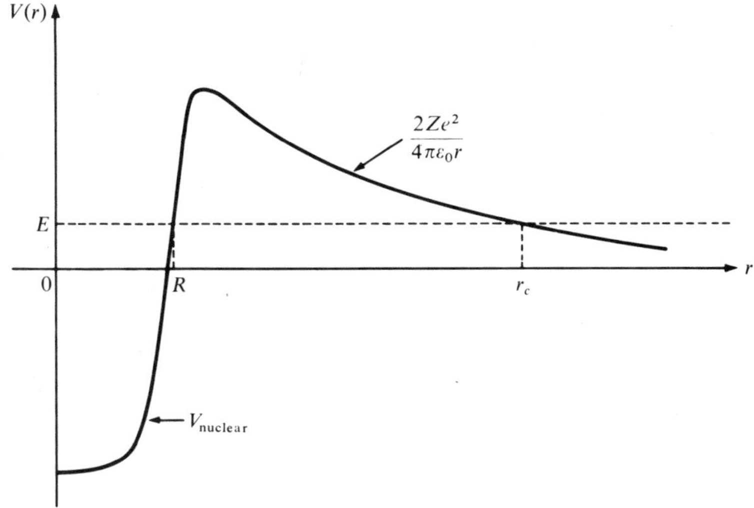

Fig. 10.3-2. The potential energy of an alpha particle near a nucleus. The behavior of $V(r)$ is not well defined for $r \lesssim R$, but is the Coulomb interaction between a charge $+2e$ and a charge $+Ze$ for r somewhat larger than R.

This useful result has been obtained here by a rather rough argument. A more careful derivation[9] yields the same conclusion.

PROBLEM 10.13

In Section 2.3 the transmission coefficient of the rectangular barrier, Figure 2.3-1, is given by (2.3-2). Compare (2.3-2) with (18), both in general and for the special case of very small transmission coefficient.

The escape of alpha particles from nuclei[10] provides a characteristic example. One can picture the particle as existing inside the nucleus, bouncing against the barrier, and occasionally tunneling out (Figure 10.3-2).

[9] E. Merzbacher, *Quantum Mechanics*, 2d ed. (New York: John Wiley & Sons, Inc., 1970), Chap. 7.

[10] K. W. Ford, *Classical and Modern Physics* (Lexington, Mass.: Xerox College Publishing, 1972) Section 26.2; E. Segre, *Nuclei and Particles* (New York, W. A. Benjamin, Inc., 1964) Chapter 7. Measured decay energies and half lives are available in the Table of Isotopes in the current edition of the CRC *Handbook of Chemistry and Physics* (Cleveland: CRC Press, Inc.). An advanced treatment that examines the formation of the alpha inside the nucleus and the effects of nonspherical potentials is M. A. Preston and R. K. Bhaduri, *Structure of the Nucleus* (Reading, Mass.: Addison-Wesley Publishing Company, Inc., 1975) Chapter 11.

For $l = 0$, since $u(r)$ satisfies an equation, (7.3-1), of the same form as (10.3-1), the estimate (18) can be taken over in the form

$$T \simeq \exp\left[-2\int_R^{r_c} \alpha(r)dr\right],$$

$$r_c = \frac{2Ze^2}{4\pi\varepsilon_0 E}, \tag{10.3-19}$$

$$\hbar\alpha(r) = \sqrt{2\mu\left(\frac{2Ze^2}{4\pi\varepsilon_0 r} - E\right)} = \sqrt{2\mu E\left(\frac{r_c}{r} - 1\right)}.$$

PROBLEM 10.14

Evaluate (19). Use the result and the parameters of typical heavy nuclei to produce a formula for the alpha decay half life. Compare the predictions of your formula with a few experimental values. The references in footnote 10 will help.

10.4 *The Variation Method*

Any reasonable potential well has a ground state ψ_0 that has a single maximum near the middle and that decreases at large distances. You are now a seasoned veteran who knows how wave functions behave, and who can, without solving the Schrödinger equation, write an expression with some adjustable parameters to approximate ψ_0. The variation method shows how to optimize such a guess and get an approximate value for the energy.

The Schrödinger equation is

$$H\psi_0(\vec{r}) = E_0\psi_0(\vec{r}).$$

Assume that ψ_0 is not degenerate, so that there is a unique ψ_0 with energy E_0. Neither $\psi_0(\vec{r})$ nor E_0 is known, but $\psi(\vec{r})$ looks like a reasonable guess that should not differ too much from ψ_0:

$$\psi(\vec{r}) = \psi_0(\vec{r}) + \Delta\psi(\vec{r}),$$

where $\Delta\psi$ is a small function if the guess is good. Write an expansion of ψ in terms of the exact and unknown ψ_n, the orthonormal eigenfunctions of H that satisfy $H\psi_n = E_n\psi_n$.

$$\psi = \sum_n C_n\psi_n.$$

The expectation value of the energy for the state ψ is

$$\begin{aligned}\langle E\rangle = \int \psi^* H\psi &= \int\Big(\sum_n C_n^*\psi_n^*\Big)H\Big(\sum_m C_m\psi_m\Big)\\ &= \sum_{n,m} C_n^* C_m E_m \int \psi_n^*\psi_m\\ &= \sum_n |C_n|^2 E_n,\end{aligned} \tag{10.4-1}$$

as in Sections 3.6 and 3.7. The lowest of the E_n is E_0, so that

$$\int \psi^* H\psi \geq \sum_n |C_n|^2 E_0.$$

Normalization requires, as in (3.7-2), that $\sum_n |C_n|^2 = 1$.

$$\int \psi^* H\psi \geq E_0. \tag{10.4-2}$$

The equality holds only if ψ happens to be exactly ψ_0. If ψ is a very good guess, then $\int \psi^* H\psi$ is only slightly greater than E_0. One obtains the best value for the energy by varying parameters until $\int \psi^* H\psi$ is minimized, because (2) shows that this quantity always lies above the exact answer E_0.

$$E_{\text{best}} = \langle E\rangle_{\text{min}} = \left(\int \psi^* H\psi\right)_{\text{min}}. \tag{10.4-3}$$

The variation must, of course, keep ψ normalized. For $n > 0$ the coefficients C_n are given by

$$\begin{aligned}C_n = \int \psi_n^*\psi &= \int \psi_n^*(\psi_0 + \Delta\psi)\\ &= \int \psi_n^*\,\Delta\psi;\end{aligned}$$

they are of the same order of smallness as the mistake $\Delta\psi$ in the wave function. Their contribution to $\langle E\rangle$ is, from (1), the square of a small quantity, so that even a guess ψ that is only fair can produce an $\langle E\rangle$ that is close to the exact E_0.

Section 2.9 examines bound states for $V(x) = gx$, $x \geq 0$; $V(x) = \infty$, $x < 0$. The ground state energy is, according to (2.9-12), $2.3381(g^2\hbar^2/2M)^{1/3}$. We shall try the variation method on this problem with the trial function

$$\psi = 2\gamma^{3/2} x e^{-\gamma x}$$

where $2\gamma^{3/2}$ normalizes ψ. This form starts out linearly from $x = 0$, reaches a maximum, and decreases at large distances, but it does not behave like the exact solution in detail. The procedure is to compute $\langle E \rangle$ as a function of the parameter γ and then to vary γ until $\langle E \rangle$ is minimized.

$$\begin{aligned}\langle E \rangle &= \int_0^\infty 2\gamma^{3/2} x e^{-\gamma x}\left(-\frac{\hbar^2}{2M}\frac{d^2}{dx^2} + gx\right)2\gamma^{3/2} x e^{-\gamma x}\, dx \\ &= 4\gamma^3 \int_0^\infty \left(\frac{\hbar^2 \gamma}{M} x - \frac{\hbar^2\gamma^2}{2M} x^2 + gx^3\right) e^{-2\gamma x}\, dx \\ &= \frac{\hbar^2\gamma^2}{2M} + \frac{3g}{2\gamma}.\end{aligned}$$

The region in which the particle is likely to be found has a size of roughly $1/\gamma$. If this size is small, the kinetic energy $\hbar^2\gamma^2/2M$ is large. If this size is large, the potential energy $3g/2\gamma$ is large. As in many other examples, the energy is determined by a balance between the outward push of the uncertainty principle and the inward pull of the attractive force; see the discussion of the hydrogen atom in Section 1.3. The minimum is determined by

$$\frac{d}{d\gamma}\langle E \rangle = \frac{\hbar^2\gamma}{M} - \frac{3g}{2\gamma^2} = 0, \qquad \gamma = \left(\frac{3gM}{2\hbar^2}\right)^{1/3},$$

so that

$$\begin{aligned}\langle E \rangle_{\min} &= \frac{\hbar^2}{2M}\left(\frac{3gM}{2\hbar^2}\right)^{2/3} + \frac{3g}{2}\left(\frac{2\hbar^2}{3gM}\right)^{1/3} \\ &= \frac{3}{2}\left(\frac{9}{2}\right)^{1/3}\left(\frac{g^2\hbar^2}{2M}\right)^{1/3} = 2.4764\left(\frac{g^2\hbar^2}{2M}\right)^{1/3},\end{aligned}$$

6% too high. As shown by (2), the result cannot lie below the true value. To do better, one must choose a trial ψ that can approach the actual ψ_0 more closely, either because its functional form is a better match, or because there are several parameters in ψ that can be varied, or both.

PROBLEM 10.15

Find the ground state energy of the harmonic oscillator by using the variation method with a Gaussian trial wave function.

The variation method is not restricted to finding ground states. For spherically symmetric potentials and any orbital angular momentum quantum number l, one can work with the separated radial equation (7.1-3) in much the same fashion as with the one-dimensional x equation.

PROBLEM 10.16

Find the energy of the lowest $l = 1$, or P state of charmonium. The equation to be solved is (7.4-2). Since $u \sim r^{l+1}$ near $r = 0$, try $u \sim r^2 e^{-\gamma r/2}$. Normalize the trial u, obtain $\langle E \rangle$, and use γ as the variation parameter. Use the M, g, and V_0 values given in Section 7.4. Spin-orbit interactions are not included, so that your theory agrees with experiment if your result falls somewhere in or near the group of three states ${}^3P_2$, ${}^3P_1$, and ${}^3P_0$ shown in Figure 7.4-1.

The variation method is of great value in atomic and molecular structure calculations. The ground state of the helium atom is a good example. There both electrons are in spatial states similar to the hydrogen-like $\psi_{100}(r, \theta, \varphi)$ for $Z = 2$, with some distortion because of the Coulomb repulsion between the two electrons. Under exchange of the two electrons,

$$\psi(\vec{r}_1, \vec{r}_2) = \psi_{100}(\vec{r}_1)\psi_{100}(\vec{r}_2)$$

is even, so that the spin state must be the odd singlet ζ_{00} of (9.1-13). The Schrödinger equation that governs the spatial behavior is, as discussed in Section 4.2,

$$\left[-\frac{\hbar^2}{2M}(\nabla_1^2 + \nabla_2^2) - \frac{Ze^2}{4\pi\varepsilon_0}\left(\frac{1}{r_1} + \frac{1}{r_2}\right) + \frac{e^2}{4\pi\varepsilon_0|\vec{r}_1 - \vec{r}_2|}\right]\psi(\vec{r}_1, \vec{r}_2) = E\psi(\vec{r}_1, \vec{r}_2). \tag{10.4-4}$$

The first term in the Hamiltonian is the total kinetic energy of the two electrons. The second term gives the Coulomb attraction of the electrons by the nucleus, with $Z = 2$ for He, $Z = 3$ for the ion Li^+, $Z = 4$ for Be^{++}, and so forth. The third term gives the Coulomb repulsion between the electrons and makes things hard. The nucleus is assumed to be fixed at the origin.[11]

Take as the trial wave function

$$\psi(\vec{r}_1, \vec{r}_2) = \left(\frac{e^{-r_1/a}}{a^{3/2}\sqrt{\pi}}\right)\left(\frac{e^{-r_2/a}}{a^{3/2}\sqrt{\pi}}\right), \qquad a \equiv \frac{4\pi\varepsilon_0\hbar^2}{ze^2M}. \tag{10.4-5}$$

If the number z were taken to be the nuclear charge number Z, then ψ would be the product of two $n = 1$ states for a hydrogen-like system without any corrections. However, z will be taken as the variation parameter. This approach makes

[11] Nuclear motion effects are discussed by D. Park, *Introduction to the Quantum Theory*, 2d ed. (New York: McGraw-Hill Book Company, 1974), Sec. 15.3.

good sense because each electron is partly shielded from the $+Ze$ nuclear charge by the other electron, and therefore sees an effective charge between $(Z - 1)e$ and Ze.

The first step is to obtain the expectation value of the Hamiltonian. The kinetic energy of each electron contributes

$$\iiint \frac{e^{-r/a}}{a^{3/2}\sqrt{\pi}}\left(-\frac{\hbar^2}{2M}\nabla^2\right)\frac{e^{-r/a}}{a^{3/2}\sqrt{\pi}}\,d\tau$$

$$= \frac{-\hbar^2}{2Ma^3\pi}\int_0^\infty e^{-r/a}\frac{1}{r^2}\frac{d}{dr}\left(r^2\frac{d}{dr}e^{-r/a}\right)4\pi r^2\,dr \qquad (10.4\text{-}6)$$

$$= \frac{\hbar^2}{2Ma^2} = \frac{e^4Mz^2}{2(4\pi\varepsilon_0\hbar)^2}$$

The Coulomb interaction of each electron with the nucleus contributes

$$\int_0^\infty \left(\frac{e^{-r/a}}{a^{3/2}\sqrt{\pi}}\right)^2\left(\frac{-Ze^2}{4\pi\varepsilon_0 r}\right)4\pi r^2\,dr = \frac{-Ze^2}{4\pi\varepsilon_0 a}$$

$$= \frac{-Ze^4Mz}{(4\pi\varepsilon_0\hbar)^2}. \qquad (10.4\text{-}7)$$

The third term is the electrostatic energy of the charge distribution of electron #1 in the potential created by the charge distribution of electron #2. View the charge distribution of #2 as a set of spherical shells of radius r_2 and thickness dr_2. Those inside r_1 contribute

$$-\frac{e}{4\pi\varepsilon_0}\frac{e^{-2r_2/a}}{\pi a^3}\frac{4\pi r_2^2\,dr_2}{r_1}$$

as if their charge were concentrated at the origin. Those outside contribute

$$-\frac{e}{4\pi\varepsilon_0}\frac{e^{-2r_2/a}}{\pi a^3}\frac{4\pi r_2^2\,dr_2}{r_2},$$

independent of r_1. The electrostatic potential is therefore

$$U_2(r_1) = \frac{-e}{\pi\varepsilon_0 a^3}\left[\frac{1}{r_1}\int_0^{r_1} r_2^2 e^{-2r_2/a}\,dr_2 + \int_{r_1}^\infty r_2 e^{-2r_2/a}\,dr_2\right]$$

$$= \frac{-e}{4\pi\varepsilon_0}\left[\frac{1}{r_1} - \left(\frac{1}{a} + \frac{1}{r_1}\right)e^{-2r_1/a}\right].$$

The potential energy contributed by the repulsion between the electrons is

$$\iiint \left(-e\frac{e^{-2r_1/a}}{\pi a^3}\right)U_2(r_1)d\tau_1 = \frac{5}{8}\frac{e^2}{4\pi\varepsilon_0 a}$$

$$= \frac{5}{8}\frac{e^4 Mz}{(4\pi\varepsilon_0 \hbar)^2}. \tag{10.4-8}$$

The sum of the three contributions is twice (6) plus twice (7) plus (8),

$$\langle E\rangle = \frac{e^4 M}{(4\pi\varepsilon_0\hbar)^2}\left(z^2 - 2Zz + \frac{5z}{8}\right), \tag{10.4-9}$$

and $d\langle E\rangle/dz = 0$ gives

$$z = Z - \tfrac{5}{16}. \tag{10.4-10}$$

On the average each electron interposes $\frac{5}{16}$ of its charge distribution between the nucleus and the other electron. The value (10) substituted into (9) gives

$$\langle E\rangle_{\min} = \frac{-e^4 M}{(4\pi\varepsilon_0\hbar)^2}(Z - \tfrac{5}{16})^2 = -2 \times 13.61\ \text{eV}(Z - \tfrac{5}{16})^2. \tag{10.4-11}$$

This result is the energy of the two K shell electrons, that is, its negative is the amount of energy needed to pull both electrons out of the atom. After one electron is removed, the remaining ion is in a simple hydrogen-like state, and has energy -13.61 eV Z^2. The magnitude of the difference,

$$I = [2(Z - \tfrac{5}{16})^2 - Z^2]13.61\ \text{eV}, \tag{10.4-12}$$

is the ionization energy for the first of the two electrons. For helium, $\langle E\rangle_{\min} = -77.50$ eV, while the experimental value for the energy needed to remove both electrons is -78.88 eV. The ionization energy of the first electron is 23.07 eV according to (12), while the experimental result is 24.48 eV.

PROBLEM 10.17

a. Obtain the energy of the two K shell electrons by a first-order perturbation method. Use the first two terms of the Hamiltonian in (4) as H_0 and the third term as H'. Use the unperturbed wave function (5) with $z = Z$. All integrals needed are done above, and this problem is very easy if you do your bookkeeping right.

b. Make a table that shows the experimental, variation, and perturbation values for I, the ionization energy of the first electron for He, Li^+, and Be^{++}. Comment on the difference between the variation and perturbation methods. (Experimental results are given in tables of *ionization potentials* in, for example, the CRC *Handbook of Chemistry and Physics.*)

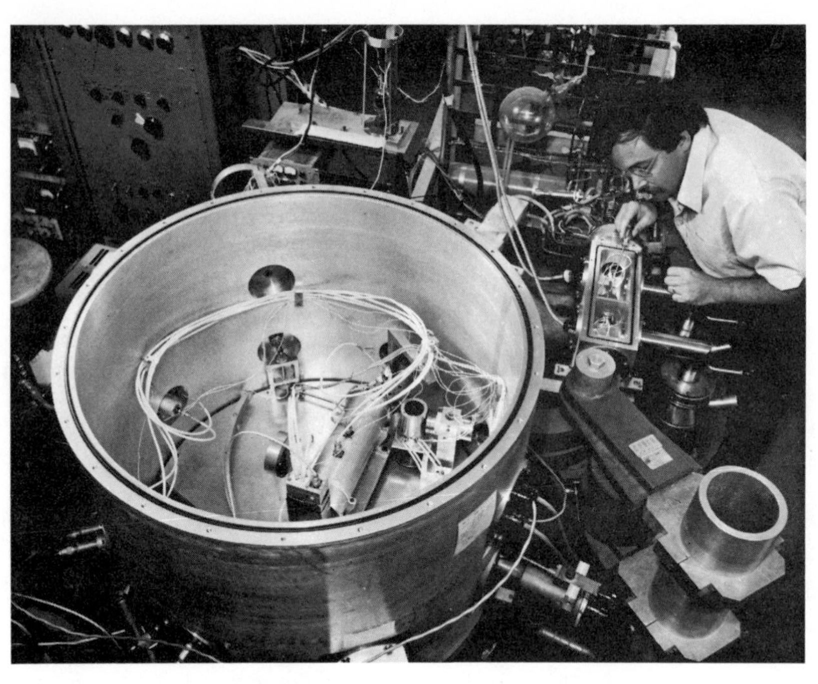

An ion–atom scattering experiment. Gas enters the apparatus at the upper right and is ionized and accelerated in the rectangular box that is being adjusted. The ions are then deflected by the field of the magnet at the lower right, so that those with the desired energy and mass enter the large cylindrical scattering chamber. The target gas is in the small cylindrical cup near the right of the scattering chamber. The scattered particles are recorded by a detector that can be moved about to determine the scattering probability as a function of angle. During the measurements, the accelerator box and the scattering chamber are sealed and evacuated. The apparatus has been used to study a variety of processes such as the scattering of H^+ by Ar and of He^{++} by He in the 2 to 300 eV region. (*Courtesy of R. L. Champion, L. D. Doverspike, and B. T. Smith, Physics Department, College of William and Mary*)

11

Scattering

With the conventional choice $V(r = \infty) = 0$, negative energy states are bound states. The particle cannot get away, its wave function must fit into some region, and there is a discrete spectrum of allowed energy eigenvalues.

If the particle has positive energy, it can come from and go to indefinitely large distances, its wave function is not forced to fit into some region, and the allowed energies form a continuum. The question "What are the possible energy eigenvalues?" has the uninteresting answer, "All energies > 0." A typical interesting question is, "If the particle comes in with initial momentum $\vec{p}_i$, what is the probability that it is scattered so that its final momentum is $\vec{p}_f$?" The $E > 0$ states are scattering states.

In Chapter 2 scattering states are considered along with bound states. However, most of the calculations there are essentially one dimensional, and deal with particles that can do nothing except continue in their original direction or scatter through 180°. This chapter deals with three-dimensional scattering that can give deflections through any angle.

11.1 Scattering Wave Functions and Cross Sections

The basic ingredients of a scattering experiment are an incident beam, a scattering center described by the $V(\vec{r})$ that acts on the incident particles, and scattered particles that can be recognized by a detector. The $V(\vec{r})$ is caused by something that recoils during the collision; perhaps $V(\vec{r})$ is the Coulomb field of a nucleus that is deflecting protons. The process is not an interaction of the incident particle with a fixed force center, but is usually an interaction between two particles that are fairly well isolated from the rest of the world. It is necessary to hedge and say "usually" because the recoil might not be enough to make negligible the chemical forces that hold the target particle in place. If the incident and target particles can be treated as an isolated system, the process is best described by the Section 4.1 formalism as the motion in relative coordinates of a single particle with reduced mass. This description is the same as that used in Chapter 7 for $E < 0$ states.

PROBLEM 11.1

Estimate what the energy must be if it is right to treat the target particle as free for (i) 180° scattering and (ii) 2° scattering of

a. a neutron by a proton.

b. an alpha particle by a gold nucleus.

In typical atomic, nuclear, and particle interactions, $V(\vec{r})$ is appreciable only in a very small region. Outside this scattering region there are no significant forces. Choose the z axis to be the direction of the incident momentum, and the wave function that describes the incident beam outside the scattering region to be

$$\psi_i = Ae^{ikz}, \qquad k = \frac{\sqrt{2\mu E}}{\hbar}. \tag{11.1-1}$$

The incident flux density $\vec{j}_i$ is then $\hat{e}_z|A|^2\hbar k/\mu$. Note that ψ_i is an eigenstate of L_z with eigenvalue zero:

$$L_z\psi_i = \left(x\frac{\hbar}{i}\frac{\partial}{\partial y} - y\frac{\hbar}{i}\frac{\partial}{\partial x}\right)Ae^{ikz} = 0. \tag{11.1-2}$$

If $V(\vec{r})$ is rotationally symmetric about the z axis, then the scattered particles will also be in an $L_z = 0$ state.

The choice (1) does not limit the accuracy of the usual scattering calculation, but it is really an approximation. According to (1), the incident particles have precisely the energy $\hbar^2k^2/2\mu$ and are described by a single coherent wave of infinite extent. As discussed in Section 2.8, any real particle beam consists of wave packets of limited extent, somewhat indefinite energy, and random relative phases. Nevertheless, (1) is adequate for most scattering computations because typically the energy is fairly well defined, the coherence length is large compared to the scattering region, and collimating apertures are large compared to the wavelength.[1]

Far from the origin a scattered particle is described by a spherical wave that is moving out to the large r region. Such a wave is best treated in spherical polar coordinates, and some of the Chapter 7 results are applicable. The radial behavior $R(r)$ is described by (7.1-1) with $E > 0$ and with $V(r)$ omitted.

$$-\frac{\hbar^2}{2\mu}\frac{d^2}{dr^2}(rR) + \frac{\hbar^2 l(l+1)}{2\mu r^2} rR = ErR.$$

The scattering is assumed to be elastic, that is, the energy E of the scattered wave is the same as that of the incident wave. Because r is large, the centrifugal potential is negligible. The equation is

$$\frac{d^2}{dr^2}(rR) + k^2(rR) = 0,$$

and is solved by $rR \sim e^{\pm ikr}$. The general solution for the wave ψ_d at the detector in the large r region is the sum of the two possibilities for R, e^{+ikr}/r and e^{-ikr}/r, each multiplied by some function of the angles θ and φ.

$$\psi_d = \frac{A}{r}[f(\theta, \varphi)e^{+ikr} + g(\theta, \varphi)e^{-ikr}]. \qquad (11.1\text{-}3)$$

The same constant A is inserted in (3) as in (1) to provide a reminder that ψ_d must be proportional to ψ_i. To understand what (3) means, evaluate the probability current density

$$\vec{j}_d = \frac{\hbar}{2\mu i}(\psi_d^* \vec{\nabla}\psi_d - \text{complex conjugate}), \qquad (11.1\text{-}4)$$

with the gradient in spherical polar coordinates

$$\vec{\nabla}\psi_d = \frac{\partial\psi_d}{\partial r}\hat{r} + \frac{1}{r}\frac{\partial\psi_d}{d\theta}\hat{\theta} + \frac{1}{r\sin\theta}\frac{\partial\psi_d}{\partial\varphi}\hat{\varphi}. \qquad (11.1\text{-}5)$$

In the large r region the only significant term comes from the derivatives $\partial e^{\pm ikr}/\partial r$; all other terms have higher powers of r in the denominator. The first

[1] Although the plane wave approximation (1) is generally adequate and mathematically simpler, it is interesting and instructive to do the calculation with wave packets. See E. Merzbacher, *Quantum Mechanics*, 2d ed. (New York: John Wiley & Sons, Inc., 1970), Sec. 11.2.

term in (3), $Af(\theta, \varphi)e^{+ikr}/r$, by itself contributes

$$\mathbf{j}_d \xrightarrow[\text{large } r]{} \frac{\hbar}{2\mu i}\left[\frac{A^*}{r} f^* e^{-ikr} \frac{A}{r} f \frac{\partial}{\partial r} e^{ikr}\hat{\mathbf{r}} - \text{complex conjugate}\right]$$

$$= \frac{\hbar k}{\mu} |A|^2 \frac{|f(\theta, \varphi)|^2}{r^2} \hat{\mathbf{r}}. \tag{11.1-6}$$

This expression makes good sense. The current density is directed outward from the scattering center, decreases $\sim 1/r^2$, is proportional to the incident flux density $j_i = |A|^2 \hbar k/\mu$, and depends on the direction through $|f(\theta, \varphi)|^2$. The second term in ψ_d makes bad sense. Because of the e^{-ikr}, it contributes a spherical wave that comes in toward the scattering center from large distances. The physical condition that the particles are scattered outward requires that $g(\theta, \varphi)$ must be set equal to zero.

PROBLEM 11.2

a. Use (5) in (4) and obtain without approximation the current densities produced by $Af(\theta, \varphi)e^{+ikr}/r$ and by $Ag(\theta, \varphi)e^{-ikr}/r$. Then show explicitly that the large r behavior is given by (6) and by a similar expression for the term proportional to $g(\theta, \varphi)$.

b. What circumstances would produce a nonzero $g(\theta, \varphi)$?

Suppose that, as shown in Figure 11.1-1, the detector presents a small area δa to the outgoing spherical wave. The count rate, the mean number of counts per second $\langle dN/dt \rangle$, equals the magnitude of the current density times δa,

$$\left\langle \frac{dN}{dt} \right\rangle = \frac{\hbar k}{\mu} |A|^2 \frac{|f(\theta, \varphi)|^2}{r^2} \delta a$$

$$= j_i |f(\theta, \varphi)|^2 \, \delta\Omega \tag{11.1-7}$$

where $\delta\Omega$ is the solid angle intercepted by the counter. The count rate is therefore the product of the incident flux density that is determined by an accelerator or other source, the solid angle that is determined by the size and placement of the detector, and the quantity $|f(\theta, \varphi)|^2$ that is determined by the nature of the interaction. This quantity has the units of an area, and is the *differential scattering cross section*:

$$\frac{d\sigma}{d\Omega}(\theta, \varphi) = |f(\theta, \varphi)|^2. \tag{11.1-8}$$

It is the mean number of particles scattered by one scattering center per second in the direction θ, φ, per unit solid angle, per unit incident flux. Its connection

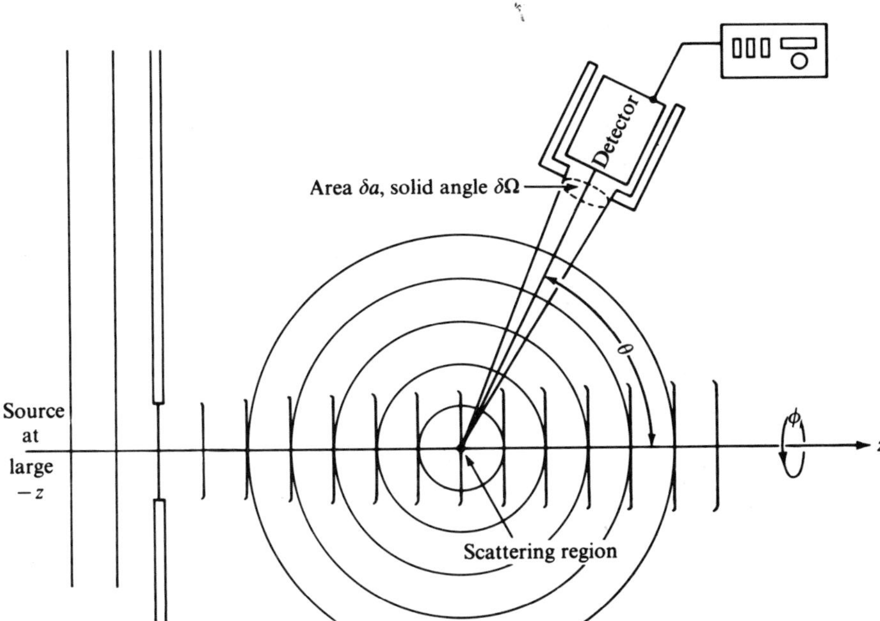

Fig. 11.1-1. A scattering experiment. The detector is at (r, θ, φ) in the relative coordinates, is shielded from the unscattered incident particles, and intercepts a solid angle $\delta\Omega$. All collimating apertures are large compared to the wavelength, so that diffraction effects due to these apertures are negligible.

with $f(\theta, \varphi)$ is quantum physics, but the concept of scattering cross section is the same as that used in classical mechanics.[2]

The total scattering cross section σ is the integral of (8) over all angles:

$$\sigma = \iint \frac{d\sigma}{d\Omega}(\theta, \varphi) d\Omega. \tag{11.1-9}$$

This quantity equals the mean number of particles scattered per second in any direction, per scattering center, per unit incident flux.

The coordinates r, θ, φ are measured in the relative coordinate system, which is at rest with respect to the center of mass of the two interacting particles. Some experiments are done with colliding beams, but the usual situation is that, before the collision, the target particle is at rest in the laboratory. It is often necessary to transform from the center-of-mass system to the laboratory system in order to compare theory with experiment. The transformation is the same in

[2] K. R. Symon, *Mechanics*, 3d ed. (Reading, Mass.: Addison-Wesley Publishing Co., Inc., 1971), Secs. 3.16, 4.6, and 4.8; J. B. Marion, *Classical Dynamics of Particles and Systems*, 2d ed. (New York: Academic Press, Inc., 1970), Chap. 9; D. Park, *Introduction to the Quantum Theory*, 2d ed. (New York: McGraw-Hill Book Company, 1974), Secs. 9.1 and 9.2.

classical and quantum mechanics. Review as needed this transformation and other classical mechanics aspects of cross-section calculations; see the references given in footnote 2.

The central job of a quantal scattering calculation is to find $f(\theta, \varphi)$. This quantity, the *scattering amplitude*, determines the cross section, and is itself determined by the energy and by the $V(\vec{r})$. Like any function of angle, it can be expanded in spherical harmonics,

$$f(\theta, \varphi) = \sum_{l,m} C_{lm} Y_{lm}(\theta, \varphi). \tag{11.1-10}$$

Each Y_{lm}, as shown in Section 6.2, is an eigenstate of L^2 with eigenvalue $\hbar^2 l(l+1)$, and of L_z with eigenvalue $\hbar m$. According to (2), the incident particles must have $m = 0$, and as observed there, the scattered particles must also have $m = 0$ if $V(\vec{r})$ is rotationally symmetric about the z axis. It then follows that only $m = 0$ terms can appear in (10). The Y_{l0} do not depend on φ. Therefore

$$f = f(\theta) \qquad \text{only if} \qquad V \neq V(\varphi). \tag{11.1-11}$$

The following sections illustrate some methods that are used to obtain $f(\theta, \varphi)$.

PROBLEM 11.3

A 2.0-cm × 2.0-cm × 0.10-cm-thick lead target is bombarded by 10^6/sec cm^2 neutrons with 0.10-MeV kinetic energy. A 10-cm^2 counter, 50 cm from the target, receives 3.0/sec elastically scattered neutrons, independent of the counter's angular location. Neglect the recoil of the lead nuclei.

a. Evaluate σ and $d\sigma/d\Omega$.

b. Write ψ_i and ψ_d, with numerical values given for all constants as far as these data permit. (Take as ψ_d the wave scattered by a single Pb nucleus.)

11.2 *The Born Approximation*

The Section 11.1 discussions of the incident wave ψ_i and detected wave ψ_d should not obscure that there is one wave function ψ which describes the entire system. In the pipe that goes from the accelerator to the scattering region, ψ is very well approximated by the ψ_i of (11.1-1). At the detector, ψ is very well approximated by the ψ_d of (11.1-3) with $g(\theta, \varphi) = 0$. In the scattering region, ψ depends on the details of the interaction, and its form there is what connects ψ_d to ψ_i. To obtain this link, it is necessary to solve the Schrödinger equation

$$-\frac{\hbar^2}{2\mu}\nabla^2\psi(\vec{r}) + V(\vec{r})\psi(\vec{r}) = E\psi(\vec{r}),$$

or

$$(\nabla^2 + k^2)\psi(\vec{r}) = w(\vec{r})\psi(\vec{r}),$$
$$k = \frac{\sqrt{2\mu E}}{\hbar}, \qquad w(\vec{r}) = \frac{2\mu V(\vec{r})}{\hbar^2}. \tag{11.2-1}$$

To solve (1), we will develop the *Green's function* method. The effort is a good investment; the method is valuable for many quantal scattering problems, and also for a variety of calculations in classical mechanics and in electricity and magnetism.[3] Set aside (1) for the moment and look at the equation

$$(\nabla^2 + k^2)G(\vec{r}, \vec{r}') = \delta(\vec{r} - \vec{r}'), \tag{11.2-2}$$

where the three-dimensional delta function is $\delta(x - x')\delta(y - y')\delta(z - z')$. It satisfies

$$\delta(\vec{r} - \vec{r}') = 0, \qquad \vec{r} \neq \vec{r}',$$

and

$$\iiint \delta(\vec{r} - \vec{r}')d\tau' = 1, \quad d\tau' \equiv dx'\,dy'\,dz',$$

if $\vec{r}' = \vec{r}$ is in the volume of integration. A solution to (2) is called a *Green's function.* The method rests on the observation that, if $\psi(\vec{r})$ satisfies

$$\psi(\vec{r}) = Ae^{ikz} + \iiint G(\vec{r}, \vec{r}')w(\vec{r}')\psi(\vec{r}')d\tau', \tag{11.2-3}$$

then $\psi(\vec{r})$ satisfies (1). To verify this statement, operate on (3) with $\nabla^2 + k^2$. The first term contributes nothing;

$$(\nabla^2 + k^2)Ae^{ikz} = 0.$$

In the second term, exchange the order of differentiation and integration. Note that ∇^2 differentiates the coordinates $\vec{r} = (x, y, z)$, not the variables of integration $\vec{r}' = (x', y', z')$.

$$\begin{aligned}(\nabla^2 + k^2)\psi(\vec{r}) &= \iiint (\nabla^2 + k^2)G(\vec{r}, \vec{r}')w(\vec{r}')\psi(\vec{r}')d\tau' \\ &= \iiint \delta(\vec{r} - \vec{r}')w(\vec{r}')\psi(\vec{r}')d\tau' \\ &= w(\vec{r})\psi(\vec{r}),\end{aligned}$$

[3] For one-dimensional classical mechanics illustrations, see K. R. Symon, *Mechanics*, 3d ed. (Reading, Mass.: Addison-Wesley Publishing Co., Inc., 1971), Sec. 2.11; J. B. Marion, *Classical Dynamics of Particles and Systems*, 2d ed. (New York: Academic Press. Inc., 1970), Sec. 4.6. Many applications to electricity and magnetism are given by J. D. Jackson, *Classical Electrodynamics*, 2d ed. (New York: John Wiley & Sons, Inc., 1975).

the same as (1), where (2) gives the first step, and the general property (2.3-4) of the delta function gives the second step.

One way to find $G(\vec{r}, \vec{r}')$ is to begin with an educated guess. At large r,

$$\psi_d = Af(\theta, \varphi)\frac{e^{ikr}}{r} \tag{11.2-4}$$

is the result of wavelets scattered outward from many points in the scattering region. Equation (2) indicates that $G(\vec{r}, \vec{r}')$ is the contribution of a point source at $\vec{r} - \vec{r}' = 0$, and it makes sense to try

$$G(\vec{r}, \vec{r}') = C\frac{e^{ikq}}{q}, \qquad \vec{q} \equiv \vec{r} - \vec{r}', \qquad q \equiv |\vec{q}|. \tag{11.2-5}$$

To see whether the guess is good and to evaluate C, substitute (5) into (2). Shift the origin to $\vec{r}'$, that is, to $q = 0$. The radial derivative in ∇^2 becomes a derivative with respect to q, and the angular derivatives give zero because (5) is spherically symmetric with respect to $q = 0$. The task then becomes to check the validity of

$$\frac{1}{q}\frac{d^2}{dq^2}\left(qC\frac{e^{ikq}}{q}\right) + k^2C\frac{e^{ikq}}{q} = \delta(\vec{q}). \tag{11.2-6}$$

For $q \neq 0$ the left side is zero and (6) is satisfied. To study the $q = 0$ behavior, integrate (6) over the volume of a very small sphere of radius ε centered at $q = 0$.

$$\int_\varphi \int_\theta \int_{q=0}^{\varepsilon} \left[\nabla^2\left(C\frac{e^{ikq}}{q}\right) + k^2\frac{Ce^{ikq}}{q}\right]d\tau = 1. \tag{11.2-7}$$

For sufficiently small ε, $kq \ll 1$, the second term becomes

$$\int_0^\varepsilon k^2\frac{C}{q}4\pi q^2\,dq = 2\pi k^2 C\varepsilon^2,$$

and vanishes as ε goes to zero. The first term is

$$\iiint \vec{\nabla}\cdot\left(\vec{\nabla}C\frac{e^{ikq}}{q}\right)d\tau = \iint\left(\vec{\nabla}C\frac{e^{ikq}}{q}\right)\cdot d\vec{S}$$

where Gauss' divergence theorem (Section 1.5) turns the volume integral into an integral over the surface of the small sphere. For $kq \ll 1$, with $\hat{q}$ a unit vector outward from $q = 0$, the integral is

$$\iint\left(\vec{\nabla}\frac{C}{q}\right)\cdot d\vec{S} = C\iint\left(-\frac{\hat{q}}{q^2}\right)\cdot\hat{q}q^2\sin\theta\,d\theta\,d\varphi = -4\pi C.$$

Equation (6) then becomes $-4\pi C + 0 = 1$, and with $C = -1/4\pi$, the guess (5) is indeed a solution to (2);

$$G(\vec{r}, \vec{r}\,') = -\frac{e^{ik|\vec{r}-\vec{r}\,'|}}{4\pi|\vec{r} - \vec{r}\,'|}. \tag{11.2-8}$$

The constant C is not arbitrary because (2) is inhomogeneous; multiplying the term on the right of (2) by three would triple the solution.

The Green's function (8) is not the most general solution to (2). A term in e^{-ikq}/q could be added, but, as observed in Section 11.1, it would describe a spherical wave coming *in* toward the scattering center from large distances. Also, any combination of plane waves with $|\vec{k}| = k$ could be added because they are solutions to the homogeneous equation $(\nabla^2 + k^2)G = 0$. The result (8) is, however, *the* solution which describes waves that spread out from the scattering region. The Green's function is determined by two kinds of requirements. First, it must satisfy (2). Second, it must obey the boundary conditions demanded by the physical meaning of the calculation.

With the Green's function (8), (3) becomes

$$\psi(\vec{r}) = Ae^{ikz} - \frac{1}{4\pi}\iiint \frac{e^{ik|\vec{r}-\vec{r}\,'|}}{|\vec{r} - \vec{r}\,'|} w(\vec{r}\,')\psi(\vec{r}\,')d\tau'. \tag{11.2-9}$$

This integral equation is equivalent to the differential equation (1) plus the boundary condition that at large distances there must be only the incident wave and the outgoing scattered wave. Assume that $w(\vec{r}\,')$ is negligible except in a small scattering region. At the detector $|\vec{r}| \gg |\vec{r}\,'|$, so that

$$\begin{aligned} |\vec{r} - \vec{r}\,'| &= \sqrt{r^2 - 2rr'\cos\gamma + r'^2} \\ &\simeq r - r'\cos\gamma, \end{aligned} \tag{11.2-10}$$

where γ is the angle between $\vec{r}$ and $\vec{r}\,'$ (Figure 11.2-1). In the denominator of (9), $|\vec{r} - \vec{r}\,'|$ can be replaced by just r. In the exponent the next term in (10) must be kept because $kr'\cos\gamma$ can easily change the phase appreciably. Then (9) becomes

$$\psi(\vec{r}) \xrightarrow[\text{large } r]{} Ae^{ikz} - \frac{e^{ikr}}{4\pi r}\iiint e^{-ikr'\cos\gamma} w(\vec{r}\,')\psi(\vec{r}\,')d\tau'. \tag{11.2-11}$$

Comparison with (4) shows that

$$f(\theta, \varphi) = \frac{-1}{4\pi A}\iiint e^{-ikr'\cos\gamma} w(\vec{r}\,')\psi(\vec{r}\,')d\tau'. \tag{11.2-12}$$

Except for the exploitation of the inequality $|\vec{r}| \gg |\vec{r}\,'|$, no approximation has been made. As it stands, (12) may not look very useful because the integrand involves the unknown exact wave function ψ. Now suppose that the distortion in the incident wave $\psi_i = Ae^{ikz}$ is modest even in the scattering region. Then

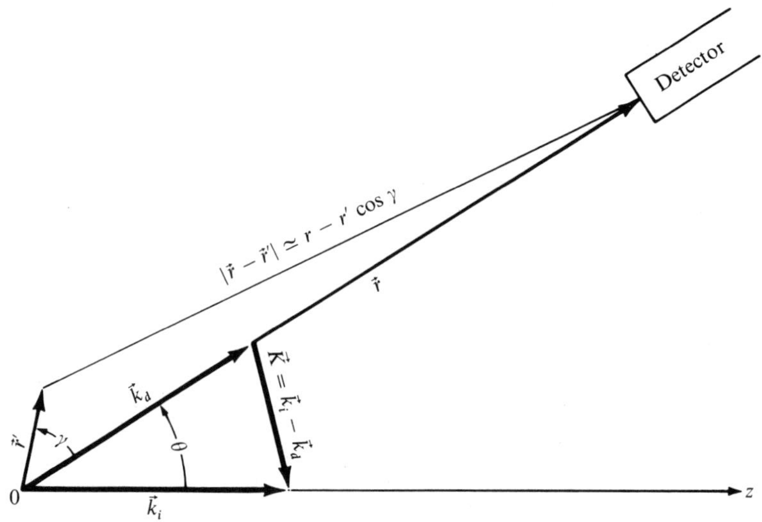

Fig. 11.2-1. The scattering geometry. At the detector $|\vec{r}| \gg |\vec{r}'|$. The scattering angle is θ. The incident momentum is $\hbar\vec{k}_i$, the momentum detected after scattering is $\hbar\vec{k}_d$, and the momentum $\hbar\vec{K}$ is transferred to the scattering center. In elastic scattering, only the direction, not the magnitude, of the momentum can change: $|\vec{k}_i| = |\vec{k}_d| \equiv k$.

$\psi(r')$ can be replaced by $Ae^{ikz'}$ in the integral;

$$f(\theta, \varphi) = \frac{-1}{4\pi} \iiint e^{ik(z' - r'\cos\gamma)} w(\vec{r}')d\tau'. \tag{11.2-13}$$

The approximation that changes (12) into (13) was first applied to quantum physics by M. Born, and is known as the Born approximation. It is a perturbation approach much like that of Section 10.1 in that the unperturbed ψ_i is used in the perturbing interaction to calculate the change in the wave function. The process can be taken to second order by using (13) to get the first-order wave function and then inserting it in the integral, to third order by using a second-order wave function, and so forth. We shall use only the first order (13). This first-order Born approximation is adequate for the scattering of high-energy particles by potential energy functions of modest magnitude and extent because there is rather little distortion of the incident wave under these conditions.

From Figure 11.2-1,

$$kz' - kr' \cos\gamma = \vec{k}_i \cdot \vec{r}' - \vec{k}_d \cdot \vec{r}' = \vec{K} \cdot \vec{r}',$$

where $\hbar\vec{K}$ is the momentum transferred by the scattered particle to the scattering center. In terms of $\vec{K}$, (13) is

$$f(\theta, \varphi) = \frac{-1}{4\pi} \iiint w(\vec{r}')e^{i\vec{K}\cdot\vec{r}'} d\tau'. \tag{11.2-14}$$

If $w(\vec{r})$ is spherically symmetric, $w(\vec{r}') = w(r')$, and the z' axis can be taken in the $\vec{K}$ direction without affecting the form of w.

$$f(\theta) = -\frac{1}{4\pi}\iiint w(r')e^{iKr'\cos\theta'}\sin\theta'\,d\theta'\,d\varphi' r'^2\,dr'$$

$$= -\frac{1}{2}\int_0^\infty w(r')\left[\int_0^\pi e^{iKr'\cos\theta'}\sin\theta'\,d\theta'\right]r'^2\,dr'$$

$$= -\frac{1}{K}\int_0^\infty w(r')\sin Kr'\,r'\,dr'. \qquad (11.2\text{-}15)$$

As anticipated in (11.1-11), (15) does not depend on φ. The incident beam is an eigenstate of L_z with eigenvalue zero, and a spherically symmetric potential cannot change the angular momentum. The angular dependence of $f(\theta)$ resides entirely in K. From Figure 11.2-1,

$$K = 2k\sin\frac{\theta}{2}. \qquad (11.2\text{-}16)$$

The integrand in (15) oscillates because of the sin Kr', and the integral converges if

$$\lim_{r'\to\infty} r'w(r') = 0, \qquad (11.2\text{-}17)$$

that is, if w decreases faster than the Coulomb potential. Throughout this discussion it is assumed that there is a scattering region where $V(r)$ is appreciable, and distant regions where the incident and detected beams are undisturbed because $V(r)$ is negligible. The condition (17) says what is required. It may seem disappointing that the important case $w \sim 1/r$ is excluded, but we shall see that Coulomb scattering can be treated if the screening by surrounding charges is considered.

PROBLEM 11.4

Obtain, as outlined below, the Born approximation for a one-dimensional reflection coefficient. Compare the steps with the development of this section.

Suppose that $V(x)$ is zero for large $|x|$ and some small function near the origin. "Small" means that $|V(x)| \ll E$ everywhere, where $E = \hbar^2k^2/2M$ is the energy of the incident beam Ae^{ikx}.

a. Write the one-dimensional analogs of (1), (2), and (3). Verify that the analog of (3) satisfies that of (1).

b. Obtain the Green's function. Not much work is required; it is the solution of (2.3-5) for $2MU/\hbar^2 = 1$ and $\psi(0) = 1$, except for a shift in origin.

The result is

$$G(x, x') = \left(1 - \frac{1}{2ik}\right)e^{+ik(x-x')} + \frac{1}{2ik}e^{-ik(x-x')}, \qquad x \le x',$$

$$= e^{+ik(x-x')}, \qquad x > x'. \tag{11.2-18}$$

c. In the analog of (3) make the Born approximation. Select the term that describes a wave scattered back from the scattering region toward $-\infty$. Then show that the reflection coefficient is

$$R = \left|\frac{1}{2ik}\int_{-\infty}^{+\infty} w(x')e^{+2ikx'}\,dx'\right|^2, \tag{11.2-19}$$

where R is the ratio of reflected to incident current at large $-x$, outside the region where $w(x) \neq 0$.

d. Suppose $V(x) = V_0, 0 \le x \le a$, and 0 otherwise. Use (19) to evaluate R. The exact answer is $1 - T$, where T is the transmission coefficient given by (2.3-1). Show that the exact calculation and the Born approximation give the same result if $E \gg V_0$.

The road to (15) is long and hard, but the result makes it possible to calculate rather easily an approximate scattering cross section for many forms $V(r)$. The integral in (15) can be a little clumsy, but it is a single definite integral that can often be evaluated analytically, and can always be evaluated numerically.

EXAMPLE

Obtain the Born approximation scattering cross section for the Gaussian potential energy function

$$V(r) = V_0 e^{-r^2/a^2}.$$

SOLUTION

From (15), with $w = 2\mu V/\hbar^2$,

$$f(\theta) = -\frac{2\mu V_0}{\hbar^2 K}\int_0^\infty e^{-r^2/a^2} \sin Kr\, r dr.$$

The integral is given in standard tables,[4] and the result is

$$f(\theta) = -\frac{\mu V_0 a^3}{2\hbar^2}\sqrt{\pi}e^{-K^2a^2/4}.$$

[4] See, for example, the very useful I. S. Gradshteyn and I. M. Ryzhik, *Table of Integrals, Series, and Products*, 4th ed., trans. edited by A. Jeffrey (New York: Academic Press, Inc., 1965), p. 495.

With $K = 2k\sin(\theta/2)$, the differential scattering cross section becomes

$$\frac{d\sigma}{d\Omega} = |f(\theta)|^2 = \left(\frac{\mu V_0 a^3}{2\hbar^2}\right)^2 \pi e^{-2k^2a^2\sin^2(\theta/2)}.$$

The differential cross section integrated over all angles gives the total cross section.

$$\begin{aligned}\sigma &= \int_0^{2\pi}\int_0^{\pi} \frac{d\sigma}{d\Omega}\sin\theta\, d\theta\, d\varphi \\ &= 2\pi^2\left(\frac{\mu V_0 a^3}{2\hbar^2}\right)^2 \int_0^{\pi} e^{-2k^2a^2\sin^2(\theta/2)}\sin\theta\, d\theta.\end{aligned}$$

$$\sin\theta\, d\theta = 2\sin\frac{\theta}{2}\cos\frac{\theta}{2}\,d\theta = 4\sin\frac{\theta}{2}\,d\left(\sin\frac{\theta}{2}\right).$$

Let $\sin^2(\theta/2) = u$.

$$\sigma = \frac{1}{2}\left(\frac{\pi\mu V_0 a^3}{\hbar^2}\right)^2\left(\frac{1 - e^{-2k^2a^2}}{k^2a^2}\right).$$

Notice that $f(\theta)$ depends on the sign of the potential energy function, but that the cross section does not. Notice also that at high energies, where $ka \gg 1$, there is little scattering except near $\theta = 0$, because the Gaussian in $d\sigma/d\Omega$ is very small if $ka\sin\theta/2 \gg 1$. At low energies, $ka \ll 1$, the scattering is nearly isotropic, and the total cross section is

$$\sigma \simeq \left(\frac{\pi\mu V_0 a^3}{\hbar^2}\right).$$

Some of these features obtain for any reasonable $V(r)$, and are discussed further at the end of this section.

The potential energy function

$$V(r) = U_0\frac{e^{-r/a}}{r} \tag{11.2-20}$$

is applicable in several contexts. With $-U_0 \simeq$ a few tens of MeV $\times\, 10^{-13}$ cm and $a \simeq 1.4 \times 10^{-13}$ cm, it describes the long-range part of the nucleon-nucleon interaction due to single pion exchange, and is called the Yukawa potential. The qualitative background is discussed in Section 1.3. With $U_0 = q_1q_2/4\pi\varepsilon_0$ and $a \simeq 0.5$ Å, it describes the *screened* Coulomb interaction between charges q_1 and q_2. In the scattering of an alpha particle by a nucleus, the Coulomb interaction is U_0/r, with $U_0 = 2Ze^2/4\pi\varepsilon_0$. This Coulomb interaction is felt by the alpha particle when it is closer to the nucleus than most of the electrons, but at distances of the order of the atomic radius, the electrons that surround the nucleus reduce the effective charge. At something like 10^{-8} cm the alpha particle sees a nucleus with charge $+Ze$ surrounded by a negative cloud of charge near $-Ze$, that is, it sees a rather small effective charge. The factor $e^{-r/a}$ provides a convenient model of this screening effect. The scattering amplitude (15) is

$$f(\theta) = -\frac{2\mu U_0}{\hbar^2 K}\int_0^\infty e^{-r'/a}\sin Kr'\,dr'$$

$$= -\frac{2\mu U_0}{\hbar^2}\frac{1}{K^2+\dfrac{1}{a^2}}. \tag{11.2-21}$$

For the screened Coulomb interaction, with expression (16) for K and $E = \hbar^2 k^2/2\mu$,

$$f(\theta) = -\frac{q_1 q_2}{4\pi\varepsilon_0}\frac{1}{4E\sin^2\dfrac{\theta}{2}+\dfrac{\hbar^2}{2\mu a^2}}. \tag{11.2-22}$$

Near $\theta = 0$ the screening term $\hbar^2/2\mu a^2$ keeps $f(\theta)$ finite. If, for sufficiently large θ and E, $\hbar^2/2\mu a^2$ can be neglected,

$$f(\theta) = \frac{-q_1 q_2}{4\pi\varepsilon_0 E}\frac{1}{\left(2\sin\dfrac{\theta}{2}\right)^2}, \tag{11.2-23}$$

$$\frac{d\sigma}{d\Omega}(\theta) = |f(\theta)|^2 = \left(\frac{q_1 q_2}{4\pi\varepsilon_0 E}\right)^2\frac{1}{\left(2\sin\dfrac{\theta}{2}\right)^4}. \tag{11.2-24}$$

This result is the same as the famous Rutherford scattering cross section obtained in classical mechanics.[5] There is another coincidence. The quantal calculation for $V \sim 1/r$ can be done exactly[6] without resort to the Born or other approximation. The scattering amplitude differs from (23), but only by a complex factor of magnitude unity, so that the exact quantum Coulomb cross section is also (24). This double coincidence has been discussed more than it has been illuminated. The $1/r$ potential, or $1/r^2$ force, matches the geometry

[5] K. R. Symon, *Mechanics*, 3d ed. (Reading, Mass.: Addison-Wesley Publishing Co., Inc., 1971), Sec. 3.16; J. B. Marion, *Classical Dynamics of Particles and Systems*, 2d ed. (New York: Academic Press, Inc., 1970), Sec. 9.5.

[6] E. Merzbacher, *Quantum Mechanics*, 2d ed. (New York: John Wiley & Sons, Inc., 1970), Sec. 11.8; L. I. Schiff, *Quantum Mechanics*, 3d ed. (New York: McGraw-Hill Book Company, 1968), Sec. 21.

of our space in a unique way and seems to have many magic properties. It is fun to speculate about the course that physics would have taken in 1911 if the quantum and Newtonian answers were different, and Rutherford's classical calculation had failed to match the alpha scattering data.

PROBLEM 11.5

Obtain the Born approximation differential scattering cross section for $V(r) = C/r^2$, that is, for a $1/r^3$ force. Note that, unlike (24), the result contains $\hbar$, so that it cannot be the same as the classical result.

PROBLEM 11.6

Because there is electronic screening, (22) rather than (23) should be used to study the scattering of alpha particles by atoms. For 5-MeV alphas scattered by gold, estimate the range of angles for which screening changes the cross section by more than 1%. Are there other effects that give deflections through angles in this range?

Suppose that alphas are scattered by ^{4}He nuclei, so that the incident and target particles are identical. In the center-of-mass system, the two particles approach each other along one line and then recede from each other along another line. If one alpha leaves the scene of the collision in the direction (θ, φ), the other must leave in the direction $(\pi - \theta, \varphi + \pi)$. The detector registers a count if either alpha hits it. What is the differential scattering cross section that describes the results? The classical answer is

$$\frac{d\sigma}{d\Omega}(\theta, \varphi) + \frac{d\sigma}{d\Omega}(\pi - \theta, \varphi + \pi) = \left(\frac{q_1 q_2}{4\pi\varepsilon_0 E}\right)^2 \left[\frac{1}{\left(2\sin\dfrac{\theta}{2}\right)^4} + \frac{1}{\left(2\cos\dfrac{\theta}{2}\right)^4}\right].$$

In classical physics it is obvious and right to say that the cross section for #1 to be scattered through θ, plus the cross section for #2 to be scattered through θ, is the cross section for either #1 or #2 to be scattered through θ. In quantum physics this answer is wrong. The incident and target particles are indistinguishable, and, as is pointed out at the end of Section 4.2, *indistinguishable possibilities can interfere.*

In the language developed in Section 4.2, alpha particles are bosons; they contain an even number of fermions and have zero spin. The total wave function must therefore be even under exchange of the two alphas. The coordinates are the relative coordinates which describe the vector $\vec{r} = (r, \theta, \varphi)$ that extends from, say, alpha #1 to alpha #2. Exchanging #1 and #2 means changing $\vec{r}$ into $-\vec{r} = (r, \pi - \theta, \varphi + \pi)$. The sum

$$\psi(r, \theta, \varphi) + \psi(r, \pi - \theta, \varphi + \pi)$$

is therefore the required symmetric wave function. At the detector the wave is the sum of the two waves that describe the two indistinguishable particles,

$$A\frac{e^{ikr}}{r}[f(\theta, \varphi) + f(\pi - \theta, \varphi + \pi)],$$

and the differential scattering cross section is

$$\frac{d\sigma}{d\Omega} = |f(\theta, \varphi) + f(\pi - \theta, \varphi + \pi)|^2, \tag{11.2-25}$$

rather than $|f(\theta, \varphi)|^2$. In classical physics, cross sections add. In quantum physics, amplitudes add, and interference effects can appear in the square of the sum. With the Born approximation $f(\theta)$ given by (23), the alpha-alpha Coulomb scattering is described by

$$\frac{d\sigma}{d\Omega} \simeq \left(\frac{q_1 q_2}{4\pi\varepsilon_0 E}\right)\left[\frac{1}{\left(2\sin\frac{\theta}{2}\right)^4} + \frac{1}{\left(2\cos\frac{\theta}{2}\right)^4} + \frac{2}{\left(2\sin\frac{\theta}{2}\right)^2\left(2\cos\frac{\theta}{2}\right)^2}\right]. \tag{11.2-26}$$

The third term is the result of interference between the two indistinguishable possibilities. This expression is written as an approximate equality because the Born approximation does not give one important aspect of the third term. The complex factor mentioned above that appears in the exact solution but not in the Born approximation does make a difference here. It is shown in the references of footnote 6 that the exact result is

$$\frac{d\sigma}{d\Omega} = \left(\frac{q_1 q_2}{4\pi\varepsilon_0 E}\right)^2\left[\frac{1}{\left(2\sin\frac{\theta}{2}\right)^4} + \frac{1}{\left(2\cos\frac{\theta}{2}\right)^4} + \cos\left(\frac{2q_1 q_2 \ln \tan^2(\theta/2)}{4\pi\varepsilon_0 \hbar v}\right)\frac{2}{\left(2\sin\frac{\theta}{2}\right)^2\left(2\cos\frac{\theta}{2}\right)^2}\right], \tag{11.2-27}$$

where v is the relative speed. This equation illustrates a feature of the relation between classical and quantum physics. The exact form (27) has an oscillatory factor in the numerator of the interference term. In the classical limit, where $\hbar$ is negligible, the oscillations are so rapid that in any realizable experiment the interference term averages to zero, so that it is observable only in the quantum domain. This fact does not follow from the approximate (26).

PROBLEM 11.7

Obtain the proton-proton Coulomb differential scattering cross section. Some of the Section 9.1 results are relevant because protons are spin $\frac{1}{2}$ fermions. Do the calculation

a. for incident and target protons both polarized in the same direction, that is, with $m_s = +\frac{1}{2}$ for both with respect to some axis.
b. for unpolarized incident and target protons.

Expression (15) can be used with a variety of potential energy functions. A typical example is the sphere of constant potential energy,

$$\begin{aligned} V(r) &= -V_0, \qquad 0 \le r \le b, \\ &= 0, \qquad r > b, \end{aligned}$$

for which Section 7.1 treats the bound states. The scattering amplitude is

$$\begin{aligned} f(\theta) &= \frac{2\mu V_0}{\hbar^2 K} \int_0^b \sin Kr\, r dr \\ &= \frac{2\mu V_0 b^3}{\hbar^2} \frac{\sin Kb - Kb \cos Kb}{(Kb)^3}, \end{aligned} \qquad (11.2\text{-}28)$$

with K given by (16). The differential scattering cross section for nonidentical incident and target particles is $|f(\theta)|^2$. This result, the one for the Gaussian potential, the one for the screened Coulomb interaction, and the Born approximation cross section for any reasonable $V(r)$ share several features.

Suppose that $w(r)$ is negligible for $r > b$. If $Kb \ll 1$, then $(\sin Kr')/K$ is near r' wherever w is appreciable, and (15) becomes

$$f(\theta) = -\frac{\mu}{2\pi\hbar^2} \iiint V(r')d\tau', \qquad (11.2\text{-}29)$$

a constant times the volume integral of V. For any energy, $Kb = 2kb \sin \theta/2$ is negligible near $\theta = 0$, so that the square of (29) gives the cross section for scattering close to the forward direction. For sufficiently small energy, $Kb \ll 1$ at all angles. Since f depends on θ only through K, (29) is then independent of θ, the cross section is isotropic in the center-of-mass system, and the total cross section is given by

$$\begin{aligned} \sigma &= 4\pi \frac{d\sigma}{d\Omega} = 4\pi |f|^2 \\ &= \frac{\mu^2}{\pi\hbar^4} \left[\iiint V(r')d\tau' \right]^2. \end{aligned} \qquad (11.2\text{-}30)$$

For example, the sphere of constant potential gives at low energies

$$\sigma = \frac{\mu^2}{\pi\hbar^4} \left(\frac{4\pi}{3} V_0 b^3 \right)^2. \qquad (11.2\text{-}31)$$

This value is given directly by (30), and also by 4π times the square of the $Kb \to 0$ limit of (28).

At high energies K is large unless θ is small. A large K in (15) makes $\sin Kr'$ oscillate with short wavelength and makes the integral small because positive and negative contributions nearly cancel each other. The main contributions come from angles such that $Kb \lesssim 1$. For example, (28) shows that very little of the scattering is contributed by $Kb \gtrsim 3$. Most of the scattering satisfies $2kb \sin \theta/2 \lesssim 1$, or $\theta \lesssim 1/kb$, and high-energy scattering occurs almost entirely at small θ.

PROBLEM 11.8

For $V(r) = V_0 e^{-r/a}$, use the Born approximation to obtain $d\sigma/d\Omega$. Examine the low-energy and high-energy limits of the result and verify that these limits have the properties discussed above.

11.3 *Low-Energy Phase Shifts*

The approach of this section is particularly useful at low scattering energies, and it can deal with strong interactions that have $|V(\vec{r})| \gg E$. It therefore complements the Born approximation, which works if $|V(\vec{r})| \ll E$.

Sections 11.1 and 11.2 show how the A in $\psi_i = Ae^{ikz}$ cancels when the cross section is obtained, and we can simplify the expressions by choosing the incident wave amplitude equal to one. Write $\psi_i = e^{ikz}$ in spherical polar coordinates;

$$\psi_i = e^{ikr \cos\theta}.$$

The orbital angular momentum eigenstates, the spherical harmonics $Y_{lm}(\theta, \varphi)$, are complete in the sense that they can be used to expand any function of the angles. From (11.1-11), or from the absence of φ in $e^{ikr\cos\theta}$, it follows that ψ_i is an eigenstate of L_z with $m = 0$, so that only the Y_{l0} are needed to expand ψ_i.

$$e^{ikr\cos\theta} = \sum_{l=0}^{\infty} R_l(r) Y_{l0}(\theta), \qquad (11.3\text{-}1)$$

where the coefficients $R_l(r)$ must contain the r dependence. It requires a small mental wrench to see a plane wave as a superposition of spherical waves, but this expansion turns out to be very useful.

Because the Y_{l0} are orthonormal, the R_l can be found by multiplying (1) by Y_{l0}^* and integrating over the sphere. The φ integration gives 2π, and the R_l are

$$R_l(r) = 2\pi \int_0^{\pi} e^{ikr\cos\theta} Y_{l0}^*(\theta) \sin\theta \, d\theta.$$

The first two coefficients are, with the Y_{l0} given in (6.2-34),

$$R_0(r) = 2\pi \int_0^\pi e^{ikr\cos\theta} \frac{1}{\sqrt{4\pi}} \sin\theta\, d\theta$$

$$= \sqrt{4\pi}\frac{\sin kr}{kr} = \sqrt{4\pi}\, j_0(kr),$$

$$R_1(r) = 2\pi \int_0^\pi e^{ikr\cos\theta} \sqrt{\frac{3}{4\pi}} \cos\theta \sin\theta\, d\theta$$

$$= 2i\sqrt{3\pi}\left(\frac{\sin kr}{(kr)^2} - \frac{\cos kr}{kr}\right)$$

$$= 2i\sqrt{3\pi}\, j_1(kr).$$

The same $j_l(kr)$ as in (7.1-9) have to appear. Both in Section 7.1 and here we are dealing with the radial dependence of angular momentum eigenstates in a region of constant potential energy. Since

$$Y_{l0} = \sqrt{\frac{(2l+1)}{4\pi}}\, P_l(\cos\theta),$$

$$e^{ikz} = \sum_{l=0}^{\infty} \sqrt{\frac{2l+1}{4\pi}}\, R_l(r) P_l(\cos\theta) \tag{11.3-2}$$

$$= j_0(kr) + 3ij_1(kr)\cos\theta + \cdots.$$

The higher terms will not be used here, but one can show that

$$e^{ikz} = \sum_{l=0}^{\infty} (2l+1) i^l j_l(kr) P_l(\cos\theta). \tag{11.3-3}$$

Suppose that there is a spherically symmetric $V(r)$ that is negligible for $r > b$. Because of the spherical symmetry, each of the angular momentum eigenstates can be treated separately. There is nothing that can take an $l = 0$ incident particle and scatter it into an $l = 1$ state. One can first discuss what happens to the $l = 0$ contribution, then to the $l = 1$ contribution, and so forth. The actual ψ is seen as a superposition of *partial waves*, one for each l, that can be treated one at a time.

For $r < b$, ψ depends on the details of $V(r)$. For $r > b$, ψ will also be changed, but some general and simple things can be said about the nature of the changes. Since this region has $V(r) = 0$, the free-particle solution must still apply, but it now has the form

$$\psi(r > b) = \sum_{l=0}^{\infty} (2l+1) i^l [A_l j_l(kr) + B_l n_l(kr)] P_l(\cos\theta), \tag{11.3-4}$$

where the A_l and B_l are constants, and the $n_l(kr)$ are given, along with the $j_l(kr)$, by (7.1-9). This solution is applicable only for $r > b$, so that the bad behavior of the $n_l(kr)$ at $r = 0$ does not exclude them.

Suppose next that the energy is low, or that the range of $V(r)$ is small, in the sense that $kb \ll 1$. Near the origin $j_0(kr) \simeq 1$, so that it is large where $V(r)$ is effective. The $l = 0$ partial wave is modified and B_0, as well as A_0, may be appreciable. The S wave radial dependence in (4) can be written in the form

$$A_0 j_0(kr) + B_0 n_0(kr) = A_0 \frac{\sin kr}{kr} - B_0 \frac{\cos kr}{kr}$$

$$= C_0 \frac{\sin(kr + \delta_0)}{kr}$$

where $C_0 = \sqrt{A_0^2 + B_0^2}$ and $\delta_0 = -\tan^{-1}(B_0/A_0)$. The modifications in the $l = 1$ and higher partial waves are negligible because their $j_l(kr)$ are small where $V(r)$ is effective. Near the origin $j_1(kr) \simeq kr/3$, and, in general, $j_l(kr) \sim (kr)^l$. This observation is the same as that made at the beginning of Chapter 7, particularly in (7.1-5). In the $kb \ll 1$ approximation, (4) becomes

$$\psi(r > b) = C_0 \frac{\sin(kr + \delta_0)}{kr} + \sum_{l=1}^{\infty} (2l + 1)i^l j_l(kr) P_l(\cos\theta). \qquad (11.3\text{-}5)$$

This equation must give the large r form of the general scattering wave function

$$\psi(r > b) \xrightarrow[\text{large } r]{} e^{ikz} + \frac{f(\theta)}{r} e^{ikr}. \qquad (11.3\text{-}6)$$

Since the contributions of the different l values are orthogonal, the $l = 0$ parts of the terms in (6) must satisfy

$$[\psi(r > b)]_{l=0} \xrightarrow[\text{large } r]{} [e^{ikz}]_{l=0} + [f(\theta)e^{ikr}/r]_{l=0}. \qquad (11.3\text{-}7)$$

The $l = 0$ part of e^{ikz} is $j_0(kr) = \sin kr/kr$, and the $l = 0$ part of $\psi(r > b)$ is $C_0 \sin(kr + \delta_0)/kr$. With the sines written in terms of exponentials, (7) is

$$C_0 \frac{e^{i(kr+\delta_0)} - e^{-i(kr+\delta_0)}}{2ikr} = \frac{e^{ikr} - e^{-ikr}}{2ikr} + \frac{f_0(\theta)e^{ikr}}{r}.$$

To make the coefficients of the incoming wave e^{-ikr} the same on both sides, $C_0 = e^{+i\delta_0}$. With this choice the coefficients of the outgoing wave e^{+ikr} are the same on both sides if the $l = 0$ part of $f(\theta)$ is

$$f_0(\theta) = \frac{e^{2i\delta_0} - 1}{2ik} = \frac{e^{i\delta_0}}{k} \sin\delta_0. \qquad (11.3\text{-}8)$$

There is no angular dependence; $l = 0$ states are spherically symmetric. Since $kb \ll 1$, there is little scattering in $l > 0$ states, f_0 is the only important contribution, and $|f_0|^2$ is the differential scattering cross section. The total cross section is

$$\sigma = 4\pi |f_0|^2 = \frac{4\pi}{k^2} \sin^2 \delta_0. \tag{11.3-9}$$

If kb is not small, then the $l = 1$ and perhaps higher partial waves will also be modified, and a rather straightforward generalization[7] of our procedure produces additional terms in the cross section. The number of partial waves that contribute can be estimated from a semiclassical argument. The particles that have angular momentum $l\hbar$ are incident along a line that passes at a perpendicular distance of about $l\hbar/p = l/k$ from the force center; that is, their *impact parameter* is about l/k. If $l/k \lesssim b$, the range of the force, there is appreciable scattering, while if $l/k \gg b$, there is not. Of course, there are no sharp trajectories, but the argument provides a good guide. The first maximum of $j_l(kr)$ is near l/k, and $j_l(kr)$ is negligible for $r \ll l/k$.

It is useful to concentrate our ignorance in the parameter δ_0. This quantity, the $l = 0$ *phase shift*, is determined by $V(r)$ in a direct way. Consider again the sphere of constant potential energy,

$$\begin{aligned} V(r) &= -V_0, \qquad 0 \le r \le b, \\ &= 0, \qquad r > b. \end{aligned} \tag{11.3-10}$$

The treatment here has similarities with that of the bound states in Section 7.1. It is again convenient to use $u(r) = rR(r)$. Give the name $u_0(r)$ to the $u(r)$ that satisfies (7.1-3) for $l = 0$. The interior $u_0(r)$ that is finite at the origin is

$$\sin Kr, \qquad K = \frac{\sqrt{2\mu(E + V_0)}}{\hbar}$$

The ratio of the derivative to the function is $K \cot Kb$ at $r = b$. The exterior $u_0(r)$, is, from (5), proportional to $\sin(kr + \delta_0)$. Its ratio of derivative to value is $k \cot(kb + \delta_0)$ at b. Continuity of this ratio requires

$$k \cot(kb + \delta_0) = K \cot Kb, \tag{11.3-11}$$

$$\delta_0 = \tan^{-1}\left(\frac{k}{K} \tan Kb\right) - kb. \tag{11.3-12}$$

This equation determines δ_0 and therefore the scattering cross section.

[7] E. Merzbacher, *Quantum Mechanics*, 2d ed. (New York: John Wiley & Sons, Inc., 1970), Sec. 11.5; L. I. Schiff, *Quantum Mechanics*, 3d ed. (New York: McGraw-Hill Book Company, 1968), Sec. 19.

PROBLEM 11.9

Show that (12) gives the same σ as (11.2-31), the Born approximation result for a sphere of constant potential, when it should. "When it should" means that the use of only the $l = 0$ phase shift is justified, and that also the conditions for the validity of the Born approximation are satisfied.

PROBLEM 11.10

a. On the same graph, sketch $u_0(r)$ for zero potential energy and for the $V(r)$ given by (10). Select the parameters so that the magnitude of the phase shift δ_0 is $\pi/4$. Is δ_0 positive or negative?

b. Repeat (**a**) for $V(r) = +V_0 > 0$, $r \leq b$; $V(r) = 0$, $r > b$.

PROBLEM 11.11

Let $V(r)$ describe an attractive well with an impenetrable core:

$$\begin{aligned} V(r) &= \infty, && r < c, \\ &= -V_0, && c \leq r \leq b \\ &= 0, && r > b. \end{aligned}$$

Suppose that if the energy of the incident particle is equal in magnitude to the depth of the well, $E = \hbar^2 k^2/2\mu = V_0$, the phase shift δ_0 is zero. The cross section is then close to zero if the scattering of the higher partial waves is small. Sketch $u(r)$. Obtain c, the radius of the impenetrable core, in terms of the other parameters.

PROBLEM 11.12

Near 1 eV there is a sharp minimum in the cross section for the scattering of electrons by argon, krypton, and xenon atoms. This phenomenon is called the Ramsauer–Townsend effect[8] in honor of its discoverers. The explanation is that the phase shift δ_0 happens to be an integral multiple of π under these circumstances.

Far from the atom the incident electron sees $V(r) \simeq 0$ because of the screening of the nucleus by its own electrons. Inside the electron cloud the incident electron sees an attractive $V(r)$. Use

$$\begin{aligned} V(r) &= -76.66 \text{ eV}, && r \leq 1.000 \text{ Å} \\ &= 0, && r > 1.000 \text{ Å}. \end{aligned}$$

Calculate the energy of the incident beam that gives $\delta_0 = \pi$. Sketch $u(r)$ for these conditions. (Work to at least four-figure accuracy. A small change in the parameters can make it impossible to find a solution.)

---

[8] H. S. W. Massey and E. H. S. Burhop, *Electronic and Ionic Impact Phenomena*, Vol. I (London: Oxford University Press, 1969), Secs. 1.6.1, 6.4.1.

The neutron-proton interaction at low energies can be represented fairly well by (10). The form of V is, of course, not really right, but it contains the two basic parameters, depth and range, and low-energy, long-wavelength scattering is not sensitive to details of shape. One hopes and expects that the same forces operate in the ground state of the deuteron as in low-energy scattering. Section 7.1 shows that the choice $b = 1.40 \times 10^{-13}$ cm, $V_0 = 67$ MeV yields the right binding energy, 2.2 MeV. Do the same parameters describe neutron-proton scattering?

At low energies, say 1000 eV, the kinetic energy is negligible compared to V_0.

$$K = \frac{\sqrt{2\mu(V_0 + E)}}{\hbar} \simeq \frac{\sqrt{2\mu V_0}}{\hbar}$$

$$= \frac{\sqrt{939\text{ MeV} \cdot 67\text{ MeV}}}{\hbar c} = \frac{1.27 \times 10^{13}}{\text{cm}},$$

$$Kb = \left(\frac{1.27 \times 10^{13}}{\text{cm}}\right)(1.40 \times 10^{-13}\text{ cm}) = 1.78\text{ radians} = 102°.$$

The interior wave function for the scattering state differs only slightly from the interior wave function for the bound state; for $r \leq b$, the bound state has only slightly less curvature than low-energy scattering states because it lies so close to the top of the well. With these numbers, (12) becomes

$$\delta_{0t} = \tan^{-1}(-3.71 \times 10^{-13}\text{ cm} \cdot k) - 1.40 \times 10^{-13}\text{ cm} \cdot k. \qquad (11.3\text{-}13)$$

The phase shift is labeled with a t because it is computed from the deuteron parameters, and in the deuteron, the two nucleons are in a *triplet* spin state.

Since $k = 4.9 \times 10^{10}$/cm for 1000-eV center-of-mass energy, 3.71×10^{-13} cm $\cdot\, k$ and 1.40×10^{-13} cm $\cdot\, k$ are small numbers. For any small angle ε, $\tan^{-1}\varepsilon \simeq n\pi + \varepsilon$. The result is

$$\delta_{0t} = n\pi - 5.11 \times 10^{-13}\text{ cm} \cdot k, \qquad (11.3\text{-}14)$$

and the cross section is

$$\sigma_t = \frac{4\pi}{k^2}\sin^2(n\pi - 5.11 \times 10^{-13}\text{ cm} \cdot k)$$

$$\simeq \frac{4\pi}{k^2}(5.11 \times 10^{-13}\text{ cm} \cdot k)^2 = 3.3 \times 10^{-24}\text{ cm}^2. \qquad (11.3\text{-}15)$$

The energy 1000 eV is a good choice for comparison with experiment. On the one hand, it is low enough to make the force range b negligible compared to the wavelength and to satisfy the condition $K \gg k$. On the other, it is high enough for the target protons to recoil freely and also to make negligible the effects of coherent scattering from two or more protons. The experimental result for *unpolarized* neutrons is 20.4×10^{-24} cm^2.

The huge difference between this result and (15) is not a discrepancy. As shown in Section 9.1, the probability of scattering in a triplet state is $\frac{3}{4}$, while the probability of scattering in the singlet state is $\frac{1}{4}$.

$$\sigma = \tfrac{3}{4}\sigma_t + \tfrac{1}{4}\sigma_s. \tag{11.3-16}$$

The 20.4×10^{-24} cm^2 value is σ. The conclusion is that the cross section σ_s for scattering in the singlet state is large; from (15) and (16), $\sigma_s \simeq 72 \times 10^{-24}$ cm^2. It follows that the nuclear force is *spin dependent*: the nucleon-nucleon interaction depends on the relative spins of the nucleons.

What singlet state interaction gives $\sigma_s = 72 \times 10^{-24}$ cm^2? To find out, we reverse the development that leads to (15) by getting the phase shift from the cross section and then learning something about $V(r)$ from the phase shift. Since here (9) is

$$\sigma_s = \frac{4\pi}{k^2}\sin^2\delta_{0s},$$

the experiment shows that

$$\sin\delta_{0s} = \pm k\sqrt{\frac{\sigma_s}{4\pi}} = \pm 2.39 \times 10^{-12}\ \text{cm}\cdot k,$$

$$\delta_{0s} = n\pi \pm \sin^{-1}(2.39 \times 10^{-12}\ \text{cm}\cdot k).$$

For small k, since $\sin^{-1}\varepsilon \simeq \varepsilon$ for small ε,

$$\delta_{0s} = n\pi \pm 2.39 \times 10^{-12}\ \text{cm}\cdot k.$$

Equation (11), which connects the phase shift with $V(r)$, can be written

$$\frac{\tan K_s b_s}{K_s} = \frac{\tan(\delta_{0s} + kb_s)}{k}.$$

Here b_s is the range of the singlet potential and, with V_{0s} as the depth of the singlet potential,

$$K_s = \frac{\sqrt{2\mu V_{0s}}}{\hbar}.$$

The low-energy cross section is a single number, and cannot by itself determine more than one parameter. Detailed studies show that the range of the singlet force is larger than that of the triplet force,[9] but we shall take the ranges to be the same, $b_s = b = 1.4 \times 10^{-13}$ cm, to simplify the job of understanding the

[9] D. Park, *Introduction to the Quantum Theory*, 2d ed. (New York: McGraw-Hill Book Company, 1974), Chap. 18.

essentials. With this choice,

$$\frac{\tan K_s b}{K_s} = \frac{\tan(n\pi \pm 2.39 \times 10^{-12}\,\text{cm} \cdot k + 1.4 \times 10^{-13}\,\text{cm} \cdot k)}{k}$$

For small k the tangent on the right can be expanded to give $(\pm 2.39 + 0.14)10^{-12}$ cm $\cdot\, k$. Then

$$\frac{\tan K_s b}{K_s} = (+2.53 \text{ or } -2.25) \times 10^{-12}\,\text{cm}.$$

If $\tan K_s b$ were negative, then $K_s b > \pi/2$, the interior $u(r) \sim \sin K_s r$ would pass its maximum before r reaches the edge of the well, and there would be a singlet bound state. But that would be wrong. The deuteron has only one bound state, the triplet state. It follows that $(\tan K_s b)/K_s = +2.53 \times 10^{-12}$ cm. The equation

$$\frac{\tan(1.40 \times 10^{-13}\,\text{cm} \cdot K_s)}{K_s} = 2.53 \times 10^{-12}\,\text{cm}$$

is solved by $K_s = 1.10 \times 10^{13}$/cm, and gives for the depth of the singlet potential

$$V_{0s} = \frac{\hbar^2 K_s^2}{2\mu} = 50\,\text{MeV},$$

rather than the 67 MeV of the triplet well. The nucleon-nucleon interaction is somewhat weaker for nucleons in a singlet state than for nucleons in a triplet state. Note that

$$1.40 \times 10^{-13}\,\text{cm} \cdot K_s = 1.53\,\text{radians} = 88°,$$

just a little less than $\pi/2$, so that the singlet state attraction is almost but not quite strong enough to produce a bound state. The low-energy singlet cross section is large because there is almost a resonance; there is almost a bound state near zero energy.

The nucleon-nucleon force is *charge independent* to good accuracy. A variety of nuclear data shows that if corrections are made for electromagnetic effects, the proton-proton, proton-neutron, and neutron-neutron interactions are very nearly the same, in the same states. Can two neutrons form a bound system, a "dineutron," that is similar to the deuteron? Neutrons are fermions. If they are in an $l = 0$, or spatially even state, they must be in a singlet spin state. The interaction in the singlet state is a little too weak to bind the neutron-proton system. Charge independence says that the same interaction applies to the dineutron, which, unlike the neutron-proton system, *must* for $l = 0$ be in a singlet state, and binding should not occur. States with $l = 1, 3, \ldots$ should not be bound because they do not occur in the deuteron. The same argument applies to the "diproton," a nucleus of helium with mass number two. In addition, the Coulomb repulsion drives the two protons apart. It makes good sense that the deuteron is the only bound two-nucleon system.

PROBLEM 11.13

a. It has not been possible to measure directly the neutron-neutron scattering cross section because there is no way to make a target with a high density of free neutrons. Think carefully about neutrons being identical fermions, and calculate the low-energy unpolarized neutron-neutron scattering cross section. Assume that the nuclear force is charge independent, that is, that *in the same state*, neutron-neutron and neutron-proton scattering have the same cross section.

b. If there are two crossed beams of slow neutrons, how many scatterings will take place per second? Use a neutron speed of 10^5 cm/sec, a flux density in each beam of 10^9 neutrons/sec cm^2, a cross-sectional area of 10 cm $\times$ 10 cm for each beam, and 90° for the angle between the beams.

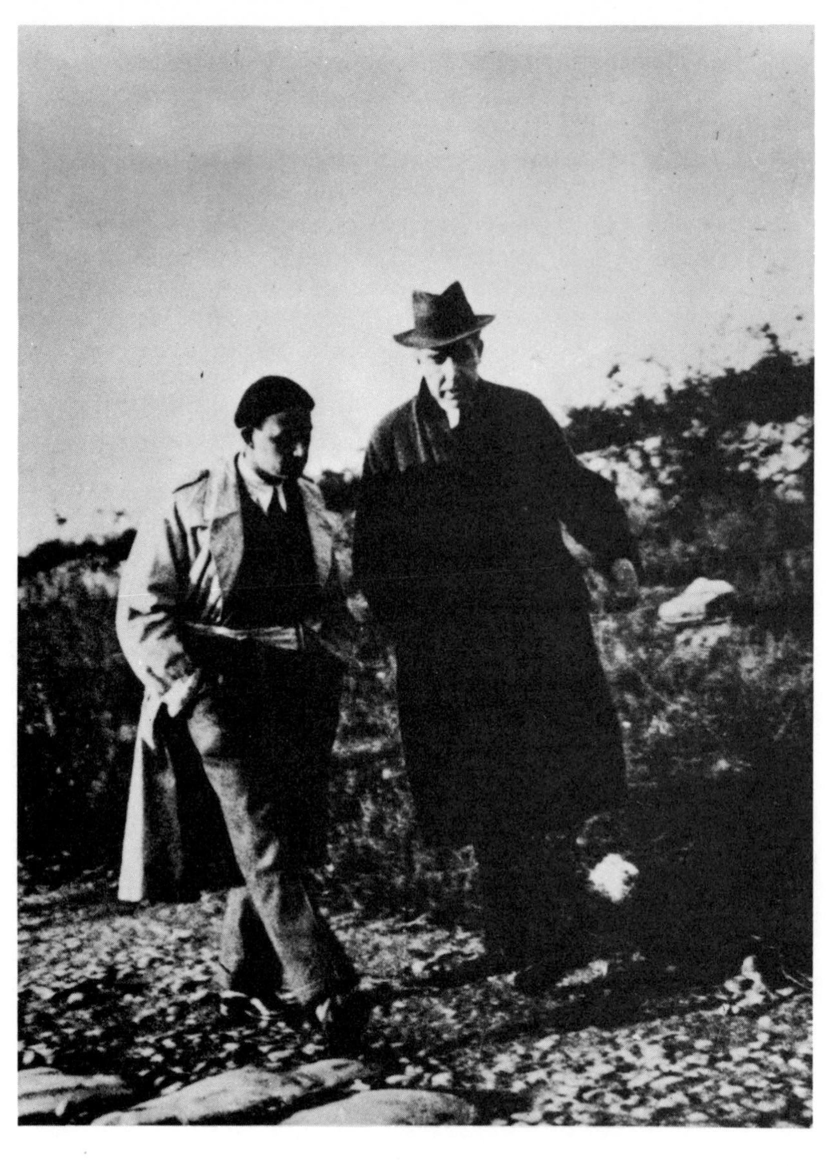

Enrico Fermi and Niels Bohr in 1931. Bohr was at that time the doyen of quantum physics, and Fermi was already known as an exceptionally able and productive scientist. W. Pauli had for several years discussed the possibility that nuclear beta decay involved the emission of a very light neutral particle that Fermi named the neutrino. *Fermi took up Pauli's suggestion and developed his theory of beta decay during 1933–34; see Section 12.3. This theory has undergone remarkably few modifications, and other theories of elementary particle interactions have been patterned after it. Two biographies of Fermi are Laura Fermi,* Atoms in the Family *(Chicago: University of Chicago Press, 1954), and Emilio Segrè,* Enrico Fermi, Physicist *(Chicago: University of Chicago Press, 1970). (Courtesy of the American Institute of Physics Niels Bohr Library, Goudsmit Collection)*

12

Transition Probabilities

Things can be perceived only if they are being altered. Particles, nuclei, atoms, and books can be detected only if their momentum is changed, or if they emit or absorb something. This chapter deals with the consequences of putting systems in a changing environment. If the Hamiltonian contains a term that depends explicitly on the time, what transitions are induced? What is the probability of these transitions?

Explicit time dependence in H makes it impossible to separate the Schrödinger equation as in Section 2.1. Exact solutions are possible for some special cases, such as the magnetic resonance example (8.4-9). Usually, however, approximation methods are needed. In the following sections, a perturbation approach is developed and applied.

12.1 | *Time-Dependent Perturbations*

The language and notation of this section are much like those of Section 10.1. There the Hamiltonian contains a small time-*in*dependent term that requires an approximation method. Here the small term *does* depend on t.

The task is to solve

$$H(x,t)\Psi(x,t) = i\hbar\frac{\partial}{\partial t}\Psi(x,t). \tag{12.1-1}$$

The x stands for all coordinates, space and spin, that the problem requires. Assume that the exact Hamiltonian can be written in the form

$$H(x,t) = H_0(x) + H'(x,t), \tag{12.1-2}$$

where the *perturbing Hamiltonian* $H'(x,t)$ contains all the time dependence, and where the unperturbed Hamiltonian $H_0(x)$ has energy eigenstates $\psi_n(x)$;

$$H_0(x)\psi_n(x) = E_n\psi_n(x). \tag{12.1-3}$$

If initially $\Psi(x,t)$ is one of the energy eigenstates,

$$\Psi(x,0) = \psi_i(x), \tag{12.1-4}$$

then $H' = 0$ lets the system stay in the state i forever, while $H' \neq 0$ causes the system to change, so that after a while there is a nonzero probability of finding it in other states.

The energy eigenstates are complete and can be used to expand Ψ at any time.

$$\Psi(x,t) = \sum_m C_m(t)\psi_m(x)e^{-iE_mt/\hbar}. \tag{12.1-5}$$

Substitute (5) into (1), with the Hamiltonian given by (2).

$$\sum_m [C_m(t)H_0\psi_m(x)e^{-iE_mt/\hbar} + C_m(t)H'\psi_m(x)e^{-iE_mt/\hbar}]$$

$$= \sum_m \left[C_m(t)E_m\psi_m(x)e^{-iE_mt/\hbar} + i\hbar\left(\frac{d}{dt}C_m(t)\right)\psi_m(x)e^{-iE_mt/\hbar}\right].$$

The first term on the left cancels the first term on the right. Multiply both sides by $\psi_n^\dagger(x)$, integrate over x and use

$$\int \psi_n^\dagger(x)\psi_m(x)dx = \delta_{nm}:$$

$$\frac{dC_n}{dt} = \frac{1}{i\hbar}\sum_m H'_{nm}(t)e^{i(E_n-E_m)t/\hbar}C_m(t). \tag{12.1-6}$$

The matrix elements of the perturbation,

$$H'_{nm}(t) \equiv \int \psi_n^\dagger(x)H'(x,t)\psi_m(x)dx \equiv \langle n|H'|m\rangle, \tag{12.1-7}$$

have the same structure as (10.1-6) and (3.8-26), but are functions of the time. The set of equations (6) is exact. For special cases, such as the two-state magnetic resonance example (8.4-9), an exact solution can be obtained, but in many important problems, (6) is an infinite set of coupled equations that can be solved only by an approximation method.

PROBLEM 12.1

Review the treatment of magnetic resonance, (8.4-9)–(8.4-11). Then write the equations (1) through (7) of this section for that problem. Show (6) with all H'_{nm} evaluated. (When comparing with Section 8.4, note that the $C_\pm(t)$ there include the $e^{-iE_nt/\hbar}$, while the $C_n(t)$ of this section do not.)

Because of (4), the initial conditions for (6) are

$$C_n(0) = \delta_{in}. \tag{12.1-8}$$

Write (6) so that $n = i$ is singled out.

$$\frac{dC_i}{dt} = \frac{H'_{ii}}{i\hbar} C_i + \sum_{m \neq i} \frac{H'_{im}}{i\hbar} e^{i(E_i - E_m)t/\hbar} C_m, \tag{12.1-9}$$

$$\frac{dC_n}{dt} = \frac{H'_{ni}}{i\hbar} e^{i(E_n - E_i)t/\hbar} C_i + \sum_{m \neq i} \frac{H'_{nm}}{i\hbar} e^{i(E_n - E_m)t/\hbar} C_m, \qquad n \neq i. \tag{12.1-10}$$

The approximation method rests on the assumption that, if t is not too large, the change from the initial conditions (8) is small. In the zeroth approximation, all H'_{rs} are neglected, all C_r are constants, and (8) is right forever. In the first approximation, products of the matrix elements and C_m are negligible second-order terms for $m \neq i$:

$$\frac{dC_i}{C_i} = \frac{H'_{ii}(t)}{i\hbar} dt, \tag{12.1-11}$$

$$dC_n = \frac{H'_{ni}}{i\hbar} e^{i(E_n - E_i)t/\hbar} C_i \, dt. \tag{12.1-12}$$

Integrating the first equation from 0 to t gives

$$C_i = \exp \int_0^t \frac{H'_{ii}}{i\hbar} dt'. \tag{12.1-13}$$

The result of this integration can then be substituted into (12), but it is not necessary to do so. The change in C_i from its initial value, unity, is small; the product of that change and H'_{ni} is a second-order quantity; and it is consistent to use simply $C_i = 1$ in (12) for a first-order calculation. In other words, in an expansion of the exponential in (13),

$$C_i(t) = 1 + \int_0^t \frac{H'_{ii}(t')}{i\hbar}\, dt' + \cdots,$$

it is sufficient to keep the first term, unity. If the approximation is to be valid, the magnitude of the perturbation must be small, and also the time elapsed since the system was definitely in state i must be short.

In this approximation the set of coupled equations is uncoupled, and (12) gives the important result

$$C_n(t) = \int_0^t \frac{H'_{ni}(t')}{i\hbar}\, e^{i(E_n - E_i)t'/\hbar}\, dt'. \tag{12.1-14}$$

For any $H'_{ni}(t)$, (14) gives $C_n(t)$, and $|C_n(t)|^2$ is the probability of finding the system in the state n at time t. The probability per second of transition from state i to state n is the rate of increase of $|C_n(t)|^2$, which is, for small t, $|C_n(t)|^2/t$.

PROBLEM 12.2

Suppose that H' is turned on at $t = 0$, and is constant in time thereafter.

$$\begin{aligned} H'(x, t) &= 0, & t &< 0, \\ &= H'(x), & t &\geq 0. \end{aligned}$$

Assume that $E_n \neq E_i$ and $|H'(x)| \ll |E_n - E_i|$ for all n.

a. Obtain the first-order C_i, C_n, $|C_i|^2$, and $|C_n|^2$ in terms of the H'_{nm}. Does the state remain normalized? What happens to the $m = i$ contribution in (5)?

b. Suppose further that H' does not depend on x. Repeat (**a**) with all matrix elements evaluated.

12.2 *The Absorption and Emission of Light*

In Section 3.11 it is pointed out that atomic states are *quasi-stationary*; they live a long time in terms of their own natural time, and their widths, $\Delta E = \hbar/\tau$, are much less than their energy differences. Their interaction with radiation is a small perturbation for which the machinery of Section 12.1 can give useful results.

The treatment in this section has important shortcomings. First, it is limited to some special cases because only the most basic ingredients of electromagnetic theory are used. Second, the radiation is taken to be a given wave, so that our approach is *semiclassical.* As indicated in Section 1.5, a completely quantal discussion of light includes from the beginning a description of the creation and destruction of photons, and requires a more elaborate formalism.[1]

We shall examine the excitation of an atom by an electromagnetic wave. The wave consists of an electric field and a magnetic field oscillating at right angles to each other and to the direction of propagation. Its magnetic force on a slow electron is much less than the electric force, and it is a good approximation to look only at the effect of the electric field. Choose the direction of propagation to be the x axis, and the direction of the electric field to be the z axis.

$$\vec{\mathscr{E}} = \mathscr{E}_0 \hat{e}_z \cos(kx - \omega t). \tag{12.2-1}$$

Excitation from the $n = 1$ to an $n = 2$ state of hydrogen requires 10.2 eV. The wavelength of 10.2 eV photons is 1220 Å. Within the $\sim 10^{-8}$ cm where the wave function of hydrogen is appreciable, $kx = 2\pi x/\lambda < 10^{-2}$, and the electric field can be approximated by $\mathscr{E}_0 \hat{e}_z \cos \omega t$. This electric field is given by minus the gradient of the electric potential $-z\mathscr{E}_0 \cos \omega t$, and the perturbing potential energy function is, for electron charge $-e$,

$$H' = e\mathscr{E}_0 z \cos \omega t. \tag{12.2-2}$$

The significant matrix element is H'_{fi}, where i describes the initial state, perhaps the ground state of the atom, and f describes the final excited state of the atom.

$$\begin{aligned} H'_{fi} &= \iiint \psi_f^*(\vec{r}) e\mathscr{E}_0 z \cos \omega t \; \psi_i(\vec{r}) d\tau \\ &= e\mathscr{E}_0 \langle f | z | i \rangle \cos \omega t. \end{aligned} \tag{12.2-3}$$

If the atom is in state i at $t = 0$ and then the wave begins to perturb it, (12.1-14) gives

$$C_f(t) = \frac{e\mathscr{E}_0}{i\hbar} \langle f | z | i \rangle \int_0^t \cos \omega t' e^{i(E_f - E_i)t'/\hbar} \, dt'.$$

With

$$\omega_{fi} \equiv \frac{E_f - E_i}{\hbar}, \tag{12.2-4}$$

[1] R. E. Leighton, *Principles of Modern Physics* (New York: McGraw-Hill Book Company, 1959), Chap. 6; E. Merzbacher, *Quantum Mechanics*, 2d ed. (New York: John Wiley & Sons, Inc., 1970), Chaps. 20 and 22; L. I. Schiff, *Quantum Mechanics*, 3d ed. (New York: McGraw-Hill Book Company, 1968), Chap. 14.

this expression becomes

$$C_f(t) = \frac{e\mathscr{E}_0}{2i\hbar}\langle f|z|i\rangle \int_0^t [e^{i(\omega_{fi}-\omega)t'} + e^{i(\omega_{fi}+\omega)t'}]dt'$$

$$= -\frac{e\mathscr{E}_0}{2\hbar}\langle f|z|i\rangle\left[\frac{e^{i(\omega_{fi}-\omega)t}-1}{\omega_{fi}-\omega} + \frac{e^{i(\omega_{fi}+\omega)t'}-1}{\omega_{fi}+\omega}\right]$$

$$= -\frac{ie\mathscr{E}_0 t}{2\hbar}\langle f|z|i\rangle\left[e^{i(\omega_{fi}-\omega)t/2}\frac{\sin(\omega_{fi}-\omega)t/2}{(\omega_{fi}-\omega)t/2}\right.$$

$$\left. + e^{i(\omega_{fi}+\omega)t/2}\frac{\sin(\omega_{fi}+\omega)t/2}{(\omega_{fi}+\omega)t/2}\right]. \qquad (12.2\text{-}5)$$

Unless ω is very close to $\pm\omega_{fi}$, the expression in the brackets becomes very small after a very short time. For 10.2-eV photons, with $\omega = 1.55 \times 10^{16}/\text{sec}$, a 1% difference between ω_{fi} and ω gives for the magnitude of the first term

$$\left|e^{i10^{14}t/\text{sec}}\frac{\sin 10^{14}t/\text{sec}}{10^{14}t/\text{sec}}\right| \leq \frac{10^{-14}\ \text{sec}}{t},$$

and $\leq 10^{-16}$ sec$/t$ for the magnitude of the second term. If, however, ω is so close to ω_{fi} that for any time of interest $(\omega_{fi} - \omega)t$ is small, then, because $\sin x/x = 1 - x^2/3! + \cdots$, the square bracket is near unity. This behavior is entirely reasonable. A photon cannot be absorbed unless its energy $h\nu = \hbar\omega$ is near $E_f - E_i = \hbar\omega_{fi}$. The uncertainty principle explains why it is not necessary for $\hbar\omega$ to equal $E_f - E_i$ precisely. For observation time t, the energy is not defined better than $\hbar/t$, and the photon energy can miss $E_f - E_i$ by this amount. At any t the main contribution to C_f therefore comes from photons with ω within $\sim \pm\hbar/t$ of ω_{fi}.

The discussion has focused on absorption, and on the sharply peaked contribution to $C_f(t)$ from $\omega \simeq \omega_{fi}$. There is also a substantial contribution if $\omega = -\omega_{fi}$, for then the second term in (5) is large. For i being the ground state and f an excited state, this condition cannot be satisfied. If, however, the system starts out in an excited state and f is a lower state, then it is a possibility, and we have a description of *stimulated emission.* This topic will be examined a little later. For excitation, that is, for the absorption of light, $\omega \simeq \omega_{fi}$, only the first term in (5) contributes significantly.

$$|C_f(t)|^2 = \frac{e^2\mathscr{E}_0^2 t^2}{4\hbar^2}|\langle f|z|i\rangle|^2 \frac{\sin^2[(\omega_{fi}-\omega)t/2]}{[(\omega_{fi}-\omega)t/2]^2}. \qquad (12.2\text{-}6)$$

The classical electromagnetic wave (1) can be translated into photons. The average energy flux density carried by such a wave[2] is $c\varepsilon_0\mathscr{E}_0^2/2$, where ε_0 is the

[2] D. Halliday and R. Resnick, *Fundamentals of Physics* (New York: John Wiley & Sons, Inc., 1974), Sec. 35–4.

permittivity constant, c is the speed of light, and $\mathscr{E}_0$ is expressed in volts per meter. The energy of each photon is $\hbar\omega$. If the flux density of photons, number per second per m^2, is $\Phi(\omega)$, then the relation

$$\Phi(\omega)\hbar\omega = \frac{c\varepsilon_0 \mathscr{E}_0^2}{2} \tag{12.2-7}$$

must be right on the average because it equates total energy with number of photons times energy per photon.

With $\mathscr{E}_0^2$ eliminated by (7), the initial rate of increase of $|C_f(t)|^2$ becomes

$$\frac{|C_f(t)|^2}{t} = \frac{2e^2}{\varepsilon_0 \hbar c}|\langle f|z|i\rangle|^2 \omega\Phi(\omega)\,\frac{\sin^2[(\omega_{fi} - \omega)t/2]}{(\omega_{fi} - \omega)^2 t}. \tag{12.2-9}$$

This expression is not quite the thing that one measures because it gives the transition probability for incident photons of definite ω, and it depends strongly on $\omega_{fi} - \omega$. Photon beams are never perfectly mono-energetic. Therefore the flux density $\Phi(\omega)$ at a definite frequency must be replaced by

$$\frac{d\Phi}{d\omega}\,d\omega,$$

the number of photons per second per m^2, with angular frequency between ω and $\omega + d\omega$. The total transition probability is then

$$W_{f\leftarrow i} = \int_0^\infty \frac{2e^2}{\varepsilon_0 \hbar c}|\langle f|z|i\rangle|^2 \omega\,\frac{d\Phi}{d\omega}\,\frac{\sin^2[(\omega_{fi} - \omega)t/2]}{(\omega_{fi} - \omega)^2 t}\,d\omega.$$

Although it is right to integrate over all ω, significant contributions come only from the neighborhood $\omega \simeq \omega_{fi}$, where $|C_f|^2$ is a sharply peaked function (Figure 12.2-1). Assume that $\omega\, d\Phi/d\omega$ varies slowly in this neighborhood, so that it can be taken out of the integral if it is evaluated at ω_{fi}.

$$W_{f\leftarrow i} = \frac{2e^2}{\varepsilon_0 \hbar c}|\langle f|z|i\rangle|^2 \omega_{fi}\left(\frac{d\Phi}{d\omega}\right)_{\omega_{fi}} \int_0^\infty \frac{\sin^2[(\omega_{fi} - \omega)t/2]}{(\omega_{fi} - \omega)^2 t}\,d\omega. \tag{12.2-10}$$

Since the only significant contributions to the integral in (10) come from the neighborhood of the peak, the integration can be taken from $-\infty$ to $+\infty$. In terms of the variable $s = (\omega - \omega_{fi})t/2$, the integral then becomes

$$\int_{-\infty}^{+\infty} \frac{\sin^2 s}{2s^2}\,ds = \frac{\pi}{2}. \tag{12.2-11}$$

The final result is

$$W_{f\leftarrow i} = \left(\frac{d\Phi}{d\omega}\right)_{\omega_{fi}} \frac{\hbar\omega_{fi}}{c}\,B_{f\leftarrow i}, \tag{12.2-12}$$

$$B_{f\leftarrow i} = \frac{\pi e^2}{\varepsilon_0 \hbar^2}|\langle f|z|i\rangle|^2. \tag{12.2-13}$$

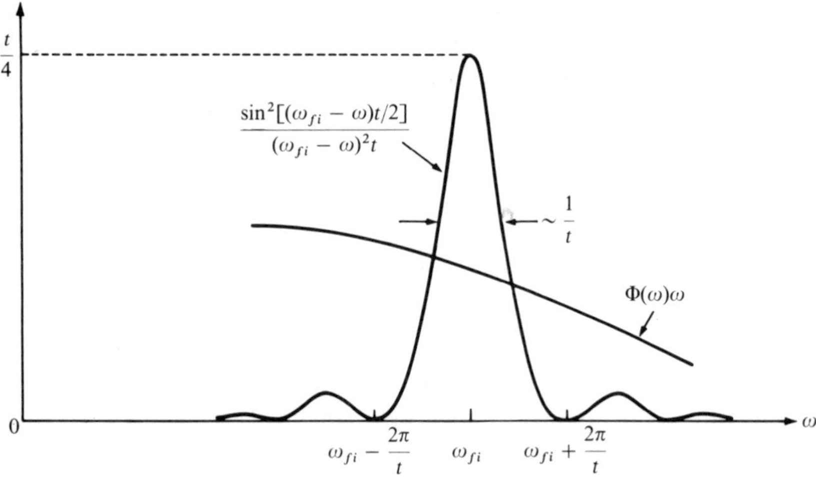

Fig. 12.2-1. The behavior of the sharply peaked integrand of (10). The height of the peak is $t/4$, its width is roughly $1/t$, and its total area is constant in time.

The *energy flux density* divided by the speed c is the *energy density*, in energy per unit volume per unit angular frequency. The quantity $B_{f\leftarrow i}$ gives the transition probability per unit energy density.

The sharply peaked integrand in (10) picks out of the distribution of incident photons those that have the right energy to induce the transition. We are, in fact, looking at another representation of the Dirac delta function:

$$\lim_{t\to\infty} \frac{2\sin^2[(\omega_{fi}-\omega)t/2]}{\pi(\omega_{fi}-\omega)^2 t} = \delta(\omega - \omega_{fi}). \tag{12.2-14}$$

The width is vanishingly small, the height at $\omega = \omega_{fi}$ is indefinitely large, and, from (11), the area under the curve is unity.

The time t does not appear in the result (12) and (13), so that $W_{f\leftarrow i}$, the probability of absorption per atom that is in state i, is a constant. This sensible conclusion emerged only after the continuous distribution of photon energies was taken into account. The limit of (9) as ω approaches ω_{fi} behaves quite differently and is time dependent because

$$\lim_{\omega\to\omega_{fi}} \frac{\sin^2[(\omega_{fi}-\omega)t/2]}{(\omega_{fi}-\omega)^2 t} = \frac{t}{4}.$$

The development (9) through (13) is right only if $d\Phi/d\omega$ varies sufficiently slowly. The state f is an excited state that decays with some mean life τ, and that therefore has a width $\hbar/\tau$ in energy and $1/\tau$ in ω. If $d\Phi/d\omega$ is nearly constant for ω in the range $\omega_{fi} \pm 1/\tau$, this refinement is unimportant. If, however, $d\Phi/d\omega$

varies appreciably in $\omega_{fi} \pm 1/\tau$, then one must recognize that there is no such thing as a single energy E_f, and a more detailed treatment is required. Such considerations are needed if the incident radiation is laser light. Lasers produce beams with a $d\Phi/d\omega$ that is very sharply peaked, and can actually be used to map out the natural widths of states.

Just as the first term in (5) accounts for absorption, the second term accounts for *stimulated emission*. As is mentioned above, if the initial state is an excited state and the final state is a lower state, say the ground state, then the emission of a photon can be induced by the radiation that is passing over the atom. The transition probability for this process is the same as that for absorption. For any two states 1 and 2,

$$(B_{2\leftarrow 1})_{\text{absorp.}} = (B_{1\leftarrow 2})_{\text{stim. em.}}, \tag{12.2-15}$$

so that, in the same radiation field,

$$(W_{2\leftarrow 1})_{\text{absorp.}} = (W_{1\leftarrow 2})_{\text{stim. em.}} \tag{12.2-16}$$

PROBLEM 12.3

Verify (15) and (16) by examining (1) through (13).

Stimulated emission is the essential process in the operation of lasers; their name stands for "light amplification by stimulated emission of radiation." Atoms are excited into a state that normally lives for a long time, radiation at the deexcitation frequency moves through these excited atoms, and they are stimulated to emit their energy. As each atom adds its contribution, the wave grows, and a large amount of coherent radiation appears.[4]

If an atom is in an excited state, it will emit a photon even if there is no stimulation. Such *spontaneous emission* is in many contexts the most familiar and important process. Atoms in a hot gas are excited by collisions and then emit their characteristic radiation, with some mean life τ, without being much influenced by the other radiation around them. As indicated at the beginning of this section, our formalism does not include a complete description of the creation and destruction of photons, and this shortcoming prevents a direct calculation of spontaneous emission. We will use a trick that was developed by Einstein in 1917, long before a modern quantum theory of radiation existed. The trick involves recognizing that there must be a relation between absorption, stimulated emission, and spontaneous emission if radiation inside a cavity is to have the equilibrium properties that are observed.

[4] For a qualitative description of lasers, see R. Eisberg and R. Resnick, *Quantum Physics of Atoms, Molecules, Solids, Nuclei, and Particles* (New York: John Wiley & Sons, Inc., 1974), Sec. 11-7.

At thermal equilibrium the radiation inside a cavity at absolute temperature T is described by the energy density distribution

$$u(\omega) = \frac{\hbar\omega^3}{\pi^2 c^3(e^{\hbar\omega/k_B T} - 1)}, \tag{12.2-17}$$

in the sense that the energy per unit volume carried by waves with angular frequency between ω and $\omega + d\omega$ is $u(\omega)d\omega$. This expression is the famous Planck's formula that began the quantum revolution in 1900.

Suppose that inside the cavity there are atoms that have levels with energies E_1 and E_2, $E_2 > E_1$, and that N_1 atoms have energy E_1, and N_2 have energy E_2. The rate of transitions from E_1 to E_2 by absorption is $N_1 B_{2\leftarrow 1} u(\omega_{12})$. The rate of transitions from E_2 to E_1 by stimulated emission is $N_2 B_{1\leftarrow 2} u(\omega_{12})$. The rate of transitions from E_2 to E_1 by spontaneous emission is N_2/τ, where τ is the mean life. When equilibrium is reached,

$$N_1 B_{2\leftarrow 1} u(\omega_{12}) = N_2 \left[B_{1\leftarrow 2} u(\omega_{12}) + \frac{1}{\tau} \right]. \tag{12.2-18}$$

Also, at equilibrium, the ratio of populations N_1 and N_2 is determined by the Boltzmann distribution

$$\frac{N_1}{N_2} = \frac{e^{-E_1/k_B T}}{e^{-E_2/k_B T}} = e^{\hbar\omega_{12}/k_B T}$$

With this ratio, (18) is

$$u(\omega_{12}) = \frac{\dfrac{1}{\tau}}{e^{\hbar\omega_{12}/k_B T} B_{2\leftarrow 1} - B_{1\leftarrow 2}}. \tag{12.2-19}$$

If (19) agrees with (17) for all T, then $B_{2\leftarrow 1} = B_{1\leftarrow 2}$, as in (15), and

$$\frac{1}{\tau} = \frac{\hbar\omega^3}{\pi^2 c^3} B_{2\leftarrow 1}. \tag{12.2-20}$$

One small repair is still needed to complete the job. The calculation of absorption and stimulated emission used the electric wave (1), which is polarized in the z direction. If (13) were substituted into (20), it would give the $1/\tau$ for an excited state with the particular orientation produced by absorption of the polarized light. In the usual light sources, atoms are jostled about and cannot know which way is up, and it is necessary to average over initial orientations. Rather than

$|\langle 1|z|2\rangle|^2$ as in (13), we write

$$\begin{aligned}\tfrac{1}{3}[|\langle 1|x|2\rangle|^2 + |\langle 1|y|2\rangle|^2 + |\langle 1|z|2\rangle|^2] \\ = \tfrac{1}{3}[\hat{e}_x\langle 1|x|2\rangle^* + \hat{e}_y\langle 1|y|2\rangle^* + \hat{e}_z\langle 1|z|2\rangle^*] \\ \cdot[\hat{e}_x\langle 1|x|2\rangle + \hat{e}_y\langle 1|y|2\rangle + \hat{e}_z\langle 1|z|2\rangle] \\ = \tfrac{1}{3}\langle 1|\vec{r}|2\rangle^* \cdot \langle 1|\vec{r}|2\rangle \\ = \tfrac{1}{3}|\langle 1|\vec{r}|2\rangle|^2.\end{aligned} \quad (12.2\text{-}21)$$

With this modification the $B_{2\leftarrow 1}$ from (13) substituted into (20) gives

$$\frac{1}{\tau} = \frac{e^2\omega_{12}^3}{3\pi\varepsilon_0\hbar c^3}|\langle 1|\vec{r}|2\rangle|^2. \quad (12.2\text{-}22)$$

Since the dimensionless *fine structure constant* is

$$\frac{e^2}{4\pi\varepsilon_0\hbar c} = \frac{1}{137.04},$$

a convenient way to write (22) is

$$\frac{1}{\tau} = \frac{1}{137.04}\frac{4\omega_{12}^3}{3c^2}|\langle 1|\vec{r}|2\rangle|^2. \quad (12.2\text{-}23)$$

The probabilities for absorption, stimulated emission, and spontaneous emission all involve the same kinds of matrix elements, the three components of $\langle 1|\vec{r}|2\rangle$. The specific functional forms of the initial and final wave functions determine the specific results, but it is useful to make some general observations.

Suppose that states 1 and 2 both have zero angular momentum. Then both wave functions are spherically symmetric, and the product $\psi_1^*(\vec{r})\psi_2(\vec{r})$ is also spherically symmetric. The matrix element

$$\langle 1|\vec{r}|2\rangle = \iiint \psi_1^*(\vec{r})\vec{r}\psi_2(\vec{r})d\tau$$

is zero because $\psi_1^*(\vec{r})\vec{r}\psi_2(\vec{r})$ is an odd function, and so there are no transitions between states 1 and 2. If a hydrogen atom is in the $n = 2$, $l = 0$ state, perhaps because of a transition from a higher state, it cannot emit a photon and go to the $n = 1$ state, because $n = 1$ implies $l = 0$. Deexcitation will usually occur through a collision with another atom rather than radiation. The restriction,

$$(j = 0) \nleftrightarrow (j = 0) \quad (12.2\text{-}24)$$

for single-photon emission, is an example of a *selection rule*. Such rules follow from the symmetries of the integrands in the matrix elements, and describe how some angular momentum changes lead to zero or very small transition probabilities. The rule (24) is absolute, in the sense that it does not depend on the approximations that have been made in this section, and it is generally applicable. It has a close analog in classical radiation theory, which shows that a spherically symmetric change in a spherically symmetric charge distribution does not cause any radiation. In some nuclei both the ground state and the first excited state have $j = 0$. Deexcitation by collision is nearly impossible because each nucleus is shielded by its electrons, and other processes remove the energy. The ^{16}O nucleus has a $j = 0$ ground state and a $j = 0$ first excited state at 6.05 MeV. It relieves itself by emitting an electron-positron pair, with 1.02 MeV going to the rest energies of the particles, and the electron and positron sharing the remaining 5.03 MeV as kinetic energy. In other cases the simultaneous emission of two photons, or the ejection of one of the atomic electrons, can occur with appreciable probability.

In the matrix elements $\langle 1|\vec{r}|2\rangle$ it is often convenient to use $\vec{r} = r\hat{r}$, with the unit vector $\hat{r}$ expressed in terms of the spherical harmonics (6.2-34).

$$\begin{aligned}\hat{r} &= \sin\theta\cos\varphi\hat{e}_x + \sin\theta\sin\varphi\hat{e}_y + \cos\theta\,\hat{e}_z \\ &= -\sqrt{\frac{2\pi}{3}}\,Y_{1+1}(\theta,\varphi)(\hat{e}_x - i\hat{e}_y) + \sqrt{\frac{2\pi}{3}}\,Y_{1-1}(\theta,\varphi)(\hat{e}_x + i\hat{e}_y) \\ &\quad + \sqrt{\frac{4\pi}{3}}\,Y_{10}(\theta,\varphi)\hat{e}_z.\end{aligned} \tag{12.2-25}$$

All contributions in $\hat{r}$ have $l = 1$. If one state has $l = 2$ and the other state has $l = 0$, the angular integration in the matrix element gives terms of the form

$$\iint Y^*_{2m'}\,Y_{1m}\,Y_{00}\,d\Omega = \frac{1}{\sqrt{4\pi}}\iint Y^*_{2m'}\,Y_{1m}\,d\Omega,$$

which are zero because of the orthogonality of the Y_{lm}. In general, the matrix elements are zero unless Δl, the change in l, is ± 1. To see that $\Delta l = 0$ gives zero, recall that, as shown in (6.2-44)), the parity of a state is $(-1)^l$. If the initial and final states have the same l, then their product $\psi_1^*\psi_2$ is necessarily an even function, $\psi_1^*\vec{r}\psi_2$ is an odd function, and the integral gives zero. Problem 12.4 deals with $|\Delta l| \geq 2$. The selection rule

$$\Delta l = \pm 1 \tag{12.2-26}$$

has a lower status than (24) because it is a consequence of our approximations. If, in the wave (1), kx is not neglected, then the formalism also describes transitions that do not obey (26). They, however, occur with much smaller probability than those that do; see the example below.

PROBLEM 12.4

Prove the statement that, if $|\Delta l| = |l_2 - l_1| \geq 2$, then $\langle 1|\vec{r}|2\rangle = 0$. If, say, $l_2 > l_1$, use the qualitative ideas of Section 9.2 to think about the angular momentum properties of $\vec{r}|1\rangle$.

If no selection rules are violated, then, for a system of size a,

$$|\langle 1|\vec{r}|2\rangle|^2 \simeq a^2, \tag{12.2-27}$$

roughly, because the matrix element is $\vec{r}$ integrated over the region in which the wave functions are appreciable. This estimate can miss the right answer by a substantial factor, but is useful for guessing magnitudes.

PROBLEM 12.5

Estimate the spontaneous emission mean life for

a. a $2p \to 1s$, that is, $(n = 2,\ l = 1) \to (n = 1,\ l = 0)$ transition in hydrogen.

b. a typical atomic transition that emits yellow light.

c. a 1.0-MeV nuclear gamma ray transition in a nucleus with $A \simeq 10$.

PROBLEM 12.6

For wavelengths around 6000 Å, the sun delivers about 0.12 watt per m^2 per Å, at noon at sea level.[5] Suppose that a typical atom has a first excited state 2.1 eV $= hc/6000$ Å above the ground state, and that the transition satisfies all selection rules. In the sun light, what is the absorption probability? If the atom is in the excited state, what are the spontaneous and stimulated emission probabilities?

EXAMPLE

In an atomic transition the initial state has $l_i = 0$, and the final state has $l_f = 2$, $m_f = +1$. Estimate roughly how the transition probability compares with that for a typical atomic transition which satisfies the selection rule $\Delta l = \pm 1$.

SOLUTION

As suggested above, retain the first power of kx in an expansion of the wave (1). Then

$$\begin{aligned} H' &= e\mathscr{E}_0 z \cos(kx - \omega t) \\ &= e\mathscr{E}_0 z[\cos kx \cos \omega t + \sin kx \sin \omega t] \\ &\simeq e\mathscr{E}_0 z[\cos \omega t + kx \sin \omega t], \\ H'_{fi} &= e\mathscr{E}_0\langle f|z|i\rangle \cos \omega t + e\mathscr{E}_0\langle f|kzx|i\rangle \sin \omega t. \end{aligned}$$

Because $\Delta l = 2$, $\langle f|z|i\rangle = 0$, and only the second term contributes. The occurrence of a $\sin \omega t$ rather than $\cos \omega t$ does not affect any conclusions because it is merely a change in the phase of the wave. Everything is the

[5] A. B. Meinel and M. P. Meinel, *Applied Solar Energy* (Reading, Mass.: Addison-Wesley Publishing Co., Inc., 1976), Sec. 2.2.

same as before except that the matrix element is $\langle f|kzx|i\rangle$ rather than $\langle f|z|i\rangle$.

The wave functions are

$$|i\rangle = R_i(r)Y_{00}(\theta, \varphi) = \frac{R_i(r)}{\sqrt{4\pi}},$$

$$|f\rangle = R_f(r)Y_{2+1}(\theta, \varphi),$$

where R_i and R_f are functions of r that are appreciable over a region of size a, about an Å for atoms.

$$zx = r^2 \cos\theta \sin\theta \cos\varphi$$

$$= r^2 \cos\theta \sin\theta \frac{e^{+i\varphi} + e^{-i\varphi}}{2}.$$

From (6.2-34),

$$Y_{2\pm1} = \mp\sqrt{\frac{15}{8\pi}} \cos\theta \sin\theta e^{\pm i\varphi}.$$

$$\therefore \quad zx = r^2\sqrt{\frac{2\pi}{15}}(Y_{2-1} - Y_{2+1}).$$

$$\therefore \quad \langle f|kzx|i\rangle = \iiint R_f^* Y_{2+1}^* kr^2 \sqrt{\frac{2\pi}{15}}(Y_{2-1} - Y_{2+1})\frac{R_i}{\sqrt{4\pi}} r^2\, dr\, d\Omega$$

$$= -\frac{1}{\sqrt{30}}\int_0^\infty R_f^* R_i kr^4\, dr,$$

because of the orthonormality of the Y_{lm}. For a rough estimate, suppose that R_f and R_i are both equal to a constant A for $r \leq a$, and zero for $r > a$. Normalization requires

$$\int_0^a A^2 r^2\, dr = 1, \qquad A = \sqrt{\frac{3}{a^3}}.$$

Then

$$\langle f|kxz|i\rangle = \frac{-1}{\sqrt{30}}\frac{3}{a^3}\int_0^a kr^4\, dr,$$

$$|\langle f|kzx|i\rangle|^2 = \frac{3k^2a^4}{250}.$$

The transition probability is proportional to the square of the magnitude of the matrix element. For transitions that satisfy $\Delta l = \pm 1$, this quantity is about a^2. The ratio is therefore about $10^{-2}k^2a^2$, or, for $\lambda = 6000$ Å,

$$10^{-2}\left(\frac{2\pi}{6000\ \text{Å}} \times 1\ \text{Å}\right)^2 = 10^{-8}.$$

The consequence is a mean life of around a second rather than the 10^{-8} sec typical for $\Delta l = \pm 1$ transitions.

This calculation is more elaborate than necessary for an estimate. The essential observation is that, to get $\Delta l = 2$ transitions, the orthogonality of the Y_{lm} requires one to keep the first power of kx in the expansion of the electromagnetic wave. Matrix elements are then of the order of $(ka)a = 2\pi a^2/\lambda$ rather than a. Furthermore, the integrations generally produce numerical factors that are less than unity. Atomic transitions that fail to satisfy $\Delta l = \pm 1$ are very improbable. For nuclei with size $a \simeq 6 \times 10^{-13}$ cm, and $\lambda/2\pi = 2 \times 10^{-11}$ cm for 1-MeV photons, $ka \simeq 0.03$. The relative suppression of $\Delta l = 2$ transitions is therefore not as great as in atoms.

Similar arguments show that $\Delta l = 3$ matrix elements are roughly k^2a^3, $\Delta l = 4$ matrix elements are roughly k^3a^4, and so forth. A full discussion of improbable transitions requires inclusion of the magnetic field in the light wave.

The 10.2-eV transitions between the $n = 1$ and $n = 2$ states of hydrogen will now be treated more carefully. The $n = 1$ state has $l = 0$ and $m = 0$ and can be written $|100\rangle$. Because of the selection rule (24), the transitions described here cannot occur between the ground state and the $|200\rangle$ state. Transitions to and from the three states $n = 2, l = 1, m = 0, \pm 1$ are possible. From Section 7.3,

$$|100\rangle = R_{10}(r)Y_{00}(\theta, \varphi), \qquad R_{10} = \frac{2}{a^{3/2}}e^{-r/a},$$

$$|21m\rangle = R_{21}(r)Y_{1m}(\theta, \varphi), \qquad R_{21} = \frac{re^{-r/2a}}{2^{3/2}a^{5/2}\sqrt{3}},$$

where the Bohr radius $a = 4\pi\varepsilon_0\hbar^2/\mu e = 0.529$ Å. The matrix elements are

$$\langle 21m|\vec{r}|100\rangle = \iiint R_{21}^* Y_{1m}^* r\hat{r} R_{10} Y_{00} r^2\, dr\, d\Omega$$

$$= \left[\iint Y_{1m}^* \hat{r} Y_{00}\, d\Omega\right]\left[\int R_{21}^* R_{10} r^3\, dr\right]. \qquad (12.2\text{-}28)$$

The radial integration is the same for all three m values.

$$\int_0^\infty R_{21}^* R_{10} r^3\, dr = \int_0^\infty \frac{r^4 e^{-3r/2a}}{a^4\sqrt{6}}\, dr = \frac{2^8 a}{3^4\sqrt{6}}.$$

The integral over angles is

$$\iint Y_{1m}^* \frac{\hat{r}}{\sqrt{4\pi}}\, d\Omega.$$

With the form (25) for $\hat{r}$, because of the orthonormality of the Y_{1m},

$$\begin{aligned}
\iint Y_{1+1}^* \frac{\hat{r}}{\sqrt{4\pi}}\, d\Omega &= \frac{-\hat{e}_x + i\hat{e}_y}{\sqrt{6}}, \\
\iint Y_{10}^* \frac{\hat{r}}{\sqrt{4\pi}}\, d\Omega &= \frac{\hat{e}_z}{\sqrt{3}}, \\
\iint Y_{1-1}^* \frac{\hat{r}}{\sqrt{4\pi}}\, d\Omega &= \frac{\hat{e}_x + i\hat{e}_y}{\sqrt{6}}.
\end{aligned} \tag{12.2-29}$$

Regardless of m,

$$\left| \iint Y_{1m}^* \frac{\hat{r}}{\sqrt{4\pi}}\, d\Omega \right|^2 = \tfrac{1}{3} \tag{12.2-30}$$

so that, for all three m values,

$$|\langle 21m|\vec{r}|100\rangle|^2 = \frac{2^{15}a^2}{3^{10}} = 0.555a^2, \tag{12.2-31}$$

in good agreement with the rough estimate (27).

It is no surprise that (31) is independent of m. The ground state is spherically symmetric, so that the mean life cannot depend on the orientation of the excited state. The angular distribution and the polarization of the radiation do depend on m, but a more detailed treatment is required to study these aspects.

From (23), with $\omega_{12} = 10.2\ \text{eV}/\hbar = 1.55 \times 10^{15}/\text{sec}$, the mean life of any of the $n = 2$, $l = 1$ states is

$$\tau = \left[\frac{1}{137.0} \frac{4\omega_{12}^3}{3c^2}\, 0.555a^2 \right]^{-1} = 1.60 \times 10^{-9}\ \text{sec}.$$

Experiments[6] of the kind described by Figure 12.2-2 give $(1.592 \pm 0.025) \times 10^{-9}$ seconds. The 10.2-eV, or 1220-Å, radiation is in the ultraviolet part of the spectrum. Because of the ω_{12}^3 factor, mean lives for transitions in the visible region are somewhat longer, typically 10^{-8} to 10^{-7} seconds. Atomic and molecular states that produce photons with wavelengths in the 10-cm region

[6] W. S. Bickel and A. S. Goodman, *Phys. Rev.* **148**, 1 (1966); H. H. Bukow et al., *Nuclear Instruments and Methods* **110**, 89 (1973).

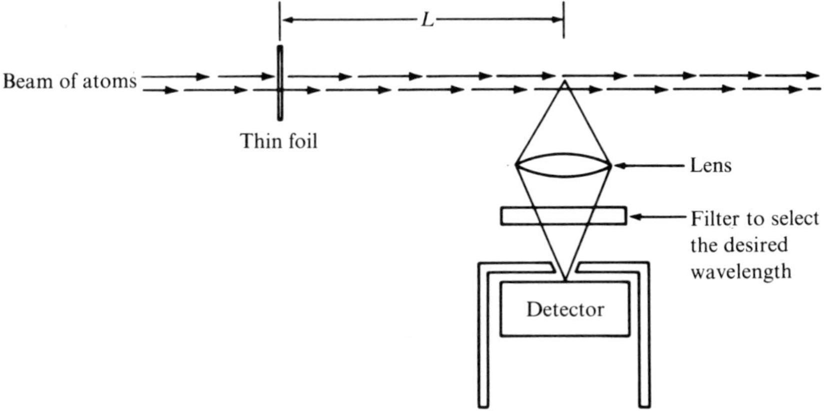

Fig. 12.2-2. A technique for measuring the mean life of atomic states. Atoms traveling at around 10^9 cm/sec are excited by collisions in a thin foil. The intensity of radiation is measured as a function of the distance L.

have mean lives in excess of

$$\left(\frac{10 \text{ cm}}{1220 \text{ Å}}\right)^3 1.6 \times 10^{-9} \text{ sec} = 30 \text{ years.}$$

In laboratory sources their spontaneous decay is unobservable, and they must be studied through absorption or stimulated emission.

PROBLEM 12.7

a. A particle of charge e is in the ($n_x = 0$, $n_y = 0$, $n_z = 1$) harmonic oscillator state. Obtain an expression for the spontaneous emission mean life for transition to the ($n_x = 0, n_y = 0, n_z = 0$) ground state. (This problem can be worked in Cartesian or in spherical polar coordinates. It is a little simpler in Cartesian coordinates.)

b. Make part (**a**) into a model of a nuclear gamma ray transition by letting the particle be a proton and choosing parameters so that the photon energy is 1.0 MeV. Obtain a number for the mean life.

12.3 *Transitions to a Continuum*

In Section 12.2 the initial and final states of the emitting or absorbing systems are presumed to have discrete energy eigenstates. The excitation of a hydrogen atom from the ground state to an $n = 2$ state requires 10.1988 eV;

photons with slightly larger or slightly smaller energy are not absorbed. A full title for Section 12.2 is *The Absorption and Emission of Light in the Transition between States That Are Well Separated in Energy from Other Possible States of the System.*

Many processes are of the kind treated in Section 12.2, and many are not. Photons with *any* energy above 13.6 eV can be absorbed by hydrogen atoms because they can induce transitions to the *continuum*, that is, to final states in which the atom has been ionized so that the electron and proton separate with positive energy. Photons with any energy above 2.23 MeV can be absorbed by a deuteron because they have enough energy to separate it into a free neutron and a free proton. Many reactions and decays that have nothing to do with electromagnetic radiation, such as beta decay, also lead to final states in which free and separated particles are flying away from each other.

Assume again that there is a single initial state ψ_i with well-defined energy E_i. A final state consists of particles with total energy E_f, which is arbitrarily close to E_i. We are, however, not interested in the transitions to just one final state, but to a group of states that differ because of slightly different energy sharing or directions of the momenta of the outgoing particles. The new thing in this section is the counting of these final states.

For any one of the final states f, (12.1-14) is

$$C_f(t) = \int_0^t \frac{H'_{fi}(t')}{i\hbar} e^{i(E_f - E_i)t'/\hbar}\, dt'. \tag{12.3-1}$$

As in Problem 12.2, suppose that at $t = 0$ the system is definitely in state i, and that then a constant perturbation begins to act on it.

$$\begin{aligned} H'(x, t) &= 0, \qquad & t < 0, \\ &= H'(x), \qquad & t \geq 0. \end{aligned} \tag{12.3-2}$$

This switching on of H' is an artifice that deserves discussion. If, for example, H' describes the nuclear beta decay interaction, then it is always present, and it seems peculiar to pretend that it begins at $t = 0$. The point is that we want to be able to say, "At $t = 0$, the system is definitely in state i." In the time between $t = -\infty$ and $t = 0$, in the absence of the perturbation, i is a stationary state, and it is possible to establish that the system is indeed described by i. There is, however, a difficulty in the turning on of the perturbation. Doing something during Δt makes the energy uncertain by $\Delta E \simeq \hbar/\Delta t$, and can therefore disturb the system badly if Δt is very small. We should not really say, "Turn on H' at $t = 0$." We should say, "Turn on H' during a short Δt, which is, however, long enough to make $\hbar/\Delta t$ negligible":

$$\Delta t \gg \frac{\hbar}{\text{characteristic energies of the process}}.$$

For the beta decay example the characteristic energies would be the energies of the emitted particles. On the other hand, Δt must be much less than the mean life τ,

$$\tau \gg \Delta t,$$

for otherwise there would be appreciable decay while H' is being increased to its final value. It follows that $\hbar/\tau$ must be small compared to the characteristic energies of the process, that is, the natural width of the state must be small. This criterion is precisely the one used in Section 3.11 to describe *quasi-stationary states*, states that decay with a mean life that is large in terms of their "natural times."

Other ways of picturing the process produce the same results. One could avoid the switching artifice by admitting that H' has always been present, while still trying to make the system be in the state i at $t = 0$. Think about the beta decay of ^{6}He to ^{6}Li, described by reaction (15) below. Since the decay interaction is weak, ^{6}He is very accurately described by the unperturbed Hamiltonian, which contains the strong nuclear and electromagnetic interactions; it is very nearly the state i. One might decide to make a chemical separation at $t = 0$ and quickly weigh the result to establish the purity and quantity of the initial sample. But every good quantum mechanic will object to the words, "at $t = 0$ and quickly." If the job is done in a time interval Δt, the energy is made uncertain by $\hbar/\Delta t$, and all of the limitations found above reappear. We now return to the calculation, which is much easier to digest than the concepts behind it.

With $\omega_i = E_i/\hbar$ and $\omega_f = E_f/\hbar$,

$$C_f = \frac{-H'_{fi}}{\hbar(\omega_f - \omega_i)}\,[e^{i(\omega_f-\omega_i)t} - 1]$$

$$= \frac{-2iH'_{fi}}{\hbar(\omega_f - \omega_i)}\,e^{i(\omega_f-\omega_i)t/2}\sin\frac{(\omega_f - \omega_i)t}{2},$$

$$|C_f|^2 = \frac{2\pi t}{\hbar^2}\,|H'_{fi}|^2\left[\frac{2\sin^2(\omega_f - \omega_i)t/2}{\pi(\omega_f - \omega_i)^2 t}\right]. \tag{12.3-3}$$

Except for the replacement of ω_{fi} by ω_f and of ω by ω_i, the expression in the last square bracket is the same as that examined in Section 12.2. Comparison with (12.2-14) shows that

$$\lim_{t\to\infty}\frac{2\sin^2(\omega_f - \omega_i)t/2}{\pi(\omega_f - \omega_i)^2 t} = \delta(\omega_i - \omega_f), \tag{12.3-4}$$

and the peak described by the expression for any reasonable t is so narrow that the delta function can be used in all of the subsequent calculations.

We want the probability of transition not to a particular state f, but to any of a dense group of states that all look about the same in the experiment.

"Looking about the same in the experiment" means that the momenta and other attributes of the emitted particles lie within an unresolvably small range. Suppose that the density in energy of such final states is $\rho(E_f)$, so that the number of these final states with energy between E_f and $E_f + dE_f$ is $\rho(E_f)dE_f$. Then the rate of transition to any of the group of states is

$$\begin{aligned} W &= \frac{1}{t}\int |C_f(E_f)|^2 \rho(E_f) dE_f \\ &= \frac{2\pi}{\hbar}\int |H'_{fi}|^2 \delta(\omega_i - \omega_f)\rho(E_f)d\omega_f, \end{aligned} \tag{12.3-5}$$

where $dE_f/\hbar$ is replaced by $d\omega_f$. Because of the delta function, the integration gives

$$W = \frac{2\pi}{\hbar}\rho(E_f)|H'_{fi}|^2. \tag{12.3-6}$$

Since $E_i = E_f$, the density of states can be written either as $\rho(E_i)$ or $\rho(E_f)$; $\rho(E_f)$ will be used here to emphasize that it is the final states that must be counted. This very important result is applicable in a great variety of situations where the perturbation leads to significant changes, but where the perturbation is small in the sense that it does not affect the energy much. It is often called the Golden Rule of quantum mechanics.

In Chapter 5 periodic boundary conditions are used to count states of particles in a large volume $\mathscr{V}$. The conclusions there can be applied directly to calculate $\rho(E_f)$. According to (5.1-23), the number of states of the form

$$\frac{e^{i\vec{p}\cdot\vec{r}/\hbar}}{\sqrt{\mathscr{V}}},$$

with magnitude of momentum below p_f, or with energy below E_f, is for each spin state

$$\frac{\mathscr{V}p_f^3}{6\pi^2\hbar^3} = \frac{\mathscr{V}}{6\pi^2\hbar^3}(2ME_f)^{3/2}. \tag{12.3-7}$$

The density $\rho(E_f)$ of such states, the number per unit energy, is the derivative of this expression with respect to E_f.

$$\rho(E_f) = \frac{\mathscr{V}(2M)^{3/2}E_f^{1/2}}{4\pi^2\hbar^3} = \frac{\mathscr{V}Mp_f}{2\pi^2\hbar^3}. \tag{12.3-8}$$

This result includes all states with magnitude of momentum p_f. The density of states with $\vec{p}_f$ pointing into a small solid angle $d\Omega$ is (8) times $d\Omega/4\pi$:

$$\frac{d\rho(E_f)}{d\Omega}d\Omega = \frac{\mathscr{V}Mp_f}{8\pi^3\hbar^3}d\Omega. \tag{12.3-9}$$

The results (6) and (9) can be used to derive from a different approach the Born approximation scattering cross section obtained in Section 11.2. Take as the initial state the wave

$$\psi_i = \frac{e^{ikz}}{\sqrt{\mathscr{V}}}$$

so that there is one incident particle in the volume $\mathscr{V}$ in which the final states are counted. Take the point of view that the scattering potential energy function $V(\vec{r})$ is the perturbation that causes transitions from ψ_i to

$$\psi_f = \frac{e^{i\vec{k}_d \cdot \vec{r}}}{\sqrt{\mathscr{V}}},$$

where, as in Section 11.2, $\vec{k}_d$ has magnitude k and points in the direction of the detector. The matrix element is

$$H'_{fi} = \iiint \psi_f^*(\vec{r}) V(\vec{r}) \psi_i(\vec{r}) d\tau$$

$$= \frac{1}{\mathscr{V}} \iiint V(\vec{r}) e^{i(kz - \vec{k}_d \cdot \vec{r})} \, d\tau.$$

The geometry is the same as before. From Figure 11.2-1,

$$\vec{k}_d \cdot \vec{r} = kr \cos \gamma, \qquad kz - kr \cos \gamma = \vec{K} \cdot \vec{r},$$

where $\hbar\vec{K}$ is the momentum transferred to the scattering center.

$$H'_{fi} = \frac{1}{\mathscr{V}} \iiint e^{ik(z - r\cos\gamma)} V(\vec{r}) d\tau$$

$$= \frac{1}{\mathscr{V}} \iiint V(\vec{r}) e^{i\vec{K} \cdot \vec{r}} \, d\tau. \tag{12.3-10}$$

The transition probability, the probability of scattering into $d\Omega$ per second, is

$$dW = \frac{2\pi}{\hbar} \frac{d\rho(E_f)}{d\Omega} d\Omega |H'_{fi}|^2$$

$$= \frac{\mu p_f \, d\Omega}{4\pi^2 \hbar^4 \mathscr{V}} \left| \iiint V(\vec{r}) e^{i\vec{K} \cdot \vec{r}} \, d\tau \right|^2.$$

The mass is again called μ because many applications deal with two-particle interactions that are best described in terms of relative coordinates and the reduced mass. With $w(r) = 2\mu V(\vec{r})/\hbar^2$ as in (11.2-1),

$$\frac{dW}{d\Omega} = \frac{p_f}{\mu \mathscr{V}} \,|\, f(\theta, \varphi)|^2,$$

$$|\, f(\theta, \varphi)|^2 = \left| \frac{1}{4\pi} \iiint w(\vec{r}) e^{i\vec{K}\cdot\vec{r}} \, d\tau \right|^2, \tag{12.3-11}$$

with the $f(\theta, \varphi)$ the same as that given by (11.2-14). The cross section is the rate $dW/d\Omega$ divided by the incident density

$$j_i = |\psi_i|^2 \times \text{speed} = \frac{p_i}{\mathscr{V}\mu}, \; p_i = p_f.$$

Therefore

$$\frac{d\sigma}{d\Omega} = |\, f(\theta, \varphi)|^2, \tag{12.3-12}$$

the same form as used throughout Chapter 11.

The meaning of the density of final states $\rho(E_f)$ is a little elusive on first acquaintance. Review the derivation of (12) with special attention to the way that the volume $\mathscr{V}$ appears in the density of states and the normalization of the wave functions, and disappears in the final expression for the differential scattering cross section.

PROBLEM 12.8

In Section 11.2, (11) emerged as an approximate result. Examine the approximations that are implicit in the treatment of Section 12.3, and show that they have the same physical content as those of Section 11.2.

Nuclear beta decay[7] provides a good example of the application of (6) to particle transmutation. The basic process is the decay of a neutron into a proton, an electron, and a neutrino[8]:

$$n \rightarrow p + e^- + \bar{\nu}_e + 0.782 \text{ MeV}, \qquad \tau = 15.3 \text{ minutes}. \tag{12.3-13}$$

The energy 0.782 MeV is the total kinetic energy of the products of the decay; it does not include rest energies. The time τ is the mean life, and must be multi-

[7] K. W. Ford, *Classical and Modern Physics* (Lexington, Mass.: Xerox College Publishing, 1972), Chaps. 25, 26, and 27, particularly Sec. 26.3; R. Eisberg and R. Resnick, *Quantum Physics of Atoms, Molecules, Solids, Nuclei, and Particles* (New York: John Wiley & Sons, Inc., 1974), Secs. 16-3, 16-4; H. Frauenfelder and E. M. Henley, *Subatomic Physics* (Englewood Cliffs, N. J.: Prentice-Hall, Inc., 1974), Chap. 11.

[8] The designation $\bar{\nu}_e$ means that the particle is actually an antineutrino of the particular kind that appears in conjunction with electrons.

plied by ln 2 = 0.693 to give the half life. Free neutrons, such as those drifting around in a nuclear reactor, undergo the decay (13), while neutrons inside a nucleus can do so only if the energy of the entire nucleus is lowered thereby. Examples include

$$^{3}\text{H} \rightarrow {}^{3}\text{He} + e^{-} + \bar{\nu}_e + 0.0186 \text{ MeV}, \qquad \tau = 17.7 \text{ years}, \tag{12.3-14}$$

$$^{6}\text{He} \rightarrow {}^{6}\text{Li} + e^{-} + \bar{\nu}_e + 3.51 \text{ MeV}, \qquad \tau = 1.17 \text{ seconds}. \tag{12.3-15}$$

In these and similar nuclear beta decays it is a good approximation to neglect the energy of recoil of the nucleus and to assume that the e^- and the $\bar{\nu}_e$ together carry off all the available kinetic energy.

To obtain the density of final states, we must violate slightly the boundaries of this book. The energies are of the order of the rest energy of the electron and, in any case, the neutrino has zero rest mass and moves with the speed of light. It is therefore necessary to use the relativistic relation between energy and momentum,[9]

$$E^2 = c^2p^2 + M^2c^4. \tag{12.3-16}$$

The relation $p = h/\lambda$ is unchanged, so that the same counting procedure as in Chapter 5 still gives, as in (7), $\mathscr{V}p^3/6\pi^2\hbar^3$ for the number of states with magnitude of momentum below p. The number of states with magnitude of momentum between p and dp is then

$$\frac{\mathscr{V}p^2\,dp}{2\pi^2\hbar^3}. \tag{12.3-17}$$

In beta decay the directions of the electron momentum and the neutrino momentum can be selected independently because the nucleus can recoil to conserve the momentum. The total number of possibilities is therefore the number of possibilities for the electron times the number of possibilities for the neutrino. The density of final states is then a term of the form (17) for the electron, multiplied by one for the neutrino, divided by dE_f, where $E_f = E_e + E_\nu$, the total energy, rest plus kinetic, of the two particles. Let $d\rho(E_f)$ be the density of final states for electron energy between E_e and $E_e + dE_e$.

$$d\rho(E_f) = \frac{1}{dE_f}\left(\frac{\mathscr{V}p_e^2\,dp_e}{2\pi^2\hbar^3}\right)\left(\frac{\mathscr{V}p_\nu^2\,dp_\nu}{2\pi^2\hbar^3}\right). \tag{12.3-18}$$

Since the neutrino has zero rest mass, (16) gives

$$p_\nu = \frac{E_\nu}{c}, \qquad dp_\nu = \frac{dE_\nu}{c}.$$

[9] K. F. Ford, *Classical and Modern Physics* (Lexington, Mass.: Xerox College Publishing, 1972) Part VI, particularly Sections 21.2–21.5; A. P. French, *Special Relativity* (New York: W. W. Norton & Co., Inc., 1968) particularly Chapter 6.

If p_e is held fixed, then any change dE_ν in neutrino energy must equal dE_f, the change in total energy. With dp_ν replaced by dE_f/c, and with p_ν replaced by $(E_f - E_e)/c$,

$$d\rho(E_f) = \frac{\mathscr{V}^2(E_f - E_e)^2 p_e^2\, dp_e}{4\pi^4\hbar^6 c^3}.$$

With

$$E_e^2 = c^2 p_e^2 + M_e^2 c^4,$$
$$E_e\, dE_e = c^2 p_e\, dp_e,$$

the density of final states is

$$d\rho(E_f) = \frac{\mathscr{V}^2}{4\pi^4\hbar^6 c^6}(E_f - E_e)^2\sqrt{E_e^2 - M_e^2c^4}\, E_e\, dE_e. \tag{12.3-19}$$

The next step is to construct the matrix element. In 1933–34 E. Fermi developed the theory of beta emission by using the theory of spontaneous photon emission as a guide. The essential element in electromagnetic transitions is an integral of the product of the initial wave function, the electromagnetic wave, the final wave function, and the charge. Fermi introduced as the analogous term in beta decay

$$H'_{fi} = \iiint \psi_{Nf}^* \frac{e^{-i\vec{k}_e\cdot\vec{r}}}{\sqrt{\mathscr{V}}}\, g(\text{OPERATORS})\, \frac{e^{i\vec{k}_\nu\cdot\vec{r}}}{\sqrt{\mathscr{V}}}\psi_{Ni}\, d\tau. \tag{12.3-20}$$

The constant g gives the strength of the interaction, just as e gives the strength of the interaction between electrons and light. The initial and final nuclear wave functions are ψ_{Ni} and ψ_{Nf}. The electron and neutrino wave functions are taken to be plane waves,[10] normalized in the same volume $\mathscr{V}$ that is used for counting states. A meaningful discussion of (OPERATORS) requires a complete relativistic description that includes spin. Much of the good physics that has been done since Fermi's formulation is buried in (OPERATORS), and it is regrettable that its treatment is beyond our scope.

For our purposes (20) can be estimated very simply. If the electron and neutrino energies are not much more than an MeV, both $k_e = p_e/\hbar$ and $k_\nu = p_\nu/\hbar$ are $\lesssim 10^{11}$/cm, $\vec{k}_e\cdot\vec{r}$ and $\vec{k}_\nu\cdot\vec{r}$ are $\lesssim 10^{-2}$ within the nuclear volume, and the exponentials can be approximated by unity. The same idea is used at the beginning of Section 12.2, where $\cos(kx - \omega t)$ is replaced by $\cos \omega t$ because the wavelength of light is large compared to the size of the atom.

$$H'_{fi} \simeq \frac{g}{\mathscr{V}} \iiint \psi_{Nf}^*(\text{OPERATORS})\psi_{Ni}\, d\tau. \tag{12.3-21}$$

[10] The distortion of the electron wave function by the Coulomb field of the nucleus is neglected here. Because the wave function is pulled in and has better overlap with the nucleus, a more accurate treatment shows that there is substantial enhancement of the probability of low-energy electron emission.

The size of H'_{fi} is sensitive to the nuclear angular momentum and parity changes, and there are selection rules with some similarities to those found for electromagnetic transitions. Again, angular momentum changes $\geq 2\hbar$ give small matrix elements. If the selection rules are obeyed, then the matrix element is large. "Large" means that the integral in (21) is near unity; if the overlap of the functions in the integrand is good, then the integral can be almost as large as the normalization integral for either ψ_{Ni} or ψ_{Nf}:

$$H'_{fi} \simeq \frac{g}{\mathscr{V}}. \tag{12.3-22}$$

With (19) and (22) the transition probability (6) can be assembled:

$$W(E_e)dE_e = \frac{g^2}{2\pi^3\hbar^7c^6} E_e(E_f - E_e)^2\sqrt{E_e^2 - M_e^2c^4}\, dE_e \tag{12.3-23}$$

is the probability per second that the nucleus emits an electron with energy between E_e and $E_e + dE_e$. The transition probability vanishes if E_e has its smallest possible value M_ec^2 or its largest possible value E_f. The electron spectrum is sketched in Figure 12.3-1. The agreement with experiment is good, except for the discrepancies caused by the Coulomb effects mentioned in footnote 10.

The behavior of the spectrum for E_e close to E_f would be different if the neutrino had a nonzero rest mass M_ν. The maximum possible E_e would be $E_f - M_\nu c^2$, and the curve in Figure 12.3-1 would cut off sharply with a vertical slope at the upper end. Studies of the $E_e \simeq E_f$ region of beta spectra show that $M_\nu < 10^{-4}M_e$.

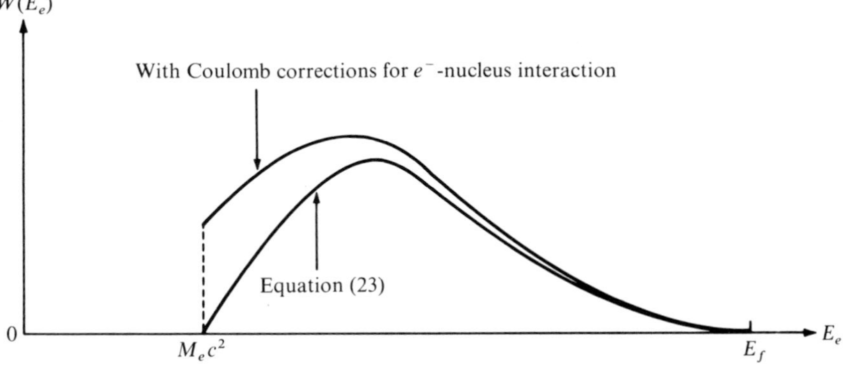

Fig. 12.3-1. The energy spectrum for electrons from nuclear beta decay.

PROBLEM 12.9

Obtain the beta transition probability under the assumption that the neutrino does have a nonzero rest mass M_ν; that is, repeat the development (18) through (23) for

$$E_\nu^2 = c^2 p_\nu^2 + M_\nu^2 c^4.$$

For the decay of tritium, reaction (14), plot on the same graph the beta spectrum for $M_\nu = 10^{-3} M_e$ and for $M_\nu = 0$.

The integral of (23) over all possibilities for E_e is the total transition probability, the reciprocal of the mean life of the beta unstable nucleus.

$$\frac{1}{\tau} = \int_{M_e c^2}^{E_f} W(E_e) dE_e. \qquad (12.3\text{-}24)$$

The result is a rather strong function of the available kinetic energy, as suggested by the data in (13), (14), and (15).

PROBLEM 12.10

Obtain τ as a function of E_f. Then use the result to make three determinations of the beta decay interaction constant g by using the data in (13), (14), and (15). The values can differ from each other because the neglect of the nucleus-electron Coulomb interaction and the estimate (22) limit our accuracy.

Our treatment of beta decay bypasses the heart of the matter by using the estimate (22) for the matrix element. It is nevertheless worth the effort because it may clarify the meaning of $\rho(E_f)$, and because it shows how $\rho(E_f)$ can affect the characteristics of a process. It is true for a variety of transitions that the shapes of particle spectra and the energy dependence of the mean life are determined by the purely kinematic state counting represented by $\rho(E_f)$, so that one can calculate these features without penetrating the mysteries of the matrix elements.

Electromagnetic processes provide another class of examples. The Section 12.2 results for the spontaneous emission of light can also be obtained from (6) if the photon is treated as the free particle for which $\rho(E_f)$ is the density of states. Here we shall not reexamine spontaneous decay but shall concentrate on photon absorption that excites a particle to a positive energy state. The calculation is in essence an adaptation of the expressions in Section 12.2 to the methods of this section.

Begin with (1), but use the perturbation (12.2-2) to describe the incident light. The matrix element is then given by (12.2-3), and (12.2-9) again describes the initial growth of $|C_f(t)|^2$:

$$\frac{|C_f(t)|^2}{t} = \frac{\pi e^2}{\varepsilon_0 \hbar c}|\langle f|z|i\rangle|^2\Phi(\omega)\omega\left\{\frac{2\sin^2[(\omega_f - \omega_i - \omega)t/2]}{\pi(\omega_f - \omega_i - \omega)^2 t}\right\}.$$

It is useful to write $\omega_f - \omega_i$ rather than ω_{fi} because now the initial state energy $\hbar\omega_i$ and the photon energy $\hbar\omega$ are taken as definite, while the final state energy $\hbar\omega_f$ lies in the continuum. The relation (12.2-14) shows that the expression in the braces becomes $\delta(\omega_f - \omega_i - \omega)$ at appreciable times. With the introduction of the density of final states $\rho(E_f)$, (5) is applicable.

$$\begin{aligned} W &= \frac{1}{t}\int |C_f(E_f)|^2\rho(E_f)dE_f \\ &= \int \frac{\pi e^2}{\varepsilon_0 \hbar c}|\langle f|z|i\rangle|^2\Phi(\omega)\omega\delta(\omega_f - \omega_i - \omega)\rho(E_f)dE_f \qquad (12.3\text{-}25) \\ &= \frac{\pi e^2}{\varepsilon_0 c}|\langle f|z|i\rangle|^2\Phi(\omega)\omega\rho(E_f), \end{aligned}$$

where $E_f = E_i + \hbar\omega$. For the final state of the ejected particle, use the plane wave

$$\frac{e^{i\vec{p}_f\cdot\vec{r}/\hbar}}{\sqrt{\mathscr{V}}}.$$

As in the other examples of this section, this choice means that the distortion of the final state wave function by the field of the atom is being neglected. The density of final states is, for nonrelativistic outgoing particles, given by (9). Since $\Phi(\omega)$ is the incident photon flux density, (25) divided by $\Phi(\omega)$ and by $d\Omega$ is the differential cross section for photodisintegration with the outgoing particle heading into $d\Omega$:

$$\frac{d\sigma}{d\Omega} = \frac{e^2 M\omega p_f}{8\pi^2\varepsilon_0\hbar^3 c}\left|\iiint e^{-i\vec{p}_f\cdot\vec{r}/\hbar}z\psi_i(\vec{r})d\tau\right|^2. \qquad (12.3\text{-}26)$$

Suppose that ψ_i is a spherically symmetric function,

$$\psi_i = Y_{00}R(r) = \frac{R(r)}{\sqrt{4\pi}}.$$

Then the integral in (26) is, with $\vec{q} = \vec{p}_f/\hbar$,

$$\frac{1}{\sqrt{4\pi}}\iiint e^{-i\vec{q}\cdot\vec{r}}zR(r)d\tau.$$

Let the direction of $\vec{q}$ be (θ_f, φ_f). The x axis was originally chosen in (12.2-1) to be the direction of the incident photon beam. However, with the $kx \ll 1$ approximation, the system knows only about the direction z of the oscillating electric field, and everything is symmetric around the z axis. Choose the x axis so that $\varphi_f = 0$. Then rotate the coordinate system through the angle θ_f so that $\vec{q}$ points in the direction of the z' axis. The electric field $\mathscr{E}_0 \hat{e}_z \cos \omega t$ becomes

$$\mathscr{E}_0(\cos \theta_f \hat{e}_{z'} - \sin \theta_f \hat{e}_{x'}) \cos \omega t.$$

This field is given by

$$-\vec{\nabla}\mathscr{E}_0(-z' \cos \theta_f + x' \sin \theta_f) \cos \omega t,$$

and the perturbation is

$$H' = e\mathscr{E}_0(z' \cos \theta_f - x' \sin \theta_f) \cos \omega t.$$

We can work in the new coordinate system if z in the matrix element is replaced by $z' \cos \theta_f - x' \sin \theta_f$. This transformation makes the integral more tractable because, with $\vec{q}$ pointing in the z' direction, $\vec{q} \cdot \vec{r}$ becomes $qr \cos \theta$.

$$\frac{1}{\sqrt{4\pi}} \int_0^\infty \int_0^\pi \int_0^{2\pi} e^{-iqr\cos\theta}(\cos \theta \cos \theta_f - \sin \theta \cos \varphi \sin \theta_f) R(r) r^3 \, dr \sin \theta \, d\theta \, d\varphi$$

$$= \sqrt{\pi} \cos \theta_f \int_0^\infty \left(\int_{-1}^{+1} e^{-iqru} u \, du \right) R(r) r^3 \, dr$$

$$= \frac{i\sqrt{\pi} \cos \theta_f}{q} \left[\int_0^\infty e^{iqr} \left(1 + \frac{i}{qr}\right) R(r) r^2 \, dr + \text{complex conjugate} \right]. \tag{12.3-27}$$

Notice the angular dependence of this term. For any $R(r)$, (27) inserted in (26) gives a differential cross section proportional to $\cos^2 \theta_f$. The particle is most likely to be thrown out in the directions $\theta_f = 0$ or π, that is, in the direction of polarization of the incident wave. The impulse given to the particle has zero component perpendicular to the electric field.

For the ground state of hydrogen, with

$$R(r) = \frac{2e^{-r/a}}{a^{3/2}},$$

the result of the integration (27) gives for the differential cross section (26)

$$\frac{d\sigma}{d\Omega} = \frac{2^9 M a^7 \omega p_f^3 \cos^2 \theta_f}{137.0\hbar^4 \left(1 + \dfrac{a^2 p_f^2}{\hbar^2}\right)^6}. \tag{12.3-28}$$

PROBLEM 12.11

Verify (28).

The total cross section is

$$\sigma = \iint \frac{d\sigma}{d\Omega}\, d\Omega = \frac{2^{11}\pi M a^7 \omega p_f^3}{3 \times 137.0\hbar^4\left(1 + \dfrac{a^2 p_f^2}{\hbar^2}\right)^6}. \tag{12.3-29}$$

Because the ionization energy is 13.6 eV, ω and p_f are related by

$$\frac{p_f^2}{2M} = \hbar\omega - 13.6\text{ eV},$$

$$\frac{a p_f}{\hbar} = \sqrt{\frac{\hbar\omega}{13.6\text{ eV}} - 1},$$

and the cross section can be written in the form

$$\sigma = \frac{2^{10}\pi a^2}{3 \times 137.0} \frac{\left(\dfrac{\hbar\omega}{13.6\text{ eV}} - 1\right)^{3/2}}{\left(\dfrac{\hbar\omega}{13.6\text{ eV}}\right)^5}. \tag{12.3-30}$$

This result is roughly right for photon energies around a few hundred eV. For photon energies $\lesssim$200 eV, there is considerable Coulomb distortion of the final state electron wave function, and the plane wave approximation is not even qualitatively right. The consequence of this distortion is the formation of an "absorption edge." The cross section rises abruptly from zero to its maximum value at the $\hbar\omega = 13.6$ eV threshold and then decreases monotonically with increasing energy. At energies much above 200 eV it is a poor approximation to assume that the wavelength of the photons is large compared to the atom.[11]

PROBLEM 12.12

Obtain, for general atomic number Z, the K shell photoelectric absorption cross section per atom by modifying the steps that lead to (30). You need to consider the two K shell electrons and use some results of Section 10.4.

---

[11] For a general summary of the data and references to experiments and detailed calculations, see J. H. Hubbell, "Survey of Photon Attenuation Coefficient Measurements, 10 eV to 100 GeV," *Atomic Data* **3**, 241 (1971). Data on the region just above threshold are given by H. P. Palenius, J. L. Kohl, and W. H. Parkinson, *Phys. Rev. A* **13**, 1805 (1976).

An estimate of the deuteron photodisintegration cross section can be obtained from (26) and (27). Neglect the few per cent D state contribution. The initial state is then described in Section 7.1 and illustrated in Figure 7.1-2. The structure of (27) shows that the principal contributions come from large r, where the deuteron wave function is the exponential tail in the classically forbidden region. To simplify the computation, assume that the deuteron is all tail and no body by taking the exponential behavior in to the origin: $\psi_i \simeq e^{-\alpha r}/r$ everywhere. In the notation of Section 7.1, the effect of this approximation is to neglect the range b of the force compared to $1/\alpha$;

$$\alpha = \frac{\sqrt{2\mu B}}{\hbar} = \frac{\sqrt{2\left(\dfrac{469.5 \text{ MeV}}{c^2}\right)(2.23 \text{ MeV})}}{\hbar}$$

$$= \frac{1}{4.32 \times 10^{-13} \text{ cm}}.$$

With

$$\psi_i(r) = \frac{R(r)}{\sqrt{4\pi}}, \qquad R(r) = \frac{\sqrt{2\alpha}\, e^{-\alpha r}}{r},$$

the approximation to the initial state is normalized;

$$\int_0^\infty |\psi_i|^2 4\pi r^2 \, dr = 1.$$

Since r is the relative coordinate and only the proton is charged, the matrix element contains not z, but $z/2$; z describes the proton-neutron separation, while $z/2$, which describes the center-of-mass to proton separation, is what we need here. These choices give for (27)

$$\frac{i\sqrt{2\pi\alpha}\cos\theta_f}{2q}\left[\int_0^\infty e^{(iq-\alpha)r}\left(1 + \frac{i}{qr}\right) r\, dr + \text{complex conjugate}\right]$$

$$= -2i\sqrt{2\pi\alpha}\cos\theta_f \frac{q}{(q^2+\alpha^2)^2},$$

where θ_f is the direction and $\hbar q$ is the final magnitude of the proton momentum. Now (26) gives

$$\frac{d\sigma}{d\Omega} = \frac{4\mu\omega\left(\dfrac{p_f}{\hbar\alpha}\right)^3 \cos^2\theta_f}{137.0\hbar\alpha^4\left(1 + \dfrac{p_f^2}{\hbar^2\alpha^2}\right)^4}, \qquad (12.3\text{-}31)$$

and the total cross section is

$$\sigma = \frac{16\pi\mu\omega\left(\dfrac{p_f}{\hbar\alpha}\right)^3}{3 \times 137.0\hbar\alpha^4\left(1 + \dfrac{p_f^2}{\hbar^2\alpha^2}\right)^4}. \tag{12.3-32}$$

With the binding energy $B = 2.23$ MeV,

$$\frac{p_f^2}{2\mu} = \hbar\omega - B,$$

$$\frac{p_f}{\hbar\alpha} = \sqrt{\frac{\hbar\omega}{B} - 1},$$

and (32) is

$$\sigma = \frac{8\pi}{3 \times 137.0\alpha^2} \frac{\left(\dfrac{\hbar\omega}{B} - 1\right)^{3/2}}{\left(\dfrac{\hbar\omega}{B}\right)^3}. \tag{12.3-33}$$

Figure 12.3-2 describes the accuracy of the result (33). The shape is nearly right, but the magnitude is low by a factor of around 2.5/1.4. About half of the discrepancy is caused by the grossly oversimplified initial state ψ_i. Most of the remainder comes from the "final state interaction," that is, from the distortion

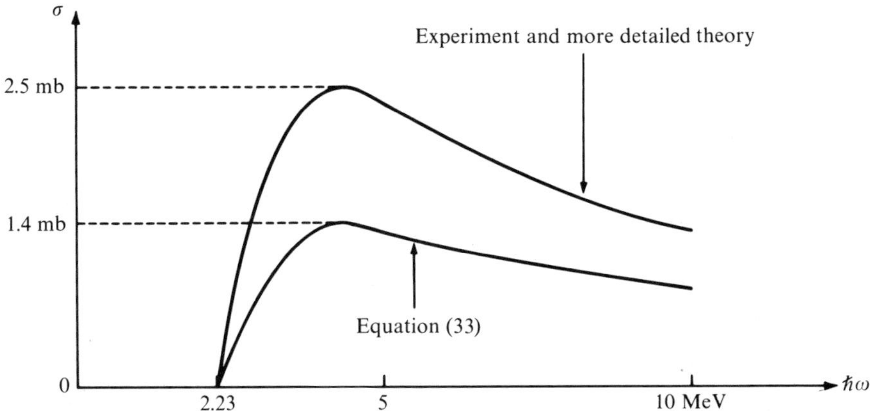

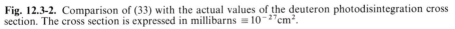

Fig. 12.3-2. Comparison of (33) with the actual values of the deuteron photodisintegration cross section. The cross section is expressed in millibarns $\equiv 10^{-27}\text{cm}^2$.

of the outgoing wave by the nuclear force between the proton and neutron. An effect of this kind is always present, but it does not do as much violence to these results as the Coulomb force does to photoionization cross sections. Other matters that must be considered to achieve detailed agreement include the *D* state contributions and, particularly near threshold, the *photomagnetic effect*. An electromagnetic wave contains an oscillating magnetic field as well as the electric field. The magnetic field is capable of flipping the spin of one of the two particles, so that there is a transition from the triplet to the singlet state. The singlet state is unbound, and therefore such a spin flip disintegrates the deuteron.[12]

[12] More detailed calculations of deuteron photodisintegration, data, and references are given by S. DeBenedetti, *Nuclear Interactions* (New York: John Wiley & Sons, Inc., 1964), Sec. 4.2; R. R. Roy and B. P. Nigam, *Nuclear Physics* (New York: John Wiley & Sons, Inc., 1967), Secs. 3.16 and 12.8.

Index

Absorption:
of light, 370–375, 379
of particles, 76–78
Airy integral, 92–96
Alpha decay, 53, 331–332
Angular distribution of gamma rays, 208
Angular momentum, 117, 209–223 (*see also* Orbital angular momentum *and* Spin)
addition of, 287–301
algebraic treatment, 213–223
commutation relations, 211–213
of composite systems, 212, 287–301
conservation, 210–211
integral or half integral values, 217
raising and lowering operators, 215, 218–219
Axioms of quantum mechanics, 131

Bands of energy in solids, 206–207
Baryons, 253
Basis functions, 137
Beta decay, 366, 384, 388–392
Binding energy of metals, 201
Bloch functions, 204
Bohr's complementarity principle, 17
Bohr-Sommerfeld-Wilson quantum condition, 329
Born approximation, 344–356, 387–388
for one-dimensional reflection coefficient, 349–350
Bose-Einstein statistics, 181
Bosons, 181, 184, 201
Bound states, 63–74
Box potential, 56–58
Bra vector, 141
Breit-Wigner resonance form, 159, 282

Canonically conjugate variables, 114
Center-of-mass coordinates, 173–174
Centrifugal potential, 226, 236
Charmonium, 254-256, 335
Chemistry, 179
Classically forbidden region, 47
localization in, 48–49
Classical trajectories as slowly varying V limit, 21
Clebsch-Gordan coefficients, 296–301
Closure, 127
Coherence length, 90–91
Collision broadening, 20
Commutator, 113–115
Commuting operators, 132–133
Complementarity, 17
Complex conjugation operator, 123, 163
Complex potential energy, 76–80, 114, 153–154
Complex vector spaces, 119, 133–134, 138
Complex waves, use of, 11–12, 47
Compressibility of metals, 200–201
Conduction electrons, 167, 186–207
Conservation:
of charge, 35–36
of probability, 25, 30–35
Correspondence principle, 13, 25
Coulomb potential:
bound states in, 18–19, 246–251
scattering by, 351–355
Current density, 32

Davisson-Germer experiment, 8–11, 23
De Broglie wavelength, 8–11
Decaying states, 79, 153–162, 385
Degeneracy:
definition, 59
harmonic oscillator, 86–87, 243
hydrogen atom, 249
one-dimensional bound states, 64–65
particle in a box, 59
and perturbation theory, 317–323
produced by rotational symmetry, 236
Delta function, 53–55, 107, 205

Delta function *(continued)*
normalization, 121-122, 157
potential energy, 55-56, 74-76
Density of states, 192-196, 386, 389-393
Deuteron, 19, 89, 239-240, 361-364
angular momentum of, 239, 301-305
magnetic moment, 284, 301-305
percent *D* state, 239, 305
photodisintegration, 384, 396-398
Diagonal matrix, 138
Diatomic molecules, 86
Diffraction:
of atoms, 11
of electrons, 8-11, 23, 91
of neutrons, 11
Dineutron, 363
Diproton, 364
Dirac bracket notation, 140-141
Dirac delta function, *see* Delta function
Displacement operator, 203-204

Ehrenfest's theorem, 149-151
Eigenfunctions, 42
expansion in terms of, 122-127
orthnormality of, 118-122
Eigenvalues, 42
as the results of measurements, 127-128
Eigenvector, 137
Electric charge, conservation of, 35-36
Electric conductivity, 206-207
Energy conservation, 149-150
for systems of particles, 171
Exchange operator, 177-180
Exclusion principle, 179-181, 189, 195, 293
Exotic atoms, 246, 316
Expansion postulate, 122-123, 155
Expectation value, 101
of functions of momentum, 108-110
of functions of position, 101-102
of operators in general, 110-114
time derivative of, 148-150
Exponential decay, 79, 161-162
classical view of, 162
Extinction length, 90

Fermi-Dirac statistics, 181
Fermi energy, 195-200, 207
Fermions, 166, 181, 183, 189, 201, 293
Fine structure, 259-260, 320
Fine structure constant, 377
Fourier integral, 107
Fourier series, 105-106
Fourier transform, 107-108
Fractional charges, 253
Free electron model of metals, 189-201
Fusion, 40

Galilean invariance, 26
Gamma ray emission, 383
Gauss' divergence theorem, 33
Golden rule for transitions, 386
Gravitational potential energy, 30
Graviton, 20, 217
Green's function, 345-347, 349-350
Group velocity, 13
Gryomagnetic ratio, 260-262

Hadrons, 253
Hamiltonian, definition, 29
Harmonic oscillator:
differential equation description, 80-87
general operator treatment, 141-145
matrix description, 146-147
in spherical polar coordinates, 242-246
^{5}He, 286
Heisenberg matrix formulation, 137-140
Heisenberg uncertainty principle, *see* Uncertainty principle
Helium, solidification of, 85
Helium atom, 168, 335-337
Hermite polynomials, 84
Hermitian conjugate, 136
Hermitian operator, 111-114, 118
Hidden variables, 37-38
Homogeneity, required of the wave equation, 23-23
Hulthén potential, 73-74
Hydrogen atom, 18-19, 167, 171, 246-251
in an electric field, 322-324
photoionization, 384, 394-395
21 cm radiation from, 306, 322
$2p$-$1s$ transition, 379, 381-383
Hydrogen-like spectra, 248-249
effect of nuclear size on, 309-310, 314-317
Hyperfine interaction, 320-322

Identical particles, 174-185
collisions between, 353-355
Inertial frame, 6, 25
Interference between indistinguishable possibilities, 184

Ket vector, 141
Kronecker delta, 106

Laguerre polynomials, 244
  associated, 250
Lasers, 375
Legendre equation and polynomials, 227–230
Leibniz' proof that atoms do not exist, 175
Linearity:
  implied by electron diffraction, 10
  of operators, 111, 163
  required of the wave equation, 23–24
Linear potential energy, 92–97, 251–256
Lithium, 176, 187–189, 200
Lorentz line shape, 159–161, 282
Lowering operator:
  for angular momentum, 215, 218–219, 231–232, 267
  for the harmonic oscillator, 143–147

Magic numbers for nuclei, 166
Magnetic field:
  energy of electrons in, 277–279
  of free electrons, 263–264
  precession of electrons in, 279–280
  precession of muons in, 280
Magnetic moment:
  deuteron, 284, 301–305
  electron, 260–265
  neutron, 283–284
  nuclei, 284
  proton, 283–284
Magnetic resonance, 280–284, 368–369
Matrix algebra, 134–137
Matrix and differential equation formulations, equivalence, 140
Mean life, 79, 161–162
Mean squared deviation, 102
Mean value, 101
Mesons, 19, 35, 253, 283–284
Minimum packet, 117
Mixture, 91
Momentum conservation for systems of particles, 171–172
Momentum operator, 109
Momentum space, 108–109
Muon, 79, 217, 316–317
  spin rotation, 280
Muonic atoms, 316–317

Natural time, 154, 385
Natural width, 20, 154, 385
Neutrino, 217, 366, 388-392
Neutron:
  beta decay of, 388–389
  diffraction, 11
  potential energy of inside nuclei, 78, 236–237
Neutron-proton scattering, 361–363
Newtonian mechanics, 5–8
Noncentral forces, 239, 302
Normalization, 31–33
  delta function, 121
Nuclear energy levels, 286
Nuclear magnetic resonance, 282–284
Nuclear interaction:
  charge independence, 363–364
  spin dependence, 362–363
Nuclear potential energy, 19, 30, 74, 78, 166, 302
Nuclear sizes from muonic atoms, 306–307
Nuclear structure, 166, 286
Number of states per unit volume, 192–196

Operator, 24, 110
Orbital angular momentum, 209, 217–218, 223–233
  effect on wave functions near the origin, 236
  eigenfunctions of, 226–233
Orthogonality, 119, 155
Orthonormality, 122, 157

Parity, 64
  of bound states in symmetric potentials, 63–65, 83, 85
  operator, 123, 163
  of spherical harmonics, 233, 302
Partial waves, 357–359
Pauli exclusion principle, 179–181, 189, 195, 293
Pauli spin matrices, 137, 269–270
Periodic boundary conditions, 193–195, 202
Periodic potential, 189, 202–206
Perturbations:
  for classical systems, 308
  of degenerate states, 317–320
  of nondegenerate states, 308–313
  time dependent, 367–370
Phase shifts, 359–362
Photodisintegration, 392–398
Photoionization of hydrogen, 384, 394–395

Pion, 19, 35
Planck's:
  constant, 11
  energy-frequency relation, 13
  formula for radiation in a cavity, 376
Poisson brackets, 114
Position space, 108–110
Positronium, 246, 254
Potential energy, meaning of in quantum physics, 29–30
Probability density, 30–31
Propagator, 126–127

Quadrupole moment, 302
Quarks, 92, 96, 252–256
Quasi-stationary states, 154–161, 385

Radiation:
  absorption, 370–375, 379
  selection rules, 377–381
  spontaneous emission, 375–377, 379, 381–383
  stimulated emission, 372, 375
Raising operator:
  for angular momentum, 215, 218–219, 231–232, 267
  for the harmonic oscillator, 143–147
Ramsauer-Townsend effect, 360
Random phases, 90–91, 272, 294–295
Range of forces, 20
Rectangular well, 60–72
Recursion relation:
  for Coulomb wave functions, 248
  for harmonic oscillator wave functions, 82
  for Legendre polynomials, 228
Reduced mass, 173
Reflection coefficient for step potential, 46–47
  Born approximation for, 349–350
Relative coordinates, 173–174
Root mean squared deviation, 102
Rotation operator, 211
Rutherford scattering cross section, 352–353

Scattering, 338–364
  amplitude, 344
  Coulomb, 351–355
  cross section, 342–343
  of identical particles, 178, 353–355
  of ions by atoms, 338
  of neutrons by neutrons, 364
  of neutrons by protons, 361–363
Schmidt orthogonalization method, 120
Schrödinger equation:
  development of, 23–29
  separation in Cartesian coordinates, 43–44
  separation in spherical polar coordinates, 224–226
  separation of time dependence, 41–43
  time independent, 42–43
Schwarz inequality, 115
Scintillant for particle detection, 100
Screened Coulomb potential, 351–352
Selection rules, 377–381, 391
Several-particle systems:
  energy conservation, 171
  momentum conservation, 171–172
  Schrödinger equation, 167–169
Shell structure of nuclei, 166
Singlet spin state, 255, 292–295
  of the hydrogen atom, 320–322
Slater determinant, 180–181, 189
Solvay conferences, *opposite* 1
Sphere of constant potential:
  bound states in, 237–242
  scattering by, 355–356, 359–360
Spherical Bessel functions, 241, 357–358
Spherical Hankel functions, 241
Spherical harmonics, 230–233
  expansion of plane waves in, 356–357
  parity of, 233, 302
Spherical Neumann functions, 241, 357–358
Spin, 102–103, 123, 167, 190, 259–284
  eigenfunctions, 266–275
  Hamiltonian, 260, 274
  and statistics, 181
Spinor, 266–277
  double valued under rotations, 277
Spin-orbit interaction, 261–262
Spins, addition of, 288–295
Standard deviation, 102
Stark effect, 322–324
Stationary state, 43
Statistics, relation between spin and, 181
Step function potential, 44–47
Stern-Gerlach experiment, 262–265, 272–273
  with free electrons, 264–265
Superposition, 91
Susceptibility, electronic paramagnetic, 197–200
Symmetry under exchange of particles, 176–184

Thomas precession, 262

Time derivatives, order of in the Schrödinger equation, 25, 33
Time derivatives of expectation values, 148–150
Transition probability, 367–398
Transitions to a continuum, 384–398
Transmission coefficient:
  delta function barrier, 56
  rectangular barrier, 51
  rectangular well, 61
  step function potential, 46–47
  WKB approximation for, 330
Triplet spin state, 255, 292–295, 301
  of the hydrogen atom, 320–322
Tunneling, 52–53

Uncertainty principle, 13–20, 58, 85, 89
  for angular momentum, 117, 220–221
  precise statement, 114–118
  special meaning for *E* and *t*, 118, 152–153
Unequal path lengths in interference experiments, 90–91
Unstable systems, 79, 153–162, 384–385

Vacuum polarization, 246
Variance, 102
Variation approximation method, 332–334
  for charmonium *P* state, 335
  for ground state of helium atom, 335–337

Wave packet, 12–14
WKB approximation, 324–326
  for bound states, 327–330
  for transmission coefficients, 330–332

Yukawa potential, 350

Zero point energy, 85
Zero-to-zero angular momentum transitions, 377–378

VARIOUS RESULTS, WITH REFERENCES TO THE SECTIONS WHERE THEY ARE DEVELOPED (Continued)

Reduced mass $= \mu = \dfrac{M_1 M_2}{M_1 + M_2}$ (4.1)

$\Psi(2, 1, t) = +\Psi(1, 2, t)$ for bosons, $= -\Psi(1, 2, t)$ for fermions (4.2)

Fermi energy $= \dfrac{\hbar^2}{2M}(3\pi^2 N)^{2/3}$, $N =$ no. spin $\frac{1}{2}$ particles/vol (5.1)

$[J_x, J_y] = i\hbar J_z, \quad [J_z, J_x] = i\hbar J_y, \quad [J_y, J_z] = i\hbar J_x$ (6.1)

$J_\pm = J_x \pm iJ_y, \quad J_\pm \psi_{jm} = N_{jm}^\pm \psi_{jm\pm 1}, N_{jm}^\pm = \hbar\sqrt{j(j+1) - m(m \pm 1)}$ (6.1)

$L^2 Y_{lm} = \hbar^2 l(l+1) Y_{lm}, \quad L_z Y_{lm} = \hbar m Y_{lm},$

$$Y_{00} = \frac{1}{\sqrt{4\pi}}, \quad Y_{10} = \sqrt{\frac{3}{4\pi}}\cos\theta, \quad Y_{1\pm 1} = \mp\sqrt{\frac{3}{8\pi}}\sin\theta e^{\pm i\varphi},$$

$$Y_{20} = \sqrt{\frac{5}{16\pi}}(3\cos^2\theta - 1), \quad Y_{2\pm 1} = \mp\sqrt{\frac{15}{8\pi}}\cos\theta\sin\theta e^{\pm i\varphi},$$

$$Y_{2\pm 2} = \sqrt{\frac{15}{32\pi}}\sin^2\theta e^{\pm 2i\varphi} \quad (6.2)$$

$$u(r) = rR(r): \frac{-\hbar^2}{2\mu}\frac{d^2u}{dr^2} + \left[V(r) + \frac{\hbar^2 l(l+1)}{2\mu r^2}\right]u = Eu,$$

$$V(r) \sim r^q, \quad q > -2: R(r) \sim r^l \quad (7.1)$$

$$V = \frac{e^2}{4\pi\varepsilon_0 r}: E_n = \frac{-\mu e^4}{2\hbar^2(4\pi\varepsilon_0)^2 n^2}, \quad a = \frac{(4\pi\varepsilon_0)\hbar^2}{\mu e^2},$$

$$\psi_{100} = \frac{2e^{-r/a}}{a^{3/2}}Y_{00}, \quad \psi_{200} = \frac{1}{\sqrt{2a^3}}\left(1 - \frac{r}{2a}\right)e^{-r/2a}Y_{00},$$

$$\psi_{21m} = \frac{1}{\sqrt{24a^3}}\frac{r}{a}e^{-r/2a}Y_{1m} \quad (7.3)$$